CW00487283

Wave Propagation

Wave Propagation

From Electrons to Photonic Crystals
and Left-Handed Materials

Peter Markoš
Costas M. Soukoulis

PRINCETON UNIVERSITY PRESS

PRINCETON AND OXFORD

Published by Princeton University Press, 41 William Street, Princeton,
New Jersey 08540
In the United Kingdom: Princeton University Press, 6 Oxford Street,
Woodstock, Oxfordshire OX20 1TW

ISBN: 978-0-691-13003-3

Library of Congress Control Number: 2007943032

British Library Cataloging-in-Publication Data is available

This book has been composed in Scala and Scala Sans

Printed on acid-free paper ∞

press.princeton.edu

Printed in the United States of America

10   9   8   7   6   5   4   3   2   1

Contents

Preface

This volume is intended to serve as a general text on wave propagation for senior under-graduates and first-year graduate students in physics, applied physics, engineering, materials sciences, optics, and other related scientific disciplines. We also hope that it will be useful for the many scientists working in different areas of wave propagation.

The importance of waves in our everyday life can hardly be overestimated. Waves are everywhere. Most of the information we receive comes to us in the form of waves. We see and hear through waves. We transmit and receive information through waves. We rely on waves to bring us music, television, Email, and wireless communications. We can cook with waves, talk to each other, and see things because of waves. Waves are also used in medicine; and in industry to examine objects, such as planes, for cracks and stress. Waves in periodic media usually exhibit some common characteristics. The most important is the possible appearance of stop bands (also called gaps) separating pass bands (also called bands). This means that the frequency axis may be divided into alternating regions of gaps and bands.

Although matrices and matrix properties are important parts of the physics and engineering curriculum, transfer matrices are rarely discussed in standard undergraduate textbooks. In this text, we introduce the transfer matrix formalism that can be used to study any one-dimensional wave propagation problem. The transfer matrix formalism will be used to solve problems about the propagation of electrons in potential barriers and wells, as well as some concerning the propagation of electromagnetic waves in photonic crystals and left-handed materials.

Chapter 1 introduces the transfer matrix method. The transfer matrix, as well as the scattering matrix, is defined. Through the transfer matrix formalism, the transmission and reflection amplitudes can be easily defined and evaluated. Both traveling and standing (bound) waves are examined.

In chapters 2 and 3, the transfer matrix method is used to calculate the transmission and reflection properties of rectangular potential wells and barriers, as well as δ-function potentials. This material can be found in any book of quantum mechanics, but the transfer matrix presentation is superior, since it can be easily extended to any number of potential barriers and wells. The ideas of resonant transmission, bound states, and bonding and antibonding states are introduced in these chapters too.

In chapter 4, the transfer matrix technique is applied to the Kronig-Penney model. The most common characteristic of periodic media, that of bands and gaps, is introduced in this chapter. The group and phase velocities, the density of states, and the effective mass are introduced and evaluated.

In chapters 5 and 6 the so-called tight binding model (or equivalently the linear combination of atomic orbitals) is introduced. The tight binding model is of central importance because it is the simplest example of wave propagation in periodic media. The transfer matrix method is used to calculate the transmission and reflection properties of a lattice of N identical atoms, as well as those with 2 or 4 atoms in the unit cell. The properties of an isolated impurity are also presented. The Brillouin zone and the Fermi energy are introduced, and metals and insulators are briefly examined.

In chapter 7, the transfer matrix formalism is used to study disordered or random systems. The transmission and reflection properties of random systems are presented. The ideas of localization and probability distributions are introduced. In one-dimensional disordered systems the eigenstates are always exponentially localized.

In chapter 8, a short introduction to the accuracy of different numerical methods is given.

In chapter 9, the transfer matrix method is used to calculate the transmission and reflection properties of EM waves for different interfaces. The interfaces examined are a dielectric to dielectric, a dielectric to metal, and a dielectric to left-handed material. The Brewster angle and evanescent waves are also discussed.

In chapter 10 the transmission and reflection properties of dielectric and metallic slabs are examined. For normal incidence, the transmission of EM waves through a dielectric slab is exactly equivalent to the problem of an electron propagating through a potential barrier. The ideas of Fabry-Pérot and resonant tunneling are introduced and discussed.

In chapter 11, the transfer matrix method is used to study surface waves at different interfaces, such as vacuum-metal, and vacuum–left-handed materials interfaces. A discussion of the experimental observation of surface waves is also given.

In chapter 12, resonant tunneling through double-layer structures of dielectrics and metals is examined. In chapter 13, photonic crystals are introduced and discussed. In addition, layered materials of metals and dielectrics and of metals and left-handed materials are examined. In chapter 14, the effective parameters of photonic crystals and left-handed materials are calculated.

Chapter 15 covers wave propagation in nonlinear structures. The ideas of bistability and gap solitons are introduced.

Chapter 16 is devoted to left-handed materials. Transmission through a left-handed slab is presented. The ideas of focusing of propagating waves, recovery of evanescent waves, and perfect lenses are discussed. A short history of the left-handed field is also presented.

Each chapter concludes with a number of problems. These consist of two types: Problems with Solutions, which are set as problems and not in the main text because they are more technical and might interfere with the flow of the material presented in the text, and Problems without Solutions, which are of medium difficulty and cover the entire chapter. All the problems are solvable on the basis of the material presented in the chapter and do not need any more advanced references.

We gratefully acknowledge fruitful discussions with our colleague and friend Thomas Koschny. A special thanks is given to Rebecca Shivvers, for her help in the first editing of this book.

P. Markoš and C. M. Soukoulis

Wave Propagation

1 Transfer Matrix

In this chapter we introduce and discuss a mathematical method for the analysis of the wave propagation in one-dimensional systems. The method uses the transfer matrix and is commonly known as the *transfer matrix* method [7,29].

The transfer matrix method can be used for the analysis of the wave propagation of quantum particles, such as electrons [29,46,49,81,82,115–117,124,103,108,131,129,141] and of electromagnetic [39,123,124], acoustic, and elastic waves. Once this technique is developed for one type of wave, it can easily be applied to any other wave problem.

First we will treat the scattering from an arbitrary one-dimensional potential. Usually, one writes the amplitudes of the waves to the left side of the potential in terms of those on the right side. This defines the *transfer matrix* **M**. Since we work in a one-dimensional system, the wave in both the left and right sides of the potential has two components, one moving to the right and one moving to the left. Therefore, the transfer matrix **M** is a 2×2 matrix. The 2×2 *scattering matrix* **S** will also be introduced; it describes the outgoing waves in terms of the ingoing waves. The relationship between the transfer and scattering matrices will be introduced. Time-reversal invariance and conservation of the current density impose strong conditions on the form of the transfer matrix **M**, regardless of the specific form of the potential. Through the transfer matrix formalism, the transmission and reflection amplitudes can easily be defined and evaluated. Both traveling and standing (bound) waves will be examined.

Once the transfer matrix is calculated for one potential, it can be easily extended to calculate analytically the transfer matrix for N identical potentials [39,165]. As the number of potentials increases, the traveling waves give rise to pass bands, while the standing or bound waves give rise to gaps in the energy spectrum of the system.

The appearance of bands and gaps is a common characteristic of wave propagation in periodic media. Bands and gaps appear in electronic systems [1,3,4,11,17,23,30], photonic crystals (electromagnetic waves) [33–36,39], phononic crystals (acoustic waves), and left-handed materials. The transfer matrix formalism is also very useful in calculating reflection and transmission properties of disordered random systems [49,59,89,90,103, 108,124,129].

1.1 A Scattering Experiment

Perhaps the simplest problem in quantum mechanics is the one-dimensional propagation of an electron in the presence of a localized potential. The motion of a quantum particle of mass m in the presence of a potential $V(x)$ in one dimension is governed by Schrödinger's equation [7,25,30]

$$-\frac{\hbar^2}{2m}\frac{\partial^2\Psi(x)}{\partial x^2} + \left[V(x) - E\right]\Psi(x) = 0. \qquad (1.1)$$

Here, $\Psi(x)$ is the *wave function* and E the energy of the electron.

In the absence of a potential, the electron is a wave that travels along in a particular direction. In the presence of a potential, we would like to know how the propagation of the electron changes. Can the electron reflect back? Can the electron pass through the potential? These questions illustrate some quantum effects not present in classical physics.

For simplicity, we assume that the potential $V(x)$ is nonzero only inside a finite region,

$$V(x) = \begin{cases} V(x) \text{ for } 0 \le x \le \ell, \\ 0 \quad \text{ for } x < 0 \text{ and } x > \ell. \end{cases} \qquad (1.2)$$

An electron approaches the sample represented by the potential $V(x)$ from either the left or the right side of the potential and is scattered by the sample. Scattering means that the electron is either reflected back or transmitted through the sample. We can measure the transmission and reflection amplitudes, t and r, respectively (they will be defined later), and from t and r we can extract information about the physical properties of the sample.

We assume that Schrödinger's equation outside the potential region is known and that it can be written as a superposition of plane waves:

$$\begin{aligned} \Psi_L(x) &= \Psi_L^+(x) + \Psi_L^-(x), \ x \le 0, \\ \Psi_R(x) &= \Psi_R^+(x) + \Psi_R^-(x), \ x \ge \ell. \end{aligned} \qquad (1.3)$$

Here, the subscripts L (Left) and R (Right) indicate the position of the particle with respect to the potential region, and the superscripts + (–) determine the direction of propagation: + means that the electron propagates in the positive direction (from left to right) and − means that the electron moves from right to left (see figure 1.1). Thus, $\Psi_L^+(x)$ is the wave function of the electron left of the sample, propagating to the right; hence it is approaching the sample. We call $\Psi_L^+(x)$ the *incident* wave, in contrast to $\Psi_L^-(x)$, which is the wave function of the electron propagating away from the sample toward the left side.

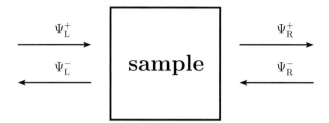

FIGURE 1.1. A typical scattering experiment. Incident waves $\Psi_L^+(x)$ and $\Psi_R^-(x)$ are scattered by the sample, characterized by the potential $V(x)$. Outgoing waves $\Psi_L^-(x)$ and $\Psi_R^+(x)$ consist of waves transmitted through the sample as well as waves reflected from the sample. Outside the sample, the wave function can be expressed as a superposition of plane waves given by equations (1.3) and (1.4).

The components of the wave function can be expressed as

$$\Psi_L^+(x) = Ae^{+iqx}, \quad \Psi_L^-(x) = Be^{-iqx},$$
$$\Psi_R^+(x) = Ce^{+iqx}, \quad \Psi_R^-(x) = De^{-iqx}. \tag{1.4}$$

Here, q is the *wave vector* related to the energy, E of the electron through the *dispersion relation*

$$E = E(q). \tag{1.5}$$

The dispersion relation (1.5) determines the physical properties of the electron in the region outside the sample ($x < 0$ and $x > \ell$). We will call these regions *leads*. To guarantee the plane wave propagation of the particle, i.e., equation (1.4), we require that both leads are translationally invariant. In the simplest cases, we will represent both leads as free space. Then $q = k$ and k is related to the energy of the free particle,

$$E = \frac{\hbar^2 k^2}{2m}. \tag{1.6}$$

More general realizations of leads, for instance consisting of periodic media, will be discussed later. In this book, we assign k to the free-particle wave vector and use q for more general cases.

1.2 Scattering Matrix and Transfer Matrix

The general solution $\Psi(x)$ of the Schrödinger equation

$$-\frac{\hbar^2}{2m}\frac{\partial^2 \Psi(x)}{\partial x^2} + \left[V(x) - E\right]\Psi(x) = 0 \tag{1.7}$$

must be a continuous function of the position x. The same must be true for the first derivative $\partial\Psi(x)/\partial x$. In particular, the requirement of the continuity of the wave function and its derivative at the boundaries of the potential $V(x)$ gives

$$\Psi_L(x=0^-) = \Phi(x=0^+), \quad \frac{\partial \Psi_L(x)}{\partial x}\bigg|_{x=0^-} = \frac{\partial \Phi(x)}{\partial x}\bigg|_{x=0^+} \tag{1.8}$$

on the left boundary of the sample, and

$$\Psi_R(x=\ell^+) = \Phi(x=\ell^-), \quad \frac{\partial \Psi_R(x)}{\partial x}\bigg|_{x=\ell^+} = \frac{\partial \Phi(x)}{\partial x}\bigg|_{x=\ell^-} \tag{1.9}$$

on the right boundary. Here, $\Phi(x)$ is the solution of Schrödinger's equation *inside* the potential region $0 \le x \le \ell$. Generally, $\Phi(x)$ cannot be expressed as a simple superposition of propagating waves.

We can, in principle, solve Schrödinger's equation (1.7) and find explicit expressions for the wave functions for any position x, including the region of the scattering potential. However, this is possible only in very few special cases, since the Schrödinger equation is not analytically solvable for a general form of the potential $V(x)$. In many cases, however, it is sufficient to know only the form of the wave function *outside* the potential region. This problem is much easier, since the wave function consists only of a superposition of plane waves, as discussed in equations (1.3) and (1.4). However, we need to estimate the coefficients A–D, defined in equation (1.4). This can be done if we know the right-hand sides of the four equations (1.8) and (1.9). Thus, the wave function outside the sample is fully determined by the four parameters that describe the scattering properties of the sample.

In general, linear relations between outgoing and incoming waves can be written as

$$\begin{pmatrix} \Psi_L^-(x=0) \\ \Psi_R^+(x=\ell) \end{pmatrix} = S \begin{pmatrix} \Psi_L^+(x=0) \\ \Psi_R^-(x=\ell) \end{pmatrix}, \tag{1.10}$$

where the matrix S,

$$S = \begin{pmatrix} S_{11} & S_{12} \\ S_{21} & S_{22} \end{pmatrix}, \tag{1.11}$$

is called the *scattering matrix*. By definition, the matrix S relates the *outgoing* waves to the *incoming* waves as shown in figure 1.1. Its elements completely characterize the scattering and transmission properties of the one-dimensional potential $V(x)$.

We can also define the *transfer matrix* M by the relation

$$\begin{pmatrix} \Psi_R^+(x=\ell) \\ \Psi_R^-(x=\ell) \end{pmatrix} = M \begin{pmatrix} \Psi_L^+(x=0) \\ \Psi_L^-(x=0) \end{pmatrix}. \tag{1.12}$$

The matrix M expresses the coefficients of the wave function on the *right-hand* side of the sample in terms of the coefficients of the wave function on the *left-hand* side.

While the representation in terms of the scattering matrix S can be easily generalized to three-dimensional systems, the transfer matrix approach is more appropriate for the analysis of one-dimensional systems and will be used frequently in the following chapters. On the other hand, physical properties of the scattering are formulated more easily by the **S** matrix.

By comparing the linear equations (1.10) and (1.12), it is easy to express the elements of the transfer matrix **M** in terms of the elements of the scattering matrix **S** (see problem 1.1):

$$\mathbf{M} = \begin{pmatrix} M_{11} & M_{12} \\ M_{21} & M_{22} \end{pmatrix} = \begin{pmatrix} S_{21} - \dfrac{S_{22} S_{11}}{S_{12}} & \dfrac{S_{22}}{S_{12}} \\ \dfrac{-S_{11}}{S_{12}} & \dfrac{1}{S_{12}} \end{pmatrix}. \tag{1.13}$$

Equivalently, we can express the elements of the scattering matrix **S** in terms of the elements of the transfer matrix:

$$\mathbf{S} = \begin{pmatrix} -\dfrac{M_{21}}{M_{22}} & \dfrac{1}{M_{22}} \\ M_{11} - \dfrac{M_{12} M_{21}}{M_{22}} & \dfrac{M_{12}}{M_{22}} \end{pmatrix}. \tag{1.14}$$

The scattering matrix **S** contains four complex parameters. In general, the matrix **S** is fully determined by eight real parameters. However, when solving a given physical problem, we can use its *physical symmetries* to reduce the number of independent parameters. Two symmetries—conservation of the current density and time-reversal symmetry—will be discussed in the following sections.

1.2.1 Conservation of the Current Density

For the time-independent problems discussed in this chapter, the total number of particles in the potential region,

$$\int_0^\ell \Psi^* \Psi \, dx, \tag{1.15}$$

is constant. For this case, in section 1.5.1 we derive the result that the current density entering the sample from one side must be equal to the current density that leaves the sample on the other side:

$$j(x=0) = j(x=\ell). \tag{1.16}$$

We remind the reader that the current density $j(x)$ is defined as

$$j(x) = \frac{\hbar i}{2m} \left[\Psi(x) \frac{\partial \Psi^*(x)}{\partial x} - \Psi^*(x) \frac{\partial \Psi(x)}{\partial x} \right]. \tag{1.17}$$

Using the definition of the current density, equation (1.17) and the expression for the current density for a plane wave, $j = (\hbar q/m)|\Psi|^2$, derived later in section 1.5.1, we can express the current density on both sides of the sample [equation (1.16)] as

$$j_L = \frac{\hbar q}{m} \left(|\Psi_L^+|^2 - |\Psi_L^-|^2 \right) = \frac{\hbar q}{m} \left(|\Psi_R^+|^2 - |\Psi_R^-|^2 \right) = j_R, \tag{1.18}$$

which are equal in magnitude, according to equation (1.16). Note that the current does not depend on x. Equation (1.18) can be rewritten in a more convenient form as

$$|\Psi_L^-|^2 + |\Psi_R^+|^2 = |\Psi_L^+|^2 + |\Psi_R^-|^2, \tag{1.19}$$

or, in vector notation

$$(\Psi_L^{-*} \; \Psi_R^{+*}) \begin{pmatrix} \Psi_L^- \\ \Psi_R^+ \end{pmatrix} = (\Psi_L^{+*} \; \Psi_R^{-*}) \begin{pmatrix} \Psi_L^+ \\ \Psi_R^- \end{pmatrix}, \tag{1.20}$$

where Ψ^* is the complex conjugate of Ψ.

Now we use equation (1.10), which relates the outgoing waves Ψ_L^- and Ψ_R^+ with the incoming waves Ψ_L^+ and Ψ_R^-. For complex conjugate waves, the relation (1.10) reads

$$(\Psi_L^{-*} \; \Psi_R^{+*}) = (\Psi_L^{+*} \; \Psi_R^{-*})S^\dagger, \tag{1.21}$$

where the conjugate matrix $S^\dagger$ is defined in appendix A as the matrix

$$S^\dagger = \begin{pmatrix} S_{11}^* & S_{21}^* \\ S_{12}^* & S_{22}^* \end{pmatrix}. \tag{1.22}$$

Inserting (1.10) and (1.21) into equation (1.20), we obtain the identity

$$(\Psi_L^{+*} \; \Psi_R^{-*})S^\dagger S \begin{pmatrix} \Psi_L^+ \\ \Psi_R^- \end{pmatrix} = (\Psi_L^{+*} \; \Psi_R^{-*}) \begin{pmatrix} \Psi_L^+ \\ \Psi_R^- \end{pmatrix}. \tag{1.23}$$

The relation (1.23) must be valid for any incoming wave. This can be guaranteed only if the scattering matrix satisfies the relation

$$S^\dagger S = 1, \tag{1.24}$$

which means that the scattering matrix is *unitary*.

The explicit form of equation (1.24) is given by

$$\begin{pmatrix} S_{11}^* & S_{21}^* \\ S_{12}^* & S_{22}^* \end{pmatrix} \begin{pmatrix} S_{11} & S_{12} \\ S_{21} & S_{22} \end{pmatrix} = 1 = \begin{pmatrix} 1 & 0 \\ 0 & 1 \end{pmatrix}. \tag{1.25}$$

After matrix multiplication, we obtain the following relationships between the matrix elements of the scattering matrix:

$$\begin{aligned} |S_{11}|^2 + |S_{21}|^2 &= 1, \quad |S_{22}|^2 + |S_{12}|^2 = 1, \\ S_{11}^* S_{12} + S_{21}^* S_{22} &= 0, \quad S_{12}^* S_{11} + S_{22}^* S_{21} = 0. \end{aligned} \tag{1.26}$$

Note, from equation (1.24) it follows also (see problem 1.2) that

$$|\det S| = 1. \tag{1.27}$$

Equation (1.24) can be also written as

$$\mathbf{S}^{\dagger} = \mathbf{S}^{-1}.$$ (1.28)

Using the expression for the inverse matrix, given by equation (A.9) we obtain

$$\begin{pmatrix} S_{11}^* & S_{21}^* \\ S_{12}^* & S_{22}^* \end{pmatrix} = \frac{1}{\det \mathbf{S}} \begin{pmatrix} S_{22} & -S_{21} \\ -S_{12} & S_{11} \end{pmatrix}.$$ (1.29)

Comparison of the matrix elements of equation (1.29) gives some additional useful relationships for the matrix elements of the scattering matrix:

$$|S_{11}| = |S_{22}|.$$ (1.30)

Then, from the third and fourth equations (1.26) we obtain that

$$|S_{12}| = |S_{21}|.$$ (1.31)

The conservation of the current density also introduces a relationship between the elements of the transfer matrix $\mathbf{M}$. We can derive them beginning with equation (1.18), describing the conservation of the current density,

$$|\Psi_L^+|^2 - |\Psi_L^-|^2 = |\Psi_R^+|^2 - |\Psi_R^-|^2.$$ (1.32)

It is easy to verify that equation (1.32) can be rewritten in the vector form as

$$(\Psi_L^{+*} \ \Psi_L^{-*}) \begin{pmatrix} 1 & 0 \\ 0 & -1 \end{pmatrix} \begin{pmatrix} \Psi_L^+ \\ \Psi_L^- \end{pmatrix} = (\Psi_R^{+*} \ \Psi_R^{-*}) \begin{pmatrix} 1 & 0 \\ 0 & -1 \end{pmatrix} \begin{pmatrix} \Psi_R^+ \\ \Psi_R^- \end{pmatrix}.$$ (1.33)

Now we use the definition of the transfer matrix, equation (1.12), and its conjugate form

$$(\Psi_R^{+*} \ \Psi_R^{-*}) = (\Psi_L^{+*} \ \Psi_L^{-*})\mathbf{M}^{\dagger}$$ (1.34)

to obtain

$$(\Psi_L^{+*} \ \Psi_L^{-*}) \begin{pmatrix} 1 & 0 \\ 0 & -1 \end{pmatrix} \begin{pmatrix} \Psi_L^+ \\ \Psi_L^- \end{pmatrix} = (\Psi_L^{+*} \ \Psi_L^{-*})\mathbf{M}^{\dagger} \begin{pmatrix} 1 & 0 \\ 0 & -1 \end{pmatrix} \mathbf{M} \begin{pmatrix} \Psi_L^+ \\ \Psi_L^- \end{pmatrix}.$$ (1.35)

Since this relation should hold for any wave function Ψ_L^+ and Ψ_L^-, we obtain the relationship [29, 103, 108, 129, 141]

$$\mathbf{M}^{\dagger} \begin{pmatrix} 1 & 0 \\ 0 & -1 \end{pmatrix} \mathbf{M} = \begin{pmatrix} 1 & 0 \\ 0 & -1 \end{pmatrix}.$$ (1.36)

The explicit form of equation (1.36) is given by

$$\begin{pmatrix} M_{11}^* & M_{21}^* \\ M_{12}^* & M_{22}^* \end{pmatrix} \begin{pmatrix} 1 & 0 \\ 0 & -1 \end{pmatrix} \begin{pmatrix} M_{11} & M_{12} \\ M_{21} & M_{22} \end{pmatrix} = \begin{pmatrix} 1 & 0 \\ 0 & -1 \end{pmatrix},$$ (1.37)

which means that the elements of the transfer matrix satisfy the following relations:

$$|M_{11}|^2 - |M_{21}|^2 = 1, \quad |M_{22}|^2 - |M_{12}|^2 = 1,$$
$$M_{11}^* M_{12} - M_{21}^* M_{22} = 0, \quad M_{12}^* M_{11} - M_{22}^* M_{21} = 0. \tag{1.38}$$

1.2.2 Time-Reversal Symmetry

Physical systems that are symmetric with respect to an inversion of time possess another symmetry which further reduces the number of independent parameters of the matrices **S** and **M**.

If the system possesses time-reversal symmetry and if $\Psi(x)$ is a solution of Schrödinger's equation, then its complex conjugate $\Psi^*(x)$ is also a solution. In our special case, the wave functions *outside* the potential region are expressed as plane waves given by equations (1.3) and (1.4). The complex conjugate of the wave

$$\phi(x) = e^{iqx} \tag{1.39}$$

is the wave

$$\phi^*(x) = e^{-iqx}, \tag{1.40}$$

which propagates in the opposite direction. This means that, after time reversal, we have the same physical system as before, but the incoming waves are Ψ_L^{-*} and Ψ_R^{+*} and the outgoing waves are Ψ_L^{+*} and Ψ_R^{-*}. Since the scattering matrix **S** relates *any* incoming waves to the outgoing waves, we obtain

$$\begin{pmatrix} \Psi_L^{+*} \\ \Psi_R^{-*} \end{pmatrix} = \mathbf{S} \begin{pmatrix} \Psi_L^{-*} \\ \Psi_R^{+*} \end{pmatrix}. \tag{1.41}$$

On the other hand, the complex conjugate of equation (1.10) reads

$$\begin{pmatrix} \Psi_L^{-*} \\ \Psi_R^{+*} \end{pmatrix} = \mathbf{S}^* \begin{pmatrix} \Psi_L^{+*} \\ \Psi_R^{-*} \end{pmatrix}. \tag{1.42}$$

Now, inserting equation (1.42) into equation (1.41), we obtain

$$\begin{pmatrix} \Psi_L^{+*} \\ \Psi_R^{-*} \end{pmatrix} = \mathbf{S}\mathbf{S}^* \begin{pmatrix} \Psi_L^{+*} \\ \Psi_R^{-*} \end{pmatrix}. \tag{1.43}$$

Since relation (1.43) must hold for any incoming waves, we conclude that

$$\mathbf{S}\mathbf{S}^* = \mathbf{S}^*\mathbf{S} = 1. \tag{1.44}$$

This condition, in conjunction with the unitary relation, equation (1.24), implies that the scattering matrix **S** must be symmetric. Indeed, in terms of matrix elements, the condition (1.44) reads

$$|S_{11}|^2 + S_{12}^* S_{21} = 1, \quad |S_{22}|^2 + S_{21}^* S_{12} = 1,$$
$$S_{11}^* S_{12} + S_{12}^* S_{22} = 0, \quad S_{21}^* S_{11} + S_{22}^* S_{21} = 0. \tag{1.45}$$

Comparison of the third equation of (1.26), $S_{11}^* S_{12} + S_{22} S_{21}^* = 0$, with the third equation of (1.45) shows that the scattering matrix **S** is a symmetric matrix when the system possesses both time-reversal symmetry and conservation of current density,

$$S_{12} = S_{21}. \tag{1.46}$$

Time-reversal symmetry also implies that

$$\begin{pmatrix} \Psi_R^{-*} \\ \Psi_R^{+*} \end{pmatrix} = \mathbf{M} \begin{pmatrix} \Psi_L^{-*} \\ \Psi_L^{+*} \end{pmatrix}. \tag{1.47}$$

The above equation follows from the definition of the transfer matrix **M** given by equation (1.12). Indeed, the wave Ψ_R^{-*} now plays the role of the *incoming* wave and the wave Ψ_R^+ is the *outgoing* wave, while Ψ_L^{-*} is the incoming wave and Ψ_L^+ is the outgoing wave. Equation (1.47) can be written as

$$\begin{pmatrix} 0 & 1 \\ 1 & 0 \end{pmatrix} \begin{pmatrix} \Psi_R^{+*} \\ \Psi_R^{-*} \end{pmatrix} = \mathbf{M} \begin{pmatrix} 0 & 1 \\ 1 & 0 \end{pmatrix} \begin{pmatrix} \Psi_L^{+*} \\ \Psi_L^{-*} \end{pmatrix}. \tag{1.48}$$

On the other hand, the complex conjugate of the relation (1.12) reads

$$\begin{pmatrix} \Psi_R^{+*} \\ \Psi_R^{-*} \end{pmatrix} = \mathbf{M}^* \begin{pmatrix} \Psi_L^{+*} \\ \Psi_L^{-*} \end{pmatrix}. \tag{1.49}$$

Comparison of the last two equations shows that, in the case of time-reversal symmetry, the transfer matrix satisfies the relationship

$$\begin{pmatrix} 0 & 1 \\ 1 & 0 \end{pmatrix} \mathbf{M} \begin{pmatrix} 0 & 1 \\ 1 & 0 \end{pmatrix} = \mathbf{M}^*. \tag{1.50}$$

With the use of symmetry (1.50), we obtain that *for systems with time-reversal symmetry*, the transfer matrix **M**, has the form

$$\mathbf{M} = \begin{pmatrix} M_{11} & M_{12} \\ M_{12}^* & M_{11}^* \end{pmatrix} \tag{1.51}$$

[29, 103, 108, 129, 141]. We also have from the expression (1.13) that $\det \mathbf{M} = S_{21}/S_{12}$. With the use of equation (1.46), we obtain, for the case of time-reversal symmetry, that

$$\det \mathbf{M} = 1. \tag{1.52}$$

From equation (1.51 it also follows that Tr $\mathbf{M} = M_{11} + M_{22} = M_{11} + M_{11}^*$ is a real number when time-reversal symmetry is preserved.

By applying both the requirement for conservation of probability flux and time-reversal symmetry, we reduce the number of independent elements of the transfer matrix to three. Indeed, $\mathbf{M}$ is given by two *complex* numbers M_{11} and M_{12}, or by four real numbers, which determine the real and imaginary parts of M_{11} and M_{12}. These four numbers are not independent, because of the constraint det $\mathbf{M} = 1$.

1.3 Transmission and Reflection Amplitudes

To find the physical meaning of the elements of the scattering matrix $\mathbf{S}$, we return to the scattering experiment described in section 1.1. Consider a particle approaching the sample from the right. As no particle is coming from the left, we have

$$\Psi_L^+ = 0. \tag{1.53}$$

We also normalize the incoming wave to unity,

$$|\Psi_R^-|^2 = 1. \tag{1.54}$$

From equation (1.10), we obtain that the transmitted wave $\Psi_L^-(x)$ is given by

$$\Psi_L^-(x=0) = S_{12}\Psi_R^-(x=\ell) \tag{1.55}$$

and the reflected wave $\Psi_L^+(x+\ell)$ is given by

$$\Psi_R^+(x=\ell) = S_{22}\Psi_R^-(x=\ell). \tag{1.56}$$

We call S_{12} the *transmission amplitude t* and S_{22} the *reflection amplitude r*:

$$t = S_{12}, \quad r = S_{22}. \tag{1.57}$$

In the same way, we consider scattering of the particle coming from the left side of the potential. We obtain $r' = S_{11}$ as the reflection amplitude, and $t' = S_{21}$ as the transmission amplitude from left to the right. Finally, in terms of transmission and reflection amplitudes, we can write the scattering matrix $\mathbf{S}$ in the form

$$\mathbf{S} = \begin{pmatrix} r' & t \\ t' & r \end{pmatrix}. \tag{1.58}$$

Using the relationship between scattering and transfer matrices, equation (1.13), we can express the transfer matrix in the form

$$\mathbf{M} = \begin{pmatrix} t' - \dfrac{rr'}{t} & \dfrac{r}{t} \\ -\dfrac{r'}{t} & \dfrac{1}{t} \end{pmatrix}. \tag{1.59}$$

For further applications, it is useful to write the transfer matrix in the form

$$\mathbf{M} = \begin{pmatrix} t' - rt^{-1}r' & rt^{-1} \\ -t^{-1}r' & t^{-1} \end{pmatrix} \tag{1.60}$$

(problem 1.3). The order of terms in expression (1.60) is important for the analysis of scattering on many scattering centers as well as for generalization of the transfer matrix to many-channel problems, discussed in section 1.5.3.

The *transmission (reflection) coefficients* are defined, respectively, as the probability that the particle is transmitted (reflected):

$$T = |t|^2 \quad \text{and} \quad R = |r|^2. \tag{1.61}$$

Using the symmetry properties of the scattering matrix, equations (1.30) and (1.31), we have

$$|r| = |r'| \quad \text{and} \quad |t| = |t'|. \tag{1.62}$$

Conservation of the current density, equations (1.26), gives

$$|t|^2 + |r|^2 = 1 \quad \text{and} \quad |t'|^2 + |r'|^2 = 1. \tag{1.63}$$

Equations (1.63) have a simple physical interpretation. When the sample contains no losses and no sources, then the electron can be either reflected back or transmitted through the sample.

Now we use the definition (1.58) and rewrite the third equations (1.26), $S_{11}^* S_{12} + S_{21}^* S_{22} = 0$, in the form

$$\frac{r}{t} = -\left(\frac{r'}{t'}\right)^*. \tag{1.64}$$

This helps us to express

$$M_{11} = t' - \frac{r}{t}r' = t' + \frac{|r'|^2}{(t')^*} = \frac{1}{(t')^*}\left(|t'|^2 + |r'|^2\right) = \frac{1}{(t')^*}, \tag{1.65}$$

and we can write the transfer matrix, given by equation (1.59) in the more symmetric form

$$\mathbf{M} = \begin{pmatrix} (t')^{*-1} & rt^{-1} \\ -t^{-1}r' & t^{-1} \end{pmatrix}. \tag{1.66}$$

When scattering is symmetric with respect to time inversion, we have also

$$t = t' \tag{1.67}$$

[see equation (1.46)]. It is also evident that for no scattering potential, $V(x) \equiv 0$, we have $T = 1$ and $R = 0$.

1.4 Properties of the Transfer Matrix

The transfer matrix enables us to study the properties of the sample through scattering experiments. Far from the sample, we prepare a plane wave with a given wave vector q and measure how this wave transmits through the potential. Of course, the transmission coefficient T, the reflection coefficient R, as well as all elements of matrices $\mathbf{S}$ and $\mathbf{M}$ are functions of q. Thus, we discover the properties of the system from its scattering response. It is evident that both the transmission and the reflection depend on the energy $E = E(q)$ of the incoming particle. In particular,

$$t = t(q) \quad \text{and} \quad r = r(q). \tag{1.68}$$

As will be shown in the next chapters, the transfer matrix is a very powerful method of analysis for one-dimensional systems.

The transfer matrix depends on the properties of the entire system represented by the potential $V(x)$ and the two leads on the left and right sides of the potential. Any change in the physical properties of the leads (regions outside the sample) also changes the transfer matrix. We also must keep in mind that, when deriving the symmetry properties of the transfer matrix, we assumed that the leads at both sides of the sample are equal to each other. We will see that this condition is not always satisfied. Although the transfer matrix method works also in the case of different leads, some symmetry relations are not satisfied. For instance, the determinant of the transfer matrix $\mathbf{M}$ can be different from one.

For completeness, we note there is also another definition of the transfer matrix used in the literature. It uses the linear relationship between coefficients A–D defined in equation (1.4), instead of wave functions:

$$\begin{pmatrix} C \\ D \end{pmatrix} = \mathbf{T} \begin{pmatrix} A \\ B \end{pmatrix}, \tag{1.69}$$

By inserting the explicit expression for the wave functions, equation (1.4), into equation (1.12), we see that the transfer matrix $\mathbf{M}$, equation (1.12), is related to $\mathbf{T}$ as

$$\mathbf{M} = \begin{pmatrix} e^{iq\ell} & 0 \\ 0 & e^{-iq\ell} \end{pmatrix} \mathbf{T}. \tag{1.70}$$

Note that, in the limit of zero potential $V \equiv 0$, the $\mathbf{T}$ matrix is the unit matrix, while the matrix $\mathbf{M}$ possesses the phase factors $e^{\pm iq\ell}$, which the particle gains as it moves between $x = 0$ and $x = \ell$. In the following we will use the transfer matrix $\mathbf{M}$. All results can be easily reformulated in terms of the matrix $\mathbf{T}$. Of course, the transmission and reflection coefficients T and R are the same in both formulations, since the transfer matrices $\mathbf{M}$ and $\mathbf{T}$ differ from each other only in the phase.

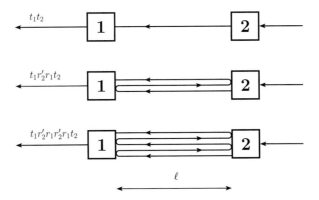

FIGURE 1.2. Schematic explanation of the calculation of the transmission through two barriers. A few paths with the electron scattered between the barriers are shown. To show different transmission paths, samples are separated by a distance ℓ. We consider $\ell \equiv 0$ in the text.

1.4.1 Multiplication of Transfer Matrices

Consider a more complicated experiment in which the particle is scattered by two individual samples. The first sample is given by the potential $V_1(x)$, located at $(a < x \leq b)$, and the second sample is determined by the potential $V_2(x)$, located at $(b \leq x < c)$. The problem of transmission and reflection through such a system can be treated in two ways: either we can use the transfer matrices $\mathbf{M}_1$ and $\mathbf{M}_2$, which determine the scattering properties of individual potentials V_1 and V_2, or we can consider the potential $V_{12}(x)$ defined on the interval $a \leq x \leq c$ and use the corresponding transfer matrix $\mathbf{M}_{12}$. Physically, it is clear that the results obtained by these two methods must be the same. This indicates that the transfer matrix $\mathbf{M}_{12}$ is completely determined by the elements of the transfer matrices $\mathbf{M}_1$ and $\mathbf{M}_2$. To derive the relationship between the transfer matrices, we express the wave function in three regions:

$$\begin{aligned}
\Psi_L(x) &= \Psi_L^+(x) + \Psi_L^-(x), & x &\leq a, \\
\Psi(x=b) &= \Psi^+(b) + \Psi^-(b), & x &\equiv b, \\
\Psi_R(x) &= \Psi_R^+(x) + \Psi_R^-(x), & x &\geq c.
\end{aligned}$$
(1.71)

Two samples are schematically shown in figure 1.2. Then, from the definition of the transfer matrix $\mathbf{M}$, equation (1.12), we have

$$\begin{pmatrix} \Psi^+(b) \\ \Psi^-(b) \end{pmatrix} = \mathbf{M}_1 \begin{pmatrix} \Psi_L^+(a) \\ \Psi_L^-(a) \end{pmatrix}$$
(1.72)

and

$$\begin{pmatrix} \Psi_R^+(c) \\ \Psi_R^-(c) \end{pmatrix} = \mathbf{M}_2 \begin{pmatrix} \Psi^+(b) \\ \Psi^-(b) \end{pmatrix}.$$
(1.73)

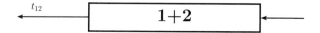

FIGURE 1.3. Transmission of the electron through the entire system consisting of two samples $1 + 2$.

By combining the equations (1.72) and (1.73), we obtain

$$
\begin{pmatrix} \Psi_R^+(c) \\ \Psi_R^-(c) \end{pmatrix} = M_2 M_1 \begin{pmatrix} \Psi_L^+(a) \\ \Psi_R^-(a) \end{pmatrix}.
\tag{1.74}
$$

As discussed above, we can consider the whole system as represented by the transfer matrix M_{12}. Then we can write

$$
\begin{pmatrix} \Psi_R^+(c) \\ \Psi_R^-(c) \end{pmatrix} = M_{12} \begin{pmatrix} \Psi_L^+(a) \\ \Psi_R^-(a) \end{pmatrix}.
\tag{1.75}
$$

A comparison of (1.74) with (1.75) gives the composition law

$$
M_{12} = M_2 M_1.
\tag{1.76}
$$

Since the matrix M_{12} is the transfer matrix of the whole system, its matrix elements determine the transmission and the reflection amplitudes [equation (1.60)] for the entire system. This enables us to determine the transmission and the reflection amplitudes of the entire system (figure 1.3) in terms of elements of the transfer matrices of the system's constituents. For example, we can calculate the transmission amplitude t_{12} for an electron approaching the system from the *right*. Using the explicit form of the transfer matrix, equation (1.60),

$$
\begin{aligned}
M &= \begin{pmatrix} t_{12}' - r_{12} t_{12}^{-1} r_{12}' & r_{12} t_{12}^{-1} \\ -t_{12}^{-1} r_{12}' & t_{12}^{-1} \end{pmatrix} \\
&= \begin{pmatrix} t_2' - r_2 t_2^{-1} r_2' & r_2 t_2^{-1} \\ -t_2^{-1} r_2' & t_2^{-1} \end{pmatrix} \begin{pmatrix} t_1' - r_1 t_1^{-1} r_1' & r_1 t_1^{-1} \\ -t_1^{-1} r_1' & t_1^{-1} \end{pmatrix},
\end{aligned}
\tag{1.77}
$$

we find by matrix multiplication that

$$
t_{12}^{-1} = t_2^{-1} t_1^{-1} - t_2^{-1} r_2' r_1 t_1^{-1},
\tag{1.78}
$$

which can be written as

$$
t_{12} = t_1 \left[1 - r_2' r_1 \right]^{-1} t_2.
\tag{1.79}
$$

The physical interpretation of formula (1.79) is more clear if we expand the right-hand side (r.h.s.) of equation (1.79) in terms of a power series:

$$
t_{12} = t_1 \left[1 + r_2' r_1 + r_2' r_1 r_2' r_1 + \cdots \right] t_2.
\tag{1.80}
$$

Table 1.1. Physical meaning of the parameters of the scattering matrix, **S** equation (1.58), and the transfer matrix, **M** equation (1.60).

t	transmission of a wave propagating from right to left
r	reflection of a wave coming from right
t'	transmission of a wave propagating from left to right
r'	reflection of a wave coming from left

Then we see that transmission amplitude t_{12} is given by the sum of the contributions of all possible paths through the two potential regions, V_1 and V_2. Three such paths are shown in figure 1.2. The first term in (1.80), $t_1 t_2$, represents the transmission through both the potentials. The second term $t_1 r_2' r_1 t_2$ corresponds to the path when an electron passes through the second sample (t_2), is reflected back from the second sample (r_1), and, after reflection from the second sample (r_2'), finally passes through the first one (t_1). Higher terms in the expansion (1.80) contain higher powers of $(r_2' r_1)^n = r_2' r_1 r_2' r_1 \ldots$. The nth term corresponds to a trajectory in which the electron is n times scattered between samples 1 and 2 before it passes through the second sample and escapes to the left.

In the same way, we can derive an expression for the reflection amplitude. From equation (1.77) we obtain that

$$-t_{12}^{-1} r_{12}' = -t_2^{-1} r_2' (t_1' - r_1 t_1^{-1} r_1') - t_2^{-1} t_1^{-1} r_1'$$
$$= -t_2^{-1} r_2' t_1' + t_2^{-1} \left[r_2' r_1 - 1 \right] t_1^{-1} r_1'. \tag{1.81}$$

Now we multiply both sides of the last equations by $-t_{12} = -t_1 [1 - r_2' r_1]^{-1} t_2$ and obtain

$$r_{12}' = r_1' + t_1 \left[1 - r_2' r_1 \right]^{-1} r_2' t_1'. \tag{1.82}$$

We remind the reader that, in agreement with our convention (table 1.1), r_{12}' is the reflection amplitude for the particle which approaches the sample from the left and is reflected back to the left.

The reflection amplitude again contains contributions of an infinite number of trajectories. The first term in equation (1.82) is just the reflection from the first barrier. All subsequent terms represent trajectories in which the electron transmits through the first barrier, then is n times reflected between both barriers, and is finally transmitted through the first sample (with transmission amplitude t_1) and leaves the system to the left.

Of special interest is the case when $V_2(x) = 0$. Then, the matrix $\mathbf{M}_2$ is the diagonal matrix and

$$\mathbf{M}_{12} = \begin{pmatrix} e^{iq\ell} & 0 \\ 0 & e^{-iq\ell} \end{pmatrix} \mathbf{M}_1, \quad \ell = c - b, \tag{1.83}$$

and consequently

$$T_{12} = \frac{1}{|t_{12}|^2} = \frac{1}{|e^{iq\ell}t_1|^2} = \frac{1}{|t_1|^2} = T_1. \tag{1.84}$$

Also, it is evident that $R_{12} = R_1$. This important result is easy to understand physically, since any reflection can appear only in the region where the potential is nonzero. The transmission and reflection coefficients through the barrier are not changed if we add a free interval of any length to the barrier. However, we must keep in mind that addition of such an interval changes the *phases* of the transmission and reflection amplitudes.

The composition relationship (1.76) can be easily generalized for the case of N barriers, resulting in the transfer matrix $\mathbf{M}$ given by

$$\mathbf{M} = \mathbf{M}_N \mathbf{M}_{N-1} \cdots \mathbf{M}_2 \mathbf{M}_1. \tag{1.85}$$

1.4.2 Propagating States

Consider a system with time-reversal symmetry. Then det $\mathbf{M} = 1$ and Tr $\mathbf{M}$ is real. The two eigenvalues λ_1 and λ_2 of the transfer matrix are related by

$$\lambda_2 = \frac{1}{\lambda_1}. \tag{1.86}$$

We will distinguish two cases. In the first case $|\lambda_1| = 1$. Then λ_1 can be written as

$$\lambda_1 = e^{iq\ell}, \tag{1.87}$$

with the wave vector q being real. As $\lambda_2 = e^{-iq\ell}$, we have Tr $\mathbf{M} = \lambda_1 + 1/\lambda_1 = 2\cos q\ell$. Note that

$$|\text{Tr } \mathbf{M}| \leq 2. \tag{1.88}$$

In the second case, we have $|\lambda_1| \neq 1$. Then we have $|\text{Tr } \mathbf{M}| = |\lambda_1 + \lambda_1^{-1}| = 2\cosh\kappa\ell > 2$. In this case, the amplitude of the transmitted wave decreases exponentially with increasing width of the potential barrier.

Thus, we conclude that equation (1.88) represents a sufficient condition for the existence of the propagating solution. Condition (1.88) is very useful in the analysis of complicated long systems. Following the composition rule (1.76) derived in section 1.4.1, we can calculate the transfer matrix as a product of transfer matrices of individual subsystems. Then, equation (1.88) allows us to determine unambiguously whether or not a given solution is propagating. In this way, we can estimate the entire spectrum of propagating solutions of the system.

1.4.3 Bound States

A one-dimensional potential well

$$V(x) < 0 \tag{1.89}$$

always has at least one bound state [11]. A bound state is characterized by a wave function that decays exponentially on both sides of the potential. We can use the transfer matrix to estimate the energy of the bound state.

The wave function of a bound state decreases exponentially for both $x > \ell$ and $x < 0$:

$$
\begin{aligned}
\Psi_R^+(x) &\propto e^{-\kappa x}, \quad x > \ell, \\
\Psi_L^-(x) &\propto e^{+\kappa x}, \quad x < 0,
\end{aligned}
\tag{1.90}
$$

where

$$q = i\kappa \quad \text{and} \quad \kappa > 0. \tag{1.91}$$

To avoid solutions that increase exponentially far from the sample, we require

$$\Psi_R^-(x) \equiv 0, \quad x > \ell, \tag{1.92}$$

and

$$\Psi_L^+(x) \equiv 0, \quad x < 0. \tag{1.93}$$

Inserting equations (1.92) and (1.93) into the transfer matrix equation (1.12), we obtain

$$\Psi_R^- = M_{21}\Psi_L^+ + M_{22}\Psi_L^-. \tag{1.94}$$

We immediately obtain the result that, for the existence of a bound state, one needs to satisfy the following equation:

$$M_{22}(i\kappa) = 0 \quad \text{for} \quad \kappa > 0. \tag{1.95}$$

The solution $\kappa_b = \kappa$ from equation (1.95) determines the energy E_b of the bound state, which is localized around the impurity. We will use this criterion to obtain different bound states for electrons and electromagnetic waves.

1.4.4 Chebyshev's Identity

A special case of the multiplication law equation (1.85) is the case when all the potential barriers are equal:

$$\mathbf{M}_1 = \mathbf{M}_2 = \cdots = \mathbf{M}_N. \tag{1.96}$$

Then the resulting transfer matrix $\mathbf{M}$ can be easily expressed in terms of the elements of the individual matrix $\mathbf{M}_1$, with the use of the Chebyshev identity [165].

Consider the transfer matrix $\mathbf{M}$,

$$\mathbf{M} = \begin{pmatrix} a & b \\ c & d \end{pmatrix}. \tag{1.97}$$

The eigenvalues λ_1 and λ_2 of the matrix $\mathbf{M}$ are

$$\lambda_1 = e^{iq\ell} \quad \text{and} \quad \lambda_2 = e^{-iq\ell}. \tag{1.98}$$

Chebyshev's identity states [39, 165] that the Nth power of the transfer matrix can be expressed as

$$\mathbf{M}^N = \begin{pmatrix} a & b \\ c & d \end{pmatrix}^N = \begin{pmatrix} a\,U_{N-1} - U_{N-2} & b\,U_{N-1} \\ c\,U_{N-1} & d\,U_{N-1} - U_{N-2} \end{pmatrix}. \tag{1.99}$$

Here, the function $U_N = U_N(q)$ is defined as

$$U_N = \frac{\sin(N+1)q\ell}{\sin q\ell}, \tag{1.100}$$

and $q\ell$ is given by the eigenvalues of the transfer matrix $\mathbf{M}$,

$$\mathrm{Tr}\,\mathbf{M} = \lambda_1 + \lambda_2 = 2\cos q\ell. \tag{1.101}$$

All the above identities are valid in the case of real q (then $|\lambda_1| = 1$) and in the case of complex $q = i\kappa$ (then $|\lambda_1| > 1$). A proof of Chebyshev's identity is given in section 1.5.2.

1.4.5 Transmission through N Identical Barriers

Chebyshev's identity allows us to derive a general expression for the transmission coefficient of N identical barriers. First, note that the transmission coefficient can be written as

$$T = |t|^2 = \frac{|t|^2}{|t|^2 + |r|^2} = \frac{1}{1 + \dfrac{|r|^2}{|t|^2}} \tag{1.102}$$

(we have used that $|t|^2 + |r|^2 = 1$). Then, comparing the matrix elements of $\mathbf{M}$ in equation (1.97) with the general expression of the transfer matrix in terms of transmission and reflection amplitudes, equation (1.60), we see that

$$M_{12} = \frac{r}{t}, \tag{1.103}$$

so that the transmission through a single barrier is

$$T_1 = \frac{1}{1 + |M_{12}|^2}. \tag{1.104}$$

Finally, from identity (1.99) we obtain the transmission for N identical barriers in the form

$$T_N = \frac{1}{1 + |M_{12}|^2 U_{N-1}^2}. \tag{1.105}$$

Using the explicit expressions for the function U_{N-1} [equation (1.100)] and for b, given by equation (1.103), we arrive at the general expression for the transmission of the particle through N identical barriers:

$$T_N = \frac{1}{1 + \left|\dfrac{r^2}{t^2}\right| \dfrac{\sin^2 Nq\ell}{\sin^2 q\ell}}, \tag{1.106}$$

where $q\ell$ is given by equation (1.101).

Relationship (1.106) plays a crucial role in the analysis of transmission through periodic systems. We only need to calculate r/t for an individual potential and we immediately obtain, with the help of equation (1.106), the transmission coefficient for any number of barriers.

1.5 Supplementary Notes

1.5.1 Current Density

We derive in this section the equation for the conservation of the particle density and show that the requirement of constant particle density in a given volume leads to the conservation of the current density.

We first multiply both sides of the Schrödinger equation

$$i\hbar \frac{\partial \Psi(x,t)}{\partial t} = -\frac{\hbar^2}{2m} \frac{\partial^2 \Psi(x,t)}{\partial x^2} + V(x)\Psi(x,t) \tag{1.107}$$

by the complex conjugate wave function Ψ^* and integrate both sides of the equation over x in the interval (x_a, x_b):

$$i\hbar \int_{x_a}^{x_b} \Psi^* \frac{\partial \Psi}{\partial t}\, dx = -\frac{\hbar^2}{2m} \int_{x_a}^{x_b} \Psi^* \frac{\partial^2 \Psi}{\partial x^2}\, dx + \int_{x_a}^{x_b} \Psi^* V(x)\Psi\, dx. \tag{1.108}$$

Then we consider the complex conjugate Schrödinger equation

$$-i\hbar \frac{\partial \Psi^*(x,t)}{\partial t} = -\frac{\hbar^2}{2m} \frac{\partial^2 \Psi^*(x,t)}{\partial x^2} + V^*(x)\Psi^*(x,t), \tag{1.109}$$

multiply it by Ψ, and again integrate from x_a to x_b,

$$-i\hbar \int_{x_a}^{x_b} \Psi \frac{\partial \Psi^*}{\partial t}\, dx = -\frac{\hbar^2}{2m} \int_{x_a}^{x_b} \Psi \frac{\partial^2 \Psi^*}{\partial x^2}\, dx + \int_{x_a}^{x_b} \Psi V^*(x)\Psi^*\, dx. \tag{1.110}$$

Now we subtract equation (1.110) from equation (1.108):

$$i\hbar \int_{x_a}^{x_b} \left[\Psi^* \frac{\partial \Psi}{\partial t} + \Psi \frac{\partial \Psi^*}{\partial t} \right] dx = -\frac{\hbar^2}{2m} \int_{x_a}^{x_b} \left[\Psi^* \frac{\partial^2 \Psi}{\partial x^2} - \Psi \frac{\partial^2 \Psi^*}{\partial x^2} \right] dx$$
$$+ \int_{x_a}^{x_b} \left[\Psi^* V(x)\Psi - \Psi V^*(x)\Psi^* \right] dx. \tag{1.111}$$

Next, we use the following identities:

$$\Psi \frac{\partial \Psi^*}{\partial t} + \Psi^* \frac{\partial \Psi}{\partial t} = \frac{\partial}{\partial t}[\Psi^* \Psi] \tag{1.112}$$

and

$$\Psi \frac{\partial^2 \Psi^*}{\partial x^2} - \Psi^* \frac{\partial^2 \Psi}{\partial x^2} = \frac{\partial}{\partial x} \left[\Psi \frac{\partial \Psi^*}{\partial x} - \Psi^* \frac{\partial \Psi}{\partial x} \right] = \frac{2m}{\hbar i} \frac{\partial}{\partial x} j(x), \tag{1.113}$$

where $j(x)$ is the current density at the point x:

$$j(x) = \frac{\hbar i}{2m} \left[\Psi \frac{\partial \Psi^*}{\partial x} - \Psi^* \frac{\partial \Psi}{\partial x} \right]. \tag{1.114}$$

Finally, we use the identity

$$\int_{x_b}^{x_a} \frac{\partial j(x)}{\partial x} dx = j(x_b) - j(x_a).$$ (1.115)

Inserting relations (1.112)–(1.115) into equation (1.111), we obtain the final result

$$\frac{\partial}{\partial t} \int_{x_a}^{x_b} \Psi^* \Psi \, dx = \left[j(x_b) - j(x_a) \right]$$
$$+ \frac{1}{i\hbar} \int_{x_a}^{x_b} \left[\Psi^* V(x) \Psi - \Psi V^*(x) \Psi^* \right] \, dx.$$ (1.116)

The physical interpretation of equation (1.116) is very simple: the term on the l.h.s. determines the change of the density of electrons in the region (x_a, x_b) versus time. This change is due either to the flux of the particle inside or outside the region (given by the first term on the r.h.s.), or to the creation (or annihilation) of particles inside the region (the last term on the r.h.s.). Note that the last term

$$\int_{x_a}^{x_b} \left[\Psi^* V(x) \Psi - \Psi V^*(x) \Psi^* \right] \, dx$$ (1.117)

is zero if the potential $V(x)$ is real:

$$V(x) = V^*(x).$$ (1.118)

The case of complex potentials corresponds to systems with absorption or gain. In this book, we will treat only real potentials.

If the last term in equation (1.116) is zero, then equation (1.116) reduces to

$$\frac{\partial}{\partial t} \int_{x_a}^{x_b} \Psi^* \Psi \, dx = j(x_b) - j(x_a).$$ (1.119)

Since we study only time-independent problems in this book, the density of particles in any region (x_a, x_b) does not change in time. Therefore the left-hand side of equation (1.116) is zero:

$$\frac{\partial}{\partial t} \int_{x_a}^{x_b} \Psi^* \Psi \, dx = 0.$$ (1.120)

This means that the number of particles in the region (x_a, x_b) does not change with time. equation (1.116) reduces to

$$j(x_a) = j(x_b),$$ (1.121)

which represents the conservation of the flux: if the number of particles inside a given region is constant, then the current flowing inward to this region must be equal to the current flowing outward.

Finally, we can express the current density for the case of a plane wave,

$$\Psi(x) = A e^{ikx}.$$ (1.122)

Inserting this into equation (1.114), we obtain the result that the current density of a plane wave is

$$j = \frac{\hbar k}{m}|\Psi|^2 = \frac{\hbar k}{m}|A|^2. \tag{1.123}$$

The current density is proportional to the wave vector k (the velocity of the particle) and to the probability density $|\Psi|^2$. Note that the current j indeed does not depend on x.

1.5.2 Proof of the Chebyshev Identity

The Chebyshev identity is used in section 1.4.4 to derive useful relations for the matrix elements of the Nth power of the transfer matrix $\mathbf{M}$. The Chebyshev identity is formulated as follows:

Consider a matrix $\mathbf{M}$

$$\mathbf{M} = \begin{pmatrix} a & b \\ c & d \end{pmatrix}. \tag{1.124}$$

Its eigenvalues λ_1 and λ_2 are given by

$$\lambda_1 = e^{iq\ell} \quad \text{and} \quad \lambda_2 = e^{-iq\ell}. \tag{1.125}$$

The Nth power of the matrix $\mathbf{M}$ can be expressed as

$$\mathbf{M}^N = \begin{pmatrix} a & b \\ c & d \end{pmatrix}^N = \begin{pmatrix} a\,U_{N-1} - U_{N-2} & b\,U_{N-1} \\ c\,U_{N-1} & d\,U_{N-1} - U_{N-2} \end{pmatrix}. \tag{1.126}$$

Here, the function $U_N = U_N(q)$ is defined as

$$U_N = \frac{\sin(N+1)q\ell}{\sin q\ell}, \tag{1.127}$$

where $q\ell$ is given by the eigenvalues of the transfer matrix $\mathbf{M}$ [equation (1.87)], and satisfies the relation

$$\mathrm{Tr}\,\mathbf{M} = \lambda_1 + \lambda_2 = 2\cos q\ell. \tag{1.128}$$

All the above identities are valid when q is real (then $|\lambda_1| = 1$) and when q is complex (then $|\lambda_1| > 1$).

We will prove the Chebyshev identity given by equation (1.126), by mathematical induction.

First, note that the identity (1.126) is satisfied for $N = 1$. Indeed, from equation (1.127) we have $U_0 \equiv 1$ and $U_{-1} \equiv 0$.

Next, assume that the identity (1.126) is valid for some $N \geq 1$. We show that then it is valid also for $N + 1$. To do so, we express

$$
\mathbf{M}^{N+1} = \mathbf{M}\mathbf{M}^N = \mathbf{M} \begin{pmatrix} a\,U_{N-1} - U_{N-2} & b\,U_{N-1} \\ c\,U_{N-1} & d\,U_{N-1} - U_{N-2} \end{pmatrix}
$$

(1.129)

$$
= \begin{pmatrix} (a^2 + bc)\,U_{N-1} - a\,U_{N-2} & b[(a+d)\,U_{N-1} - U_{N-2}] \\ c[(a+d)\,U_{N-1} - U_{N-2}] & (d^2 + bc)\,U_{N-1} - d\,U_{N-2} \end{pmatrix}.
$$

We calculate the matrix element $(M^{N+1})_{11}$:

$$
\begin{aligned}
(a^2 + bc)\,U_{N-1} - a\,U_{N-2} &= [a(a+d) - ad + bc]U_{N-1} - a\,U_{N-1} \\
&= a[(a+d)\,U_{N-1} - U_{N-2}] - U_{N-1} \\
&= a\,U_n - U_{N-1},
\end{aligned}
$$

(1.130)

where we have used the fact that $a + d = 2\cos q\ell$ and $ad - bc = \det \mathbf{M} = 1$. We also used the identity

$$
U_N \equiv 2\cos q\ell \; U_{N-1} - U_{N-2},
$$

(1.131)

which can be verified with the use of the trigonometric relations (the proof is given in problem 1.4). All the other matrix elements of the matrix $\mathbf{M}$ can be calculated in the same way. Finally, we derive

$$
\mathbf{M}^{N+1} = \begin{pmatrix} a\,U_N - U_{N-1} & b\,U_N \\ c\,U_N & d\,U_N - U_{N-1} \end{pmatrix},
$$

(1.132)

which is the relation obtained from (1.126) by the substitution $N \to N + 1$.

Starting with $N = 1$, we have just proven that relation (1.126) holds also for $N = 2$. Then, starting with $N = 2$, we find that (1.126) holds also for $N = 3$. By induction. we conclude that (1.126) is valid for any integer N.

Another Proof of the Chebyshev Identity

Another way to prove the Chebyshev identity (1.126) is the following. To get the Nth power of the matrix $\mathbf{A}$, we first find its eigenvalues and eigenvectors. We write

$$
\mathbf{M} = \mathbf{R} \begin{pmatrix} \lambda_1 & 0 \\ 0 & \lambda_2 \end{pmatrix} \mathbf{L}
$$

(1.133)

where λ_1 and λ_2 are eigenvalues, and $\mathbf{L}$ ($\mathbf{R}$) is the matrix of left (right) eigenvectors, respectively. They can be found with the help of the formulas given below in appendix A.

Then it is easy to find that

$$\mathbf{M}^N = \mathbf{R} \begin{pmatrix} \lambda_1 & 0 \\ 0 & \lambda_2 \end{pmatrix} \mathbf{L}\mathbf{R} \begin{pmatrix} \lambda_1 & 0 \\ 0 & \lambda_2 \end{pmatrix} \mathbf{L} \cdots = \mathbf{R} \begin{pmatrix} \lambda_1^N & 0 \\ 0 & \lambda_2^N \end{pmatrix} \mathbf{L}, \qquad (1.134)$$

since $\mathbf{L}\mathbf{R} = 1$. Next, one easily finds that

$$\lambda_{1,2} = e^{\pm iq\ell} \qquad (1.135)$$

with q given by (1.101). After some algebra (do it!), one finds that $\mathbf{M}^N$ is given by equation (1.99).

1.5.3 Quasi-One-Dimensional Systems

The transfer matrix can be easy generalized for the case of *quasi-one-dimensional* systems, which are finite in the direction perpendicular to the direction of propagation of the particle, and infinite in the x direction [29,47,103,52,115,131,118]. A particle propagates along the x direction, but, due to the finite size of the system in the transverse direction, it possess also transverse momentum $k_\perp$. The energy of the particle is then

$$E = \frac{\hbar^2}{2m}[k_\parallel^2 + k_\perp^2]. \qquad (1.136)$$

In the one-dimensional case $k_\perp \equiv 0$ and $k = \sqrt{k_\parallel^2 + k_\perp^2} = k_\parallel$. In quasi-one-dimensional systems, different values of $k_\perp$ are allowed, as determined by the structure of the system. If there are N allowed values of $k_{\perp n}$, $n = 1, 2, \ldots, N$, then the functions $\Psi_L^+(x)$ and $\Psi_L^-(x)$ consist of superpositions of N plane waves,

$$\Psi_L^+(x) = \sum_{n=1}^N A_n(z)e^{+ik_{\perp n}x},$$
$$\Psi_L^-(x) = \sum_{n=1}^N B_n(z)e^{-ik_{\perp n}x}. \qquad (1.137)$$

The right wave function $\Psi_R^+(x)$ and $\Psi_R^-(x)$ can be expressed in a similar way. The wave vectors $k_{\perp n}$ are given by the boundary conditions in the transverse direction. Together with the energy E they determine the nth value of $k_\parallel$.

We can introduce the $N \times N$ matrices of the transmission amplitudes t and t'. Their matrix elements t_{nm} and t'_{nm} give the transmission amplitude of the process in which an electron passes through the sample from the channel n on the left side of the sample to the channel m on the right side of the sample. In the same way, we introduce $N \times N$ matrices of the reflection amplitudes r and r'. Then the transfer matrix can be expressed as a $2N \times 2N$ matrix,

$$\mathbf{M} = \begin{pmatrix} t' - rt^{-1}r' & rt^{-1} \\ -t^{-1}r' & t^{-1} \end{pmatrix}. \qquad (1.138)$$

Note that $\mathbf{M}$ is formally identical to the one-dimensional transfer matrix given by equation (1.60). However, in this case we have to take care about the order of the matrices in the matrix products. For instance, $rt^{-1} \neq t^{-1}r$, since r and t are noncommutative matrices.

The transmission coefficient T is given as

$$T = \mathrm{Tr}\, t^\dagger t = \sum_{nm}^{N} t_{nm}^* t_{nm} \tag{1.139}$$

where t is the $N \times N$ transmission matrix. More detailed information about the transmission properties of the system can be obtained if one measures also the following transmission parameters [83, 151, 156]:

$$T_{nm} = |t_{nm}|^2 \tag{1.140}$$

and

$$T_n = \sum_m T_{nm}. \tag{1.141}$$

The matrix elements T_{nm} define the transmission amplitude from the state with $k_{\parallel n}$ to the state $k_{\parallel m}$. The transmission coefficient T_n is the transmission through the sample from the state n to all possible states m on the other side of the sample.

If there is no absorption in the system (the potential V is real) then the reflection R can be found from the conservation of the current density,

$$R = N - T \tag{1.142}$$

Note that relations (1.36) and (1.50) are valid also for the general case $N > 1$. The 2×2 matrices

$$\begin{pmatrix} 1 & 0 \\ 0 & -1 \end{pmatrix} \quad \text{and} \quad \begin{pmatrix} 0 & 1 \\ 1 & 0 \end{pmatrix} \tag{1.143}$$

are replaced by $2N \times 2N$ matrices

$$\begin{pmatrix} 1 & 0 \\ 0 & -1 \end{pmatrix} \quad \text{and} \quad \begin{pmatrix} 0 & 1 \\ 1 & 0 \end{pmatrix}, \tag{1.144}$$

where 1 is a unity $N \times N$ matrix.

1.6 Problems

Problems with Solutions

Problem 1.1
Derive the relationships (1.13) and (1.14) between the transfer matrix and scattering matrix.

Solution. The matrix equations (1.11) and (1.12) can be written explicitly as

$$\Psi_L^- = S_{11}\Psi_L^+ + S_{12}\Psi_R^-, \quad \Psi_R^+ = S_{21}\Psi_L^+ + S_{22}\Psi_R^- \tag{1.145}$$

and

$$\Psi_R^+ = M_{11}\Psi_L^+ + M_{12}\Psi_L^-, \quad \Psi_R^- = M_{21}\Psi_L^+ + M_{22}\Psi_L^-. \tag{1.146}$$

We express Ψ_R^- from the first equation (1.145):

$$\Psi_R^- = \frac{1}{S_{12}}\Psi_L^- - \frac{S_{11}}{S_{12}}\Psi_L^+ \tag{1.147}$$

and insert it into the second equation (1.145). We obtain

$$\Psi_R^+ = \left[S_{21} - \frac{S_{11}S_{22}}{S_{12}} \right]\Psi_L^+ + \frac{S_{22}}{S_{12}}\Psi_L^-. \tag{1.148}$$

Now, compare equations (1.146) with equations (1.147) and (1.148) and get

$$M_{11} = S_{21} - \frac{S_{11}S_{22}}{S_{12}}, \quad M_{12} = \frac{S_{22}}{S_{12}}$$

$$M_{21} = -\frac{S_{11}}{S_{12}}, \quad M_{22} = \frac{1}{S_{12}}. \tag{1.149}$$

The relations (1.149) are equivalent to the matrix equation (1.13). In the same way, we can express elements of the scattering matrix $\mathbf{S}$ in terms of elements of the transfer matrix $\mathbf{M}$ to obtain expression (1.14).

Problem 1.2
Prove that $|\det \mathbf{S}| = 1$ [equation (1.27)].

Solution. Since the determinant of the product of two matrices $\mathbf{AB}$ equals the product of their determinants,

$$\det \mathbf{AB} = \det \mathbf{A} \det \mathbf{B}, \tag{1.150}$$

we have for a unitary matrix $\mathbf{S}$ that

$$\det \mathbf{S} \det \mathbf{S}^\dagger = 1. \tag{1.151}$$

On the other hand,

$$\det \mathbf{S}^\dagger = S_{11}^* S_{22}^* - S_{12}^* S_{21}^* = [\det \mathbf{S}]^*. \tag{1.152}$$

By combining the previous equations we obtain

$$|\det \mathbf{S}| = 1. \tag{1.153}$$

Problem 1.3
Derive the expression (1.60) for the transfer matrix.

Solution. From the definition of the scattering matrix, equation (1.58), we have

$$\begin{pmatrix} \Psi_L^- \\ \Psi_R^+ \end{pmatrix} = \begin{pmatrix} r' & t \\ t' & r \end{pmatrix} \begin{pmatrix} \Psi_L^+ \\ \Psi_R^- \end{pmatrix}, \tag{1.154}$$

which can be written in the form

$$\Psi_L^- = r'\Psi_L^+ + t\Psi_R^-, \quad \Psi_R^+ = t'\Psi_L^+ + r\Psi_R^-. \tag{1.155}$$

From the first equation (1.155) we express $\Psi_R^- = t^{-1}\Psi_L^- - t^{-1}r'\Psi_L^+$. Inserting this expression into the second equation (1.155), we obtain

$$\begin{pmatrix} \Psi_R^+ \\ \Psi_R^- \end{pmatrix} = \begin{pmatrix} t' - rt^{-1}r' & rt^{-1} \\ -t^{-1}r' & t^{-1} \end{pmatrix} \begin{pmatrix} \Psi_L^+ \\ \Psi_L^- \end{pmatrix}. \tag{1.156}$$

Comparing this with the definition of transfer matrix, equation (1.12), we obtain expression (1.60).

Problem 1.4
Prove the identity (1.131).

Solution. To prove the relation (1.131), we start from the definition of the function U_N, given by equation (1.127), and use the relation $\sin(x \pm y) = \sin x \cos y \pm \cos x \sin y$. We obtain

$$U_N(q\ell) = \frac{\sin(N+1)q\ell}{\sin q\ell} = \frac{\sin Nq\ell \cos q\ell + \cos Nq\ell \sin q\ell}{\sin q\ell} \tag{1.157}$$

and

$$U_{N-2}(q\ell) = \frac{\sin(N-1)q\ell}{\sin q\ell} = \frac{\sin Nq\ell \cos q\ell - \cos Nq\ell \sin q\ell}{\sin q\ell}. \tag{1.158}$$

Now we sum both equations to obtain

$$U_N + U_{N-2} = \frac{2\cos q\ell \sin Nq\ell}{\sin q\ell} = 2\cos q\ell U_n, \tag{1.159}$$

which is already the required identity, equation (1.131).

Problems without Solutions

Problem 1.5
We can also define the transfer matrix $\tilde{\mathbf{M}}$ by the relation

$$\begin{pmatrix} \Psi_L^+(x) \\ \Psi_L^-(x) \end{pmatrix} = \tilde{\mathbf{M}} \begin{pmatrix} \Psi_R^+(x+\ell) \\ \Psi_R^-(x+\ell) \end{pmatrix}. \tag{1.160}$$

$\tilde{\mathbf{M}}$ expresses the wave function on the *left* side of the potential region in terms of the wave function on the *right* side. Show that $\tilde{\mathbf{M}} = \mathbf{M}^{-1}$. Derive the explicit form of the transfer matrix $\tilde{\mathbf{M}}$:

$$\tilde{\mathbf{M}} = \begin{pmatrix} \dfrac{1}{S_{21}} & -\dfrac{S_{22}}{S_{21}} \\ \dfrac{S_{11}}{S_{21}} & S_{12} - \dfrac{S_{11}S_{22}}{S_{21}} \end{pmatrix}. \tag{1.161}$$

Problem 1.6

Using the multiplication law for transfer matrices, equation (1.76), as well as the physical arguments explained in section 1.4.1, derive the composition laws for the transmission amplitude t'_{12} and reflection amplitude r_{12} of a system that consists of two samples.

Problem 1.7

Write the transmission and reflection amplitudes as follows:

$$t = |t|e^{i\phi_t}, \qquad r = |r|e^{i\phi_r}. \tag{1.162}$$

In section 1.4.1 we showed that the addition of a free space to the sample does not change transmission and reflection coefficients [equations (1.83) and (1.84)]. Show how the additional free space influences the phases ϕ_t and ϕ_r. Analysis of the phases of the transmission and reflection amplitudes is very important for inverse problems (see, for instance, section 2.6).

Problem 1.8

Repeat the analysis of section 1.4.3 with $\kappa < 0$. Show that in this case the bound state is given as a solution of the equation

$$M_{11}(i\kappa) = 0 \quad \text{for} \quad \kappa < 0. \tag{1.163}$$

2 Rectangular Potentials

The rectangular potential barrier, as shown in figure 2.1, represents one of the simplest quantum mechanical problems. We will use our transfer matrix formalism, developed in the previous chapter, to determine the transmission and reflection coefficients. Our transfer matrix results will be compared with those obtained with more traditional methods [7,15,25,30]. We will show that the transfer matrix is easy to use and can be readily extended to more complicated shapes of potentials and to disordered systems.

Schrödinger's equation is given by

$$-\frac{\hbar^2}{2m}\frac{\partial^2 \Psi}{\partial x^2} + \left[V(x) - E\right]\Psi = 0, \tag{2.1}$$

with a potential

$$V(x) = \begin{cases} 0, & x < -a, \\ V_0, & -a < x < a, \\ 0, & a < x, \end{cases} \tag{2.2}$$

and can be solved analytically; the solution can be found in many textbooks of quantum mechanics. We derive analytical formulas for the transmission and reflection coefficients for a particle approaching the potential either from the left or from the right, and find the transmission and reflection.

For the case of a potential barrier ($V_0 > 0$), it will be very useful to distinguish cases with $E > V_0$ and $E < V_0$. In the latter case, only tunneling through the barrier is possible.

For the case of a potential well ($V_0 < 0$), we calculate both the transmission coefficient (if $E > 0$) and the bound states for energies $V_0 < E < 0$. In the latter case, there is always at least one bound state when $E < 0$, and the wave function is spatially localized.

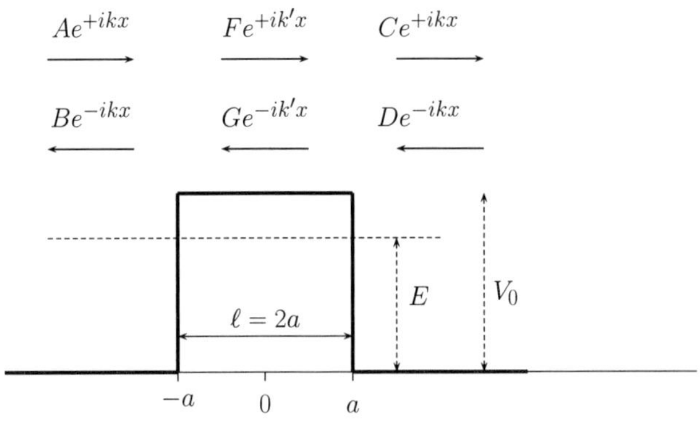

FIGURE 2.1. Rectangular potential barrier of width $\ell = 2a$ and height V_0. The solution of Schrödinger's equation consists of plane waves propagating either to the left or to the right.

When the energy $E > V_0$, we find special energies for which the transmission $T = 1$. That means that the barrier is totally transparent for these energies. Transparency of the barrier and well is due to the wave character of the electron, which is a quantum particle.

The methods developed and results obtained could easily be generalized for the analysis of the transmission of electromagnetic waves through thin slabs of various materials, which will be discussed in chapter 10.

We solve in section 2.6 an inverse problem. Suppose we obtained from experiment the results for the transmission and reflection. Can we use these data to reconstruct the scattering potential V_0 ?

Chapter 8 is related to the present chapter. We show there how to solve Schrödinger's equation (2.1) numerically. Of course, a numerical solution is not necessary for problems that can be solved exactly, but a comparison of the analytical and numerical results allows us to estimate the accuracy of the numerical procedure. This is important because many of the interesting scattering problems cannot be solved analytically.

2.1 Transfer Matrix

The solution of Schrödinger's equation (2.1) in three different regions shown in figure 2.1 can be expressed as a superposition of plane waves:

$$\Psi(x) = \begin{cases} \Psi_{\mathrm{L}}(x) = Ae^{ikx} + Be^{-ikx}, & x < -a, \\ \Psi_i(x) = Fe^{ik'x} + Ge^{-ik'x}, & -a < x < a, \\ \Psi_{\mathrm{R}}(x) = Ce^{ikx} + De^{-ikx}, & a < x. \end{cases} \tag{2.3}$$

Here k is given as

$$k = \sqrt{2mE}/\hbar \tag{2.4}$$

and

$$k' = \sqrt{2m(E - V_0)}/\hbar. \tag{2.5}$$

Note that k' is *real* for $E > V_0$ and *imaginary* for $E < V_0$. If $E > V_0$, then $k' < k$ in the case of a potential barrier ($V_0 > 0$) and $k' > k$ in the case of a potential well ($V_0 < 0$).

It is useful to characterize the potential by a *dimensionless* parameter β:

$$\beta^2 = \frac{2ma^2|V_0|}{\hbar^2}. \tag{2.6}$$

Using the dimensionless parameter β, we can express

$$2ka = 2\beta\sqrt{\frac{E}{V_0}} \tag{2.7}$$

and

$$2k'a = \begin{cases} 2\beta\sqrt{E/V_0 - 1}, & V_0 > 0, \\ 2\beta\sqrt{E/|V_0| + 1}, & V_0 < 0. \end{cases} \tag{2.8}$$

We introduce also the electron wavelength λ by the equation

$$k = \frac{2\pi}{\lambda}, \tag{2.9}$$

so that $2ka$ in equation (2.7) determines the ratio of the barrier width to the electron wavelength *outside* the barrier. Similarly, $2k'a$ gives the same ratio *inside* the barrier.

According to our definition of the transfer matrix, equation (1.12), the wave functions on both sides of the potential are connected by the equation

$$\begin{pmatrix} Ce^{+ika} \\ De^{-ika} \end{pmatrix} - \mathbf{M} \begin{pmatrix} Ae^{-ika} \\ Be^{+ika} \end{pmatrix}, \tag{2.10}$$

or, written in explicit form,

$$\begin{aligned} Ce^{ika} &= M_{11} Ae^{-ika} + M_{12} Be^{ika}, \\ De^{-ika} &= M_{21} Ae^{-ika} + M_{22} Be^{ika}. \end{aligned} \tag{2.11}$$

To find the elements of the transfer matrix, we need to express coefficients F, G, C, and D, defined in equation (2.3) in terms of A and B. To do so, we use the continuity of the wave function and its first derivative at the boundaries of the potential barrier, $x = -a$ and $x = a$. We obtain four linear relations for coefficients A–G, two for $x = -a$,

$$\begin{aligned} e^{-ika} A + e^{+ika} B &= e^{-ik'a} F + e^{+ik'a} G, \\ ke^{-ika} A - ke^{+ika} B &= k'e^{-ik'a} F - k'e^{+ik'a} G, \end{aligned} \tag{2.12}$$

and the other two for $x = a$,

$$
e^{+ika}C + e^{-ika}D = e^{+ik'a}F + e^{-ik'a}G,
$$

$$
ke^{+ika}C - ke^{-ika}D = k'e^{+ik'a}F - k'e^{-ik'a}G.
$$

(2.13)

We multiply the first equation of (2.12) by k'. Then the sum and the difference of both equations (2.12) give us the amplitudes F and G in terms of A and B:

$$
2k'e^{-ik'a}F = (k' + k)e^{-ika}A + (k' - k)e^{+ika}B,
$$

$$
2k'e^{+ik'a}G = (k' - k)e^{-ika}A + (k' + k)e^{+ika}B.
$$

(2.14)

In the same way we multiply the first equation (2.13) by k. The sum and the difference of both equations (2.13) give us the amplitudes C and D in terms of F and G:

$$
2ke^{+ika}C = (k + k')e^{+ik'a}F + (k - k')e^{-ik'a}G,
$$

$$
2ke^{-ika}D = (k - k')e^{+ik'a}F + (k + k')e^{-ik'a}G.
$$

(2.15)

Next, we insert F and G from equations (2.14) into equations (2.15) and obtain

$$
Ce^{ika} = M_{11}Ae^{-ika} + M_{12}Be^{ika},
$$

$$
De^{-ika} = M_{21}Ae^{-ika} + M_{22}Be^{ika}
$$

(2.16)

with the explicit form of the matrix elements of $\mathbf{M}$,

$$
M_{11} = \cos 2k'a + \frac{i}{2}\left(\frac{k}{k'} + \frac{k'}{k}\right)\sin 2k'a,
$$

$$
M_{12} = \qquad + \frac{i}{2}\left(\frac{k'}{k} - \frac{k}{k'}\right)\sin 2k'a,
$$

(2.17)

and

$$
M_{22} = \cos 2k'a - \frac{i}{2}\left(\frac{k}{k'} + \frac{k'}{k}\right)\sin 2k'a,
$$

$$
M_{21} = \qquad - \frac{i}{2}\left(\frac{k'}{k} - \frac{k}{k'}\right)\sin 2k'a.
$$

(2.18)

Equations (2.17) and (2.18) determine our transfer matrix (2.10). We can easily verify that $\mathbf{M}$ satisfies the requirement of the conservation of the current, equation (1.36). Also, we see that

$$
M_{22} = M_{11}^*, \qquad M_{21} = M_{12}^*,
$$

(2.19)

so that the transfer matrix has the form given by equation (1.51):

$$
\mathbf{M} = \begin{pmatrix} M_{11} & M_{12} \\ M_{12}^* & M_{11}^* \end{pmatrix}.
$$

(2.20)

This symmetry of the transfer matrix $\mathbf{M}$ follows from the symmetry of our model, defined by equation (2.1), with respect to time reversal, defined by equation (1.50).

From the expressions for the elements of the transfer matrix, equations (2.17) and (2.18), we immediately see that for the case of $V_0 = 0$ ($k' = k$) the transfer matrix $\mathbf{M}$ becomes a diagonal matrix given by

$$\mathbf{M} = \begin{pmatrix} e^{+i2ka} & 0 \\ 0 & e^{-i2ka} \end{pmatrix}, \tag{2.21}$$

as should be the case for a free electron propagating a distance of $2a$.

2.2 Transmission Coefficient: $E > V_0$

The transmission coefficient $T = |t|^2$ can be obtained either from the scattering matrix $\mathbf{S}$, equation (1.58), or from the transfer matrix, equation (1.60), as

$$T = |t|^2 = |S_{12}|^2 = \frac{1}{|M_{22}|^2}. \tag{2.22}$$

Using the expression for M_{22} given in equation (2.18), we obtain that T is equal to

$$T = \frac{1}{1 + \frac{1}{4}\left[\dfrac{k}{k'} - \dfrac{k'}{k}\right]^2 \sin^2 2k'a}. \tag{2.23}$$

From equations (2.7) and (2.8), we see that the wave vectors k and k' are related as

$$(ka)^2 - (k'a)^2 = \beta^2, \qquad E > V_0 > 0, \tag{2.24}$$

and

$$(k'a)^2 - (ka)^2 = \beta^2, \qquad E > 0 > V_0. \tag{2.25}$$

We can also derive the transmission coefficient T by using the expression given by equation (1.102),

$$T = \frac{1}{1 + \left|\dfrac{r}{t}\right|^2} = \frac{1}{1 + |M_{12}|^2}, \tag{2.26}$$

where we use the fact that $|r/t| = |M_{12}|$, as discussed in equation (1.60). By substituting the expression for M_{12} obtained in equation (2.17) into equation (2.26), we derive the same expression for T as the one given by equation (2.23).

Note that formula (2.22) is applicable to both V_0 positive and V_0 negative, *provided that* $E > V_0$. The only difference between these two cases is that $k' < k$ for the case of the potential barrier ($V_0 > 0$), while $k' > k$ for the case of the potential well ($V_0 < 0$). This can be easily verified from equation (2.5) or equations (2.24) and (2.25).

2.2.1 Resonant Transmission

For any rectangular potential barrier, there are special energies for the transmission coefficient $T = 1$. Indeed, from equation (2.23) we see that $T = 1$ when

$$\sin 2k'a = 0. \tag{2.27}$$

The last condition means that

$$2k'a = n\pi, \tag{2.28}$$

where $n = 1, 2, \ldots$. The physical meaning of the condition (2.28) is more obvious when it is expressed in terms of the wavelength $\lambda' = 2\pi/k'$. We obtain that the transmission coefficient is equal to one when the barrier width $\ell = 2a$ equals an integer multiple of half of the wavelength of the electron *inside* the barrier,

$$\ell = \frac{\lambda'}{2}n, \quad n = 1, 2, \ldots. \tag{2.29}$$

The maxima in the transmission occur when the distance 2ℓ that a particle travels *inside* the potential barrier equals an integer number of wavelengths.

We will derive a similar formula in chapter 10 when we analyze the transmission of electromagnetic waves through a dielectric slab. The resonant condition (2.29) plays an important role in the theory of Fabry-Pérot resonators.

To estimate the values of energy for which $T = 1$, we insert equation (2.28) into equation (2.8) and obtain

$$\frac{E_n}{V_0} = 1 + \frac{\pi^2 n^2}{4\beta^2} \quad \text{for} \quad V_0 > 0 \tag{2.30}$$

and

$$\frac{E_n}{|V_0|} = \frac{\pi^2 n^2}{4\beta^2} - 1 \quad \text{for} \quad V_0 < 0. \tag{2.31}$$

While n acquires all values $1, 2, \ldots$ in the case of $V_0 > 0$, n must be larger than a certain minimum value $n_{\min}^2 \geq 4\beta^2/\pi^2$ when $V_0 < 0$, because the energy E_n given by equation (2.31) must be positive.

The conditions (2.30) and (2.31) can be satisfied for any rectangular potential, independent of its height V_0 and width ℓ. What is important is not the parameters of the potential themselves, but the ratio λ'/ℓ of the wavelength of the particle *inside* the barrier, λ', to the barrier width $\ell = 2a$.

Note that, although $\sin 2k'a = 0$ also for $k' = 0$, the transmission coefficient T is less than one in this case. This is because there is also the term proportional to $(k/k') \sin 2k'a$ in the denominator of (2.23) and

$$\lim_{k' \to 0} \frac{\sin 2k'a}{k'} = 2a. \tag{2.32}$$

In this special case, we obtain a very simple expression for the transmission coefficient,

$$T = \frac{1}{1 + k^2 a^2} = \frac{1}{1 + \beta^2}, \qquad E = V_0. \tag{2.33}$$

From equation (2.5) we see that the case $k' = 0$ corresponds to the energy $E = V_0$.

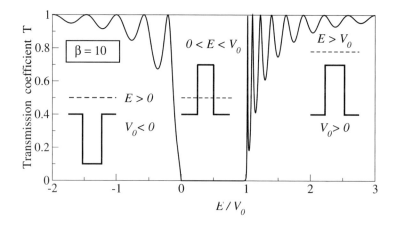

FIGURE 2.2. Transmission coefficient T as a function of dimensionless energy E/V_0 for a rectangular potential with $\beta = 10$. Negative values of E/V_0 correspond to a potential well with $V_0 < 0$, and positive values of E/V_0 correspond to a potential barrier with $V_0 > 0$. In the interval $0 < E/V_0 < 1$, only tunneling through the barrier is possible. The transmission coefficient is very small in this interval because β is large.

In figure 2.2 we plot the transmission coefficient as a function of dimensionless energy E/V_0 for $\beta = 10$. In agreement with our analytical results, we observe a set of transmission peaks $T = 1$ for both the potential barrier and the potential well. In both cases, the special energies where the transmission coefficient $T = 1$ are clearly visible. For $E/V_0 > 1$, these maxima correspond to energies E_n given by equation (2.30) with $n = 1, 2, \ldots$. In particular, the first maximum appears for $E_1 = 1.024674\,V_0$. For the case of a potential well, $V_0 < 0$, the first maximum in the transmission coefficient appears at $E = -0.209\,V_0$, which corresponds to the energy E_n given by equation (2.31) with $n = 7$.

To see how the parameters of the potential influence the transmission, we plot in figure 2.3 the transmission coefficient T for four values of the parameter β. Since $\beta^2 = 2ma^2|V_0|/\hbar^2$ [equation (2.6)], small values of β correspond either to a narrow barrier or to small potential strengths V_0. In the limit of $\beta \to 0$, both the potential barrier and the potential well are almost transparent for any energy of the quantum particle, as shown in the upper left panel of figure 2.3. Only when the energy of the electron is very small, $E \ll |V_0|$, does the particle reflect back. When β is increased, the transmission becomes smaller also for higher energies. We remind the reader that, for $E = V_0$, the transmission coefficient is $T = 1/(1 + \beta^2)$, as given by equation (2.33).

2.2.2 Wave Function

To calculate the wave function in all three regions given by equation (2.3), we need to express the different coefficients A–G. We will consider only the case when an electron is coming from the left. Then, we have $D = 0$ because no wave is coming from the right-hand side of the barrier. The wave on the left-hand side is then given by the superposition of the incoming wave Ae^{ikx} and the reflected wave Be^{-ikx}. On the right-hand side of the barrier, we have only a wave which travels through the barrier, $\Psi_R = Ce^{ikx}$. For a given

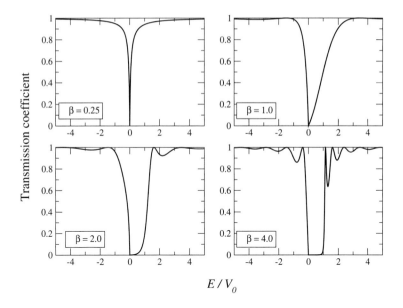

FIGURE 2.3. Transmission coefficient as a function of E/V_0 for four values of $\beta = \sqrt{2ma^2 V_0/\hbar^2}$. The small values of β correspond to a very narrow barrier in comparison with the electron wavelength ($\ell/\lambda \ll 1$; see also figure 2.9). As in figure 2.2, we plot both the cases of the potential well ($E/V_0 < 0$) and of the potential barrier ($E/V_0 > 0$). In the interval of energies $E/V_0 < 1$, only tunneling through the barrier is possible. The special case of $E = V_0$ gives $k' = 0$, and the transmission coefficient $T = 1/[1 + (ka)^2] = 1/[1 + \beta^2]$.

rectangular potential and a given energy E, the coefficients A, B, F, and G can be easily calculated with respect to C by using the formulas presented in section 2.1. For instance, we can express the coefficients F and G from equations (2.13) as follows:

$$F = \frac{k' + k}{2k'} e^{+i(k-k')a} C,$$

$$G = \frac{k' - k}{2k'} e^{+i(k+k')a} C. \tag{2.34}$$

The coefficients A and B can be found from equations (2.12). In this way, we obtain all coefficients with respect to C. The free parameter C is given by the requirement of normalization of the wave function.

Once the coefficients A, B, F, G, and C are obtained for a given energy E and a given potential V_0, we substitute them into equation (2.3) and obtain the length dependence of $\Psi(x)$ in the three regions. In figure 2.4 we present the absolute value of $\Psi(x)$ and the real part of $\Psi(x)$ for the case of resonant transmission. Notice that the width of the potential barrier is indeed an integer number of half wavelengths of the quantum particle. In figure 2.4 one clearly sees that, for $n = 1$, $\ell = 2a = \lambda'/2$, while for $n = 5$, $\ell = 2a = 5\lambda'/2$. Outside the regions of the potential barrier, the absolute value $|\Psi(x)| \equiv 1$, which means that $T = 1$. Indeed, in this special case the reflection is zero so that not only $D = 0$, but also $B = 0$. The wave function outside the potential region is $\Psi_R = \Psi_L = e^{ikx}$ and its absolute value $|\Psi_R| = |\Psi_L| = 1$. In figure 2.4 we also plot the real part of $\Psi(x)$. We see

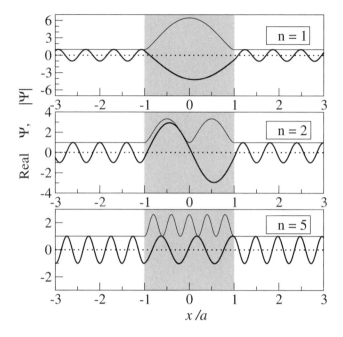

FIGURE 2.4. Real part of the wave function Ψ of an electron scattered by a potential barrier with $\beta = 10$. The energy of the electron is chosen such that the transmission coefficient $T = 1$: $E = E_n$ with E_n given by equation (2.30), and $n = 1$, 2, and 5. We clearly see that the wavelength of the electron *inside* the barrier satisfies the condition $2a = n\lambda'/2$ (2.28). Thin lines show the absolute value of the wave function $|\Psi(x)|$. Outside the barrier, $|\Psi| \equiv 1$, since $\Psi(x)$ is the plane wave traveling to the right. In contrast to this, $\Psi(x)$ consists of both left- and right-going waves inside the barrier, and $|\Psi(x)|$ oscillates as a function of x.

that Re $\Psi(x)$ oscillates as a function of position, both inside and outside the potential barrier.

In figure 2.5 we plot the absolute value of $\Psi(x)$ and Re $\Psi(x)$ for the case of a potential well with $V_0 < 0$. The energy of the incident particle is chosen in such a way as to obtain transmission equal to one. Notice also that in this case $|\Psi(x)| \equiv 1$ outside the potential well, while inside the well, $|\Psi(x)|$ oscillates and satisfies the resonant condition that the width of the well $\ell = 2a = n\lambda'/2$. In the case shown in figure 2.5 we have chosen that the energy of the incident particle is given by equation (2.31) with $n = 7$: $E_7 = 0.209|V_0|$.

The main difference between the wave functions shown in figures 2.4 and 2.5 is that the amplitude of the wave function is larger inside the barrier than inside the well. This is easy to understand from the known coefficients F and G, given in equation (2.34). The wave function inside the barrier is

$$\Psi_i(x) = Fe^{+ik'x} + Ge^{-ik'x}. \tag{2.35}$$

Therefore, the absolute value squared, $|\Psi(x)|^2$, is given by

$$|\Psi_i(x)|^2 = |F|^2 + |G|^2 + 2|F||G|\cos 2k'x, \tag{2.36}$$

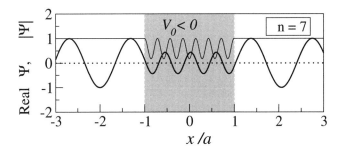

FIGURE 2.5. Real part of the wave function $\Psi(x)$ of an electron scattered from a potential well with $\beta = 10$. The energy of the electron $E_7 = 0.209|V_0|$ is given by equation (2.31) so that the transmission coefficient $T = 1$. The thin line shows the absolute value of the wave function $|\Psi(x)|$, which is unity on both sides of the potential barrier. Note that the electron wavelength is *shorter* inside the barrier than in the free space outside the potential barrier.

and oscillates between a maximum and a minimum value, $|\Psi_i^{\max}|^2 = (|F| + |G|)^2$ and $|\Psi_i^{\min}|^2 = (|F| - |G|)^2$. With the use of equations (2.34) we have

$$(|F| + |G|)^2 = \begin{cases} 1, & k' > k, \\ (k/k')^2, & k' < k, \end{cases} \tag{2.37}$$

and

$$(|F| - |G|)^2 = \begin{cases} (k/k')^2, & k' > k, \\ 1, & k' < k. \end{cases} \tag{2.38}$$

In the case of the potential barrier, when $V_0 > 0$, we have that $k > k'$, so that $|\Psi_i^{\max}| = (k/k')$ and $|\Psi_i^{\min}| = 1$. The absolute value of the wave function, $|\Psi_i(x)|^2$, oscillates between these two values. For the case of a potential well, $V_0 < 0$, we have that $k' > k$ and $|\Psi(x)|$ oscillates between the maximum value $|\Psi_i^{\max}| = 1$ and the minimum value $|\Psi_i^{\min}| = (k/k')$.

2.2.3 Maximum Reflection

Now we will look at the condition for obtaining a minimum in the transmission coefficient. From equation (2.23) we obtain a minimum of the transmission coefficient when

$$\sin 2k'a = 1. \tag{2.39}$$

We expect that in this case the transmission coefficient is close to a minimum, since the denominator in the expression (2.23) is large. Equation (2.39) is equivalent to

$$2k'a = \left(\frac{1}{2} + n\right)\pi, \tag{2.40}$$

with $n = 0, 1, \ldots$.

The corresponding energy E_n can be found from equation (2.8) as

$$\frac{E_n}{V_0} = 1 + \frac{\pi^2}{4\beta^2} \left(\frac{1}{2} + n \right)^2 \tag{2.41}$$

for the potential barrier with $V_0 > 0$, and

$$\frac{E_n}{|V_0|} = \frac{\pi^2}{4\beta^2} \left(\frac{1}{2} + n \right)^2 - 1 \tag{2.42}$$

for the potential well with $V_0 < 0$. Note, however, that the energies given by equations (2.41) and (2.42) do not determine the exact position of the minimum of the transmission coefficient, since there is another function of k and k' in the denominator in equation (2.23).

The physical meaning of the condition (2.40) is clearer when it is expressed in terms of the wavelength $\lambda' = 2\pi/k'$ inside the potential,

$$\ell = 2a = (2n+1)\frac{\lambda'}{4}. \tag{2.43}$$

So one obtains a minimum in the transmission, or, equivalently, a maximum in the reflection coefficient when the width of the potential is an odd number of $\lambda'/4$, where λ' is the wavelength of the particle inside the potential. Remember that the transmission is equal to one when the width of the potential is an integer number of $\lambda'/2$.

Figure 2.6 shows the real part of the wave function, Re $\Psi(x)$, as a function of the distance, ploted for three cases. The top two panels of figure 2.6 are for a potential barrier with $V_0 > 0$, while the lowest panel is for a potential well with $V_0 < 0$. The energy of the incident particle for the case of V_0 is given by equation (2.41) with $\beta = 10$ and $n = 1$ and 2. For the case of $V_0 < 0$, the energy of the incident particle is given by equation (2.42) with $\beta = 10$ and $n = 7$. Notice that the amplitude of the wave function on the right-hand side of the potential is smaller than that of n the left-hand side.

2.3 Tunneling: $0 < E < V_0$

In the case of $E > 0$ but $E < V_0$, the electron tunnels through the barrier. The wave vector *inside* the barrier is imaginary,

$$k' \to i\kappa', \quad \kappa' = \beta\sqrt{1 - E/V_0}. \tag{2.44}$$

From equations (2.7) and (2.8) we obtain κ', related to the wave vector of the incident electron, k, as follows:

$$(\kappa'a)^2 + (ka)^2 = \beta^2. \tag{2.45}$$

With the use of the formulas

$$\cos(ix) = \cosh x \quad \text{and} \quad \sin(ix) = i \sinh x, \tag{2.46}$$

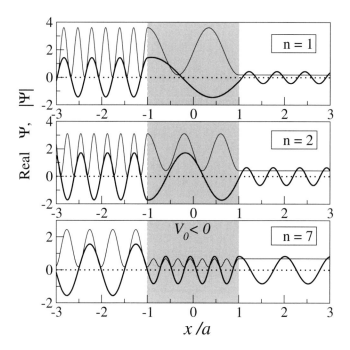

FIGURE 2.6. The real part of the wave function $\Psi(x)$ of the electron scattered by a potential barrier and well with $\beta = 10$. The energy is chosen close to a transmission minimum, and the wave is normalized in such a way that the amplitude of the incident wave $|A| = 1$. Then the amplitude of the wave function on the right side of the barrier gives the transmission coefficient $T = |t|^2 < 1$. The ratio of the barrier width to the wavelength *inside* the barrier is $2a/\lambda' = (1/4 + n/2)$. The thin line is the absolute value $|\Psi(x)|$.

we can rewrite the transfer matrix elements of the matrix **M** given by equations (2.17) and (2.18) as

$$M_{11} = \cosh 2\kappa'a + \frac{i}{2}\left(\frac{k}{\kappa'} - \frac{\kappa'}{k}\right)\sinh 2\kappa'a,$$

$$M_{12} = -\frac{i}{2}\left(\frac{\kappa'}{k} + \frac{k}{\kappa'}\right)\sinh 2\kappa'a, \qquad (2.47)$$

$$M_{22} = M_{11}^*, \qquad M_{21} = M_{12}^*.$$

By using equation (2.23) we calculate the transmission coefficient, given by

$$T = \frac{1}{1 + \frac{1}{4}\left[\frac{k}{\kappa'} + \frac{\kappa'}{k}\right]^2 \sinh^2 2\kappa'a}. \qquad (2.48)$$

It is evident from equation (2.48) that T is always smaller than unity and that T decreases as the barrier width increases. When the barrier width $\ell = 2a$ exceeds the tunneling length $\xi = 1/\kappa'$, i.e., $2\kappa'a \gg 1$, we have that $\sinh 2\kappa'a \approx \frac{1}{2}e^{2\kappa'a}$ and the

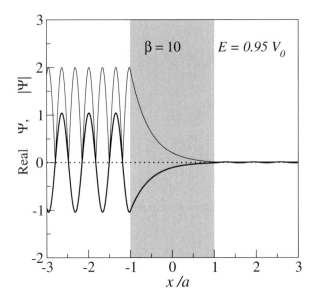

FIGURE 2.7. The real part of the wave function for the case $0 < E < V_0$. The parameter $\beta = 10$. The amplitude of the wave function is very small on the right side of the barrier. On the left side, the wave function is a superposition of the incoming wave and the reflected wave. Since the transmission is almost negligible, both incoming and reflected waves have the same amplitude. The thin line is $|\Psi|$.

transmission coefficient decreases exponentially as a function of the barrier width,

$$T \approx \left(\frac{4\kappa' k}{(\kappa')^2 + k^2} \right) e^{-4\kappa' a}. \tag{2.49}$$

There is transmission, even though the energy is lower than the height of the potential barrier. This is a wave phenomenon, and in quantum mechanics it is also exhibited by particles. In classical physics, the particles will be reflected back, if the energy $E < V_0$. The tunneling of a particle through a barrier is frequently encountered in physics. One example of tunneling is the scanning tunneling microscope and another is nuclear decay with the emission of α-particles.

In the case of tunneling, the wave function decreases inside the barrier and the transmission is always smaller than unity. We show in figure 2.7 the wave function for $\beta = 10$ with an energy only slightly smaller than V_0. We see that the wave function decreases exponentially inside the barrier and the transmission is very small, in agreement with equation (2.49). This is more clearly visible in figure 2.8, which plots the absolute value $|\Psi(x)|$ in a logarithmic scale.

In figure 2.8 we see that in the region left of the potential barrier $|\Psi(x)|$ oscillates as a function of x. This is because the wave function is a superposition of the incoming and the reflected waves:

$$\Psi_L(x) = e^{ikx} + r e^{-ikx}, \tag{2.50}$$

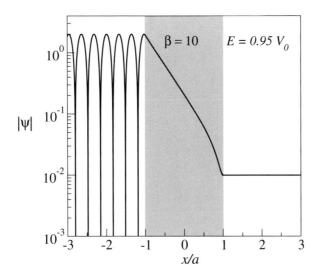

FIGURE 2.8. The same as figure 2.7, but now we plot $|\Psi(x)|$ in a logarithmic scale. We see that the wave function indeed decreases exponentially inside the barrier. Deviation from the exponential decrease is visible in the neighborhood of the right barrier, because of the interference of the right-going wave with the wave reflected from the right interface. The huge oscillations of the wave function in the region left of the barrier are explained in the text.

so that

$$|\Psi_L(x)|^2 = 1 + |r|^2 + 2|r|\cos(\phi_r + 2kx), \tag{2.51}$$

where we have expressed the complex reflection amplitude in terms of its absolute value and phase:

$$r = |r|e^{i\phi_r}. \tag{2.52}$$

Indeed, $|\Psi_L(x)|^2$ oscillates as a function of x. The maximum value

$$|\Psi_L|^2_{\max} = (1 + |r|)^2, \tag{2.53}$$

and the minimum

$$|\Psi_L|^2_{\min} = (1 - |r|)^2. \tag{2.54}$$

In the present case, $|r|$ is very close to 1 so that $|\Psi_L(x)|$ oscillates between 0 and 2.

Finally, we discuss how the transmission in the tunneling regime depends on the parameters of the potential barrier. We see from equation (2.7) that

$$ka = 2\pi\frac{a}{\lambda} = \beta\sqrt{\frac{E}{V_0}}. \tag{2.55}$$

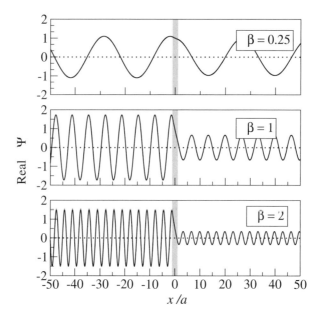

FIGURE 2.9. The real part of the wave function for three different potential barriers. The energy of the electron is $E = 0.9 V_0$. The position of the barrier is marked by vertical dashed lines. For a fixed ratio E / V_0, the transmission is determined only by the ratio of the barrier width to the electron wavelength. The potential barrier is almost transparent if its width is much smaller than the electron wavelength.

Thus, for a fixed ratio E / V_0, the parameter β determines the ratio of the electron wavelength λ to the width of the barrier, $\ell = 2a$. In figure 2.9 we see that the transmission is high when the wavelength of the electron is much larger than the barrier width. When β increases, the wave function inside the barrier decreases exponentially and, consequently, the transmission coefficient also decreases. This result again states that for a given ratio E / V_0 the only important parameter, which determines transmission properties, is the ratio of the electron wavelength to the width of the barrier.

Similar results are shown in figure 2.10. Here, however, we fix the parameter $\beta = 2$ and vary the energy of the electron. For a given β the transmission increases when E / V_0 increases, in agreement with our intuition.

2.4 Current Density

In section 1.5.1 the current density is given by the expression [equation (1.114)],

$$j = \frac{i\hbar}{2m}\left[\Psi\frac{\partial\Psi^*}{\partial x} - \Psi^*\frac{\partial\Psi}{\partial x}\right]. \tag{2.56}$$

It can be easily calculated for the wave function given by equation (2.3) for the rectangular potential. For the case of $E > V_0$, we insert the wave functions $\Psi(x)$, given by

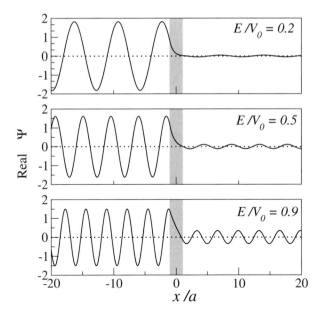

FIGURE 2.10. The real part of the wave function for three different energies of the electron. The parameter $\beta = 2$ in all three cases. As expected, the transmission is higher when the electron energy increases.

equation (2.3), in the definition of the current, equation (2.56), and we obtain

$$j = \begin{cases} j_L = \hbar k/m \left(|A|^2 - |B|^2\right), & x < -a, \\ j_{\text{barrier}} = \hbar k'/m \left(|F|^2 - |G|^2\right), & -a < x < a, \\ j_R = \hbar k/m \left(|C|^2 - |D|^2\right), & a < x. \end{cases} \tag{2.57}$$

The current density in all three regions is the same,

$$j_L = j_{\text{barrier}} = j_R, \tag{2.58}$$

since

$$k \left[|A|^2 - |B|^2\right] = k' \left[|F|^2 - |G|^2\right] = k \left[|C|^2 - |D|^2\right]. \tag{2.59}$$

To prove the first relation in equation (2.59), we use equations (2.14), which express the coefficients F and G in terms of coefficients A and B. From the first equation (2.14), we obtain the expression

$$4k'^2|F|^2 = (k'+k)^2|A|^2 + (k'-k)^2|B|^2 + (k'+k)(k'-k)\left[e^{-i2ka}AB^* + e^{+i2ka}A^*B\right], \tag{2.60}$$

and from the second equation (2.14), we derive

$$4k'^2|G|^2 = (k'-k)^2|A|^2 + (k'+k)^2|B|^2 + (k'+k)(k'-k)\left[e^{-i2ka}AB^* + e^{+i2ka}A^*B\right]. \tag{2.61}$$

Now we subtract equation (2.61) from equation (2.60). We obtain

$$4k'^2 \left[|F|^2 - |G|^2\right] = 4kk' \left[|A|^2 - |B|^2\right], \tag{2.62}$$

which is the expression we wanted to prove. In the same way, we can use equation (2.15) and prove the validity of the second relation in equation (2.59). However, we do not need to do it, since the identity $|A|^2 - |B|^2 = |C|^2 - |D|^2$ follows directly from the form of the transfer matrix $\mathbf{M}$, given by equation (2.20) which guarantees the conservation of the current density.

It is interesting to calculate the current density *inside* the barrier for the case of tunneling, $E < V_0$. In this case, k' is imaginary, $k' = i\kappa'$. Using the definition of the current density given by equation (2.56), we obtain that

$$j_{\text{barrier}} = \frac{i\hbar\kappa'}{m} \; (G^*F - F^*G). \tag{2.63}$$

The derivation of equation (2.63) is given in problem 2.2. Again, we can use equations (2.14) and show that j_{barrier} must be equal to both the current densities on the left and on the right side of the barrier. Here we show this equality for the special case $D = 0$, when the incident wave comes from the left side of the barrier. Then, from equation (2.34), we have

$$F = \frac{i\kappa' + k}{2i\kappa'} e^{+\kappa'a + ika} C \tag{2.64}$$

and

$$G = \frac{i\kappa' - k}{2i\kappa'} e^{-\kappa'a + ika} C. \tag{2.65}$$

Inserting these expressions for F and G into equation (2.63), after some algebra, we indeed obtain that

$$j_{\text{barrier}} = \frac{\hbar k}{m} |C|^2 = j_R. \tag{2.66}$$

The nonzero value of the current density *inside* the barrier means that the current and the energy transfer through the barrier in the process of tunneling. At first sight, this is surprising because there are no propagating waves in the barrier. From equation (2.63) we see that the current density is nonzero only if we have two solutions in the barrier, one decaying from the first interface and the other decaying from the second interface. Thus, the current density is coming from the "interference" of the two exponentially decaying solutions. If the right side of the potential barrier moves to infinity, we have only one interface and in this case the current density is zero.

2.5 Bound States: $V_0 < E < 0$

In the case of a potential well, $V_0 < 0$, bound states with *negative* energy $E < 0$ exist. Since $k = \sqrt{2mE}/\hbar$, we have that k is imaginary when the energy E is negative,

$$k = i\kappa, \qquad \kappa = \sqrt{\frac{2m|E|}{\hbar^2}}. \tag{2.67}$$

This means that no propagating solution exists outside the potential well. The wave function outside the barrier given by equation (2.3) either decreases or increases exponentially. To keep the solution of Schrödinger's equation normalizable, the coefficients A and D in equation (2.3) must be zero for $\kappa > 0$. On the other hand, we have from the second equation of (2.16) that

$$De^{\kappa a} = M_{21} Ae^{-\kappa a} + M_{22} Be^{\kappa a}. \tag{2.68}$$

For $D = 0$ and $A = 0$, equation (2.68) can be satisfied either by $B = 0$ or by $M_{22} = 0$. The first possibility, $B = 0$, is not interesting, since the wave function is identically equal to zero in this case. The second possibility leads to the equation

$$M_{22}(k = i\kappa) \equiv 0. \tag{2.69}$$

which is the equation for obtaining a bound state. The solution κ_b of equation (2.69) determines the energy of the bound state as

$$E_b = -\frac{\hbar^2 \kappa_b^2}{2m}. \tag{2.70}$$

Using the explicit expression (2.18) for the matrix element M_{22}, we calculate from equation (2.69)

$$\cot 2k'a = -\frac{1}{2}\left[\frac{\kappa}{k'} - \frac{k'}{\kappa}\right], \quad k' = \sqrt{2m(E - V_0)}/\hbar. \tag{2.71}$$

By use of the relation

$$\cot 2x = \frac{1}{2}(\cot x - \tan x), \tag{2.72}$$

we obtain the result that equation (2.71) has two branches of solutions:

$$\tan k'a = \frac{\kappa}{k'}, \quad \tan k'a = -\frac{k'}{\kappa}. \tag{2.73}$$

With the use of the relation

$$(k'a)^2 + (\kappa a)^2 = \beta^2, \tag{2.74}$$

which follows from equations (2.7) and (2.8), we rewrite equations (2.73) in the form

$$\tan k'a = \frac{1}{k'a}\sqrt{\beta^2 - (k'a)^2},$$

$$\tan k'a = -\frac{k'a}{\sqrt{\beta^2 - (k'a)^2}}. \tag{2.75}$$

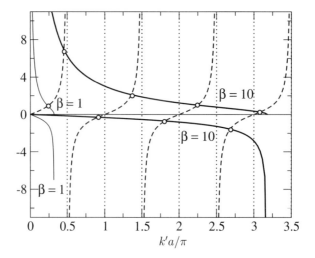

FIGURE 2.11. Graphical solution of equation (2.75) for the existence of bound states in a rectangular potential well. The dashed line is $\tan k'a$, the solid lines represent functions $\sqrt{\beta^2 - (k'a)^2}/(k'a)$ and $-k'a/\sqrt{\beta^2 - (k'a)^2}$ for $\beta = 1$ and 10, and the dotted lines show the values of $k'a = n\pi/2$, $n = 1, 2, \ldots$. The bound state is determined by the cross section of dashed and solid lines. We see that there is at least one solution for any β.

Using equation (2.74), we express the energy of the bound state (2.70) as

$$E = |V_0|\left[-1 + \frac{1}{\beta^2}(k'a)^2\right],$$ \hfill (2.76)

where $k'a$ is the solution of equations (2.75).

The graphical solution of equations (2.75) is shown in figure 2.11. We see that for any value of β there is always at least one solution which corresponds to the first equation (2.75). This means there is always at least one bound state. This result, obtained here for the special case of a rectangular potential, holds in general. In one dimension, any potential well creates at least one bound state, independent of how small is the strength of the potential well.

The number of bound states, N_b, increases when β increases. From figure 2.11, we see that for

$$N_b^+ \pi > \beta > (N_b^+ - 1)\pi,$$ \hfill (2.77)

the first equation (2.75) has exactly N_b^+ solutions. Equivalently, for

$$\left(N_b^+ + \frac{1}{2}\right)\pi > \beta > \left(N_b^- - \frac{1}{2}\right)\pi,$$ \hfill (2.78)

the second equation (2.75) has exactly N_b^- solutions. The total number of bound states $N_b = N_b^+ + N_b^-$ is therefore given as

$$N_b = \text{Int}\left[\frac{2\beta}{\pi}\right] + 1.$$ \hfill (2.79)

We see in figure 2.11 that for large β the solutions of equations (2.75) converge to

$$k'_{2n+1}a \approx (2n+1)\frac{\pi}{2} \tag{2.80}$$

and to

$$k'_{2n}a \approx 2n\frac{\pi}{2}. \tag{2.81}$$

For the limit of $\beta \to \infty$, we have an infinite number of bound states, $k'_n a = n\pi/2$, $n = 1, 2, \ldots$. Their energy, measured from the bottom of an infinitely deep potential well, reads

$$E_n = \frac{\hbar^2}{2ma^2}\frac{\pi^2 n^2}{4}, \qquad V_0 \to \infty, \tag{2.82}$$

which are well-known results from quantum mechanics for an infinite potential well.

The eigenfunctions of the potential well problem can be easily calculated, using, for instance, equations (2.34) with constant C, such that the wave function is normalized,

$$\int_{-\infty}^{+\infty} dx\, |\Psi(x)|^2 = 1. \tag{2.83}$$

For example, we show in figure 2.12 the seven eigenfunctions of the potential well with $\beta = 10$. Notice that the wave function of the bound state is symmetric $[\Psi_n(x) = \Psi_n(-x)]$ for n odd and antisymmetric $[\Psi_n(x) = -\Psi_n(-x)]$ for n even. In the limit of an infinite potential well, the wave functions possess the simple form

$$\Psi_n(x) = C \sin k'_n(x - a), \qquad |x| \leq a, \tag{2.84}$$

and $\Psi_n(x) \equiv 0$ outside the infinite potential well.

2.6 Inverse Problem for Rectangular Potential

In previous sections we calculated the transmission and reflection amplitudes for a given rectangular potential V_0. Now we solve the inverse problem. Given the transmission and reflection amplitudes t and r, we need to find the parameters of the rectangular potential.

First, from formulas (2.17) we derive an expression for the wave vector inside the barrier, k'. We start with the expressions for the elements of the transfer matrix $\mathbf{M}$,

$$M_{11} - \cos 2k'a = \frac{i}{2}\left[\frac{k}{k'} + \frac{k'}{k}\right]\sin 2k'a,$$
$$M_{12} = \frac{i}{2}\left[\frac{k'}{k} - \frac{k}{k'}\right]\sin 2k'a. \tag{2.85}$$

Square both equations and subtract the second equation from the first one to obtain

$$(M_{11} - \cos 2k'a)^2 - M_{12}^2 = -\sin^2 2k'a = \cos^2 2k'a - 1, \tag{2.86}$$

which can be solved for $\cos 2k'a$,

$$\cos 2k'a = \frac{1}{2M_{11}}\left[1 + M_{11}^2 - M_{12}^2\right]. \tag{2.87}$$

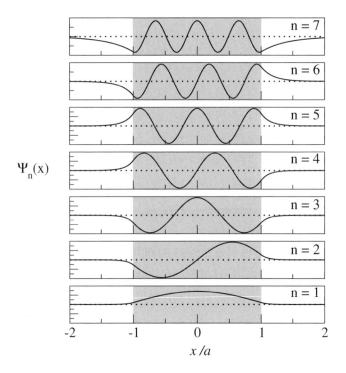

FIGURE 2.12. Wave function $\Psi_n(x)$ of bound states for the potential well with $\beta = 10$. The eigenenergies are given by equation (2.76), in which $k'a$ is the nth solution of the equation (2.75). Note the exponentially decreasing tail of the wave functions Ψ_n in the regions *outside* the potential well. The tail is longer for higher energy levels. As n increases, the wave function becomes appreciable outside the boundaries of the potential well.

Next, express $M_{11} \pm M_{12}$ as

$$M_{11} - \cos 2k'a - M_{12} = \frac{k}{k'} \sin 2k'a \tag{2.88}$$

and

$$M_{11} - \cos 2k'a + M_{12} = \frac{k'}{k} \sin 2k'a. \tag{2.89}$$

For $2k'a \neq m\pi$ (m integer) we have $\sin 2k'a \neq 0$. Then, using equation (2.87), we obtain from equations (2.88) and (2.88) the following expression for the ratio of wave vectors:

$$\frac{k'}{k} = \sqrt{\frac{(M_{11} + M_{12})^2 - 1}{(M_{11} - M_{12})^2 - 1}}. \tag{2.90}$$

The sign of the square root is positive to keep the left-hand side of equation (2.90) *positive*.

Equations (2.87) and (2.90) express k and k' in terms of transmission and reflection *amplitudes*. Note that not only the absolute values $|t|$ and $|r|$ but also the phases of the transmission and reflection are necessary to get k and k'.

Note also that equations (2.90) and (2.87) do not have a unique solution. For a given M_{11} and M_{12}, we can find several combinations of potentials V_0 and energies E. Indeed, equation (2.90) determines only the ratio E/V_0 but does not fix the actual values of E and of V_0. Equation (2.87) has an infinite number of solutions which differ by $\pi/a \times m$ with m integer. The ambiguity of the solution is not surprising. We have seen already in figure 2.3 that, for a given potential V_0, we can obtain a given value of the transmission for many energies of the electron.

2.7 Problems

Problems with Solutions

Problem 2.1
For the wave function

$$\Psi(x) = Ae^{ikx} + Be^{-ikx}, \tag{2.91}$$

calculate the current density, given by equation (2.56).

Solution. First we calculate

$$\begin{aligned}
\Psi(x)\frac{\partial \Psi^*(x)}{\partial x} &= \left[Ae^{ikx} + Be^{-ikx}\right]\left[-ikA^*e^{-ikx} + ikB^*e^{ikx}\right] \\
&= -ik\left\{|A|^2 - |B|^2 - AB^*e^{i2kx} + A^*Be^{-i2kx}\right\}.
\end{aligned} \tag{2.92}$$

In the same way, we obtain

$$\Psi^*(x)\frac{\partial \Psi(x)}{\partial x} = ik\left\{|A|^2 - |B|^2 + AB^*e^{i2kx} - A^*Be^{-i2kx}\right\}, \tag{2.93}$$

so that

$$\begin{aligned}
j &= \frac{i\hbar}{2m}\left[\Psi(x)\frac{\partial \Psi^*(x)}{\partial x} - \Psi^*(x)\frac{\partial \Psi(x)}{\partial x}\right] \\
&= \frac{\hbar k}{m}\left[|A|^2 - |B|^2\right].
\end{aligned} \tag{2.94}$$

Problem 2.2
Calculate the expression for the current density if the wave function consists of two exponentially decaying functions,

$$\Psi(x) = Fe^{-\kappa'x} + Ge^{+\kappa'x} \tag{2.95}$$

Solution. Since κ' is real, the calculation of the current is even simpler than in the case when the wave function consists of the superposition of the propagating waves. We have

$$\frac{\partial \Psi(x)}{\partial x} = -\kappa'\left[Fe^{-\kappa'x} - Ge^{\kappa'x}\right]. \tag{2.96}$$

Now, we calculate the first term in the expression for the current density,

$$\Psi(x)\frac{\partial \Psi^*(x)}{\partial x} = \left[Fe^{-\kappa'x} + Ge^{+\kappa'x}\right]\left[-\kappa'F^*e^{-\kappa'x} + \kappa'G^*e^{+\kappa x}\right]$$
$$= \kappa'\left\{-|F|^2 e^{-2\kappa'x} + |G|^2 e^{+2\kappa'x} + FG^* - F^*G\right\}, \tag{2.97}$$

and the second term,

$$\Psi^*(x)\frac{\partial \Psi(x)}{\partial x} = \left[F^*e^{-\kappa'x} + G^*e^{+\kappa'x}\right]\left[-\kappa'Fe^{-\kappa'x} + \kappa'Ge^{+\kappa'x}\right]$$
$$= \kappa'\left\{-|F|^2 e^{-2\kappa'x} + |G|^2 e^{+2\kappa'x} - FG^* + F^*G\right\}, \tag{2.98}$$

so that

$$j = \frac{i\hbar\kappa'}{m}\left[G^*F - F^*G\right]. \tag{2.99}$$

Problem 2.3
If $\mathbf{M}_{\text{box}}(\ell)$ is the transfer matrix for a rectangular potential, show that

$$\mathbf{M}_{\text{box}}(\ell_1 + \ell_2) = \mathbf{M}_{\text{box}}(\ell_2)\mathbf{M}_{\text{box}}(\ell_1). \tag{2.100}$$

Solution. Consider two rectangular potentials of widths ℓ_1 and ℓ_2, separated by a distance ℓ_h. The transfer matrix of the entire system is

$$\mathbf{M} == \mathbf{M}_{\text{box}}(\ell_2)\begin{pmatrix} e^{ik\ell_h} & 0 \\ 0 & e^{-ik\ell_h} \end{pmatrix}\mathbf{M}_{\text{box}}(\ell_1). \tag{2.101}$$

In the limit of $\ell_h \to 0$, we obtain equation (2.100). Matrix multiplication of the two transfer matrices, $\mathbf{M}_{\text{box}}(\ell_2)\mathbf{M}_{\text{box}}(\ell_1)$, indeed confirms that $\mathbf{M}_{\text{box}}(\ell_1 + \ell_2)$ has again the form of the transfer matrix for a rectangular potential, given by equations (2.17) and (2.18). For instance,

$$M_{11}(\ell) = M_{11}(\ell_1)M_{11}(\ell_2) + M_{12}(\ell_1)M_{21}(\ell_2)$$
$$= \cos k'\ell_1 \cos k'\ell_2 - \frac{1}{4}\left(\frac{k}{k'} + \frac{k'}{k}\right)^2 \sin k'\ell_1 \sin k'\ell_2$$
$$+ \frac{i}{2}\left(\frac{k}{k'} + \frac{k'}{k}\right)(\sin k'\ell_1 \cos k'\ell_2 + \cos k'\ell_1 \sin k'\ell_2) \tag{2.102}$$
$$+ \frac{1}{4}\left(\frac{k}{k'} - \frac{k'}{k}\right)^2 \sin k'\ell_1 \sin k'\ell_2.$$

The 2nd and the 4th terms together give us $-\sin k'\ell_1 \sin k'\ell_2$. Then, with the use of the relations $\cos x \cos y - \sin x \sin y = \cos(x+y)$ and $\sin x \cos y + \cos x \sin y = \sin(x+y)$, we obtain from equation (2.102) that

$$M_{11}(\ell) = \cos k'\ell + \frac{i}{2}\left(\frac{k}{k'} + \frac{k'}{k}\right)\sin k'\ell, \quad \ell = \ell_1 + \ell_2, \tag{2.103}$$

which is the first relation (2.17).

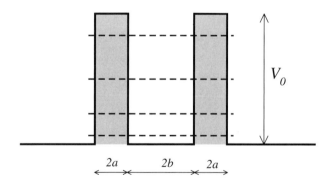

FIGURE 2.13. Two identical barriers.

Problem 2.4

Consider two identical rectangular barriers shown in figure 2.13. The energy of the incident electron is smaller than the barrier height, $E < V_0$. Show that there are special values of energy for which this system is transparent, which means that the transmission coefficient $T = 1$.

Solution. To calculate the transmission coefficient, we need the transfer matrix

$$\mathbf{M}^{(1)} = \begin{pmatrix} e^{i2kb} & 0 \\ 0 & e^{-i2kb} \end{pmatrix} \mathbf{M}_{\text{box}}, \tag{2.104}$$

where the transfer matrix $\mathbf{M}_{\text{box}}$ is the transfer matrix for the rectangular barrier, given by equation (2.47). Having an explicit form of the matrix $\mathbf{M}^{(1)}$, we immediately obtain the transmission coefficient for the system of two barriers, using the Chebyshev formula equation (1.106):

$$T = \frac{1}{1 + 4|\mathbf{M}_{12}^{(1)}|^2 \cos^2 q\ell}, \tag{2.105}$$

where $\ell = 2a + 2b$ and $2 \cos q\ell = \text{Tr}\,\mathbf{M}^{(1)}$. We can obtain resonant transmission only in the case when $\cos q\ell = 0$. Calculating the trace of the transfer matrix $\mathbf{M}^{(1)}$, we obtain

$$\text{Tr}\,\mathbf{M}^{(1)} = 2 \cos q\ell = 2 \cosh 2\kappa'a \cos 2kb$$
$$-\frac{1}{2}\left(\frac{k}{\kappa'} - \frac{\kappa'}{k}\right) \sinh 2\kappa'a \sin 2kb = 0. \tag{2.106}$$

We recall that κ' is related to k by $(\kappa'a)^2 + (ka)^2 = \beta^2$, given by equation (2.45).

Equation (2.106) can be written in the form

$$\cot 2kb = \frac{1}{2}\left(\frac{k}{\kappa'} - \frac{\kappa'}{k}\right) \tanh 2\kappa'a, \tag{2.107}$$

which is similar to the equation for the bound state of a rectangular potential, given by equation (2.71). The only difference between these two equations is in an additional term, $\tanh 2\kappa'a$, on the r.h.s. of equation (2.107). This means that the system of two barriers allows resonant transmission with the transmission coefficient $T = 1$, although only decaying solutions are allowed inside the individual barriers.

Note that, in the limit of $a \to \infty$, $\tanh 2\kappa' a \to 1$ and equation (2.107) reduces to the equation for the bound energies inside the potential barrier, equation (2.71).

Problem 2.5

For the case of total transmission through the system of two barriers, discussed in the previous problem, find the wave function in the region between the two barriers.

Solution. Since $T = 1$, the wave function both left and right of the system is e^{ikx}. The wave function between the barriers consists of both left- and right-going plane waves,

$$\Psi(x) = Ae^{ikx} + Be^{-ikx}. \tag{2.108}$$

Using the transfer matrix for the rectangular barrier, we find the coefficients A and B as

$$\begin{pmatrix} A \\ B \end{pmatrix} = \mathbf{M}_{\text{box}} \begin{pmatrix} 1 \\ 0 \end{pmatrix} = \begin{pmatrix} M_{11} \\ M_{21} \end{pmatrix}, \tag{2.109}$$

where the elements of the matrix $\mathbf{M}_{\text{box}}$ are given by equations (2.47) with the values of k and κ' given by the solution of equations (2.107) and (2.45).

We immediately see that the current density in the region between the two barriers,

$$j = |A|^2 - |B|^2 = |M_{11}|^2 - |M_{12}|^2 = \cosh^2 2\kappa' a - \sinh^2 2\kappa' a, \tag{2.110}$$

is equal to 1. However, the absolute values of the amplitudes A and B, $|A| = |M_{11}| \sim e^{2\kappa' a}$ and $|B| = |M_{12}| \sim e^{2\kappa' a}$, are large and increase exponentially as functions of the width of the barrier. Therefore, resonant tunneling requires exponentially high density of particles in the region between the barriers, which makes the process of tunneling very slow. A detailed description of how the tunneling through a double-barrier system proceeds in time is given in [150].

Problem 2.6

Consider the potential step shown in figure 2.14. Using the relations between wave functions on both sides of a potential step, given by equations (2.12), show that the transfer matrix for the interface of the potential step is given by [7]

$$\mathbf{M}_{\text{step}} = \frac{1}{2k'} \begin{pmatrix} k' + k & k' - k \\ k' - k & k' + k \end{pmatrix}, \tag{2.111}$$

where k and k' are given by equations (2.4) and (2.5), respectively. Note that

$$\det \mathbf{M}_{\text{step}} = \frac{k}{k'} \neq 1, \tag{2.112}$$

because the wave vectors differ on the two sides of the interface.

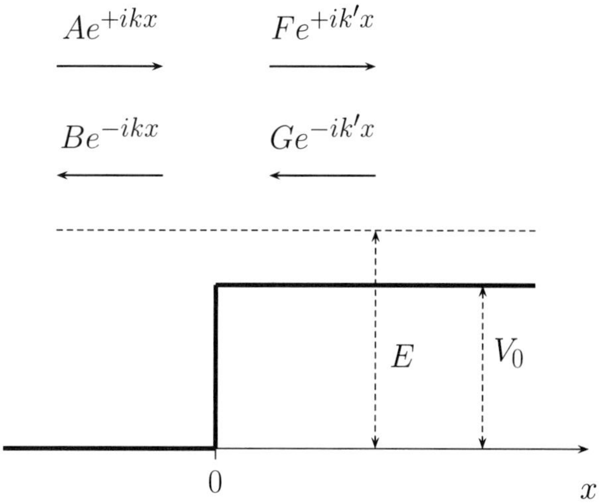

FIGURE 2.14. Potential step of height V_0.

Solution. The transfer matrix $\mathbf{M}_{\text{step}}$ is defined by the relation

$$\begin{pmatrix} F \\ G \end{pmatrix} = \mathbf{M}_{\text{step}} \begin{pmatrix} A \\ B \end{pmatrix}, \tag{2.113}$$

where the coefficients A, B, F, and G are defined in figure 2.14. From the continuity of the wave function and its first derivative at $x = 0$ we obtain

$$A + B = F + G,$$
$$ik(A - B) = ik'(F - G). \tag{2.114}$$

We solve the above system of linear equations and obtain

$$F = \frac{k' + k}{2k'} A + \frac{k' - k}{2k'} B \tag{2.115}$$

and

$$G = \frac{k' - k}{2k'} A + \frac{k' + k}{2k'} B. \tag{2.116}$$

By comparing equations (2.115) and (2.116) with equation (2.113), we obtain an explicit form of the transfer matrix $\mathbf{M}_{\text{step}}$ given by equation (2.111).

Physically, this problem can be treated as a problem of scattering of a quantum particle on an infinitesimally thin barrier located between two semi-infinite media. In contrast to the transfer matrices discussed so far, the right medium differs from the left one. Therefore, the symmetry relations derived in chapter 1 may not be valid in this case.

Problem 2.7
Using the transfer matrix $\mathbf{M}_{\text{step}}$, calculate transmission and reflection coefficients for the potential step.

Solution. We start with equations (2.111) and (2.113). Consider now a wave incident from the left side with amplitude $A \equiv 1$. Then $B = r'$ and $F = t'$, and from equation (2.113) we obtain that

$$t' = M_{11} + M_{12}r',$$
$$0 = M_{21} + M_{22}r'. \qquad (2.117)$$

Solving this linear system, we obtain

$$t' = \frac{2k'}{k+k'} \quad \text{and} \quad r' = \frac{k-k'}{k+k'}. \qquad (2.118)$$

Problems without Solutions

Problem 2.8

Consider again the potential step shown in figure 2.14. Show that for propagation in the opposite direction (the incident wave comes from the right side of the step), the transmission amplitudes t and r are given by the following expressions:

$$t = \frac{2k}{k+k'} \quad \text{and} \quad r = \frac{k'-k}{k+k'}. \qquad (2.119)$$

Note that $t \neq t'$. This is a consequence of having different wave vectors on the two sides of the potential step. Note also that neither r nor r' is zero. This is in contrast to classical mechanics, where a particle going over a potential step would slow down (to conserve energy) but would never be reflected back. In our case, we have partial reflection of the particle, even in the case when the potential decreases. This is because in quantum mechanics a particle has wave properties.

Problem 2.9

Consider now a potential step as shown in figure 2.14, but with $V_0 > E$. Intuitively, we believe that the wave incoming from the left must be totally reflected back. From equation (2.118) we indeed see that $R' = |r'|^2 = 1$, since k' is imaginary, $k' = i\kappa$. Note, however, that the transmission amplitude t' is different from zero. Why?

To solve this problem, remember that t' represents the change of amplitude of the wave in the vicinity of the step. The wave function inside the step is

$$\Psi_R = t'e^{-\kappa x}. \qquad (2.120)$$

It decreases exponentially when the distance from the step, x, increases. Although Ψ_R is nonzero inside the step, the transmission is zero, since nothing is observed far away from the step.

Problem 2.10

Calculate the current density for the case of a potential step $V_0 > E$.

Inserting the wave function (2.120) into the formula for the current density, equation (2.56), we immediately see that $j = 0$. This confirms that an exponentially decreasing wave cannot transfer any energy.

Problem 2.11
Calculate the scattering matrix $\mathbf{S}$ for a rectangular potential barrier and well.

Problem 2.12
Use the explicit expressions for the elements of the transfer matrix $\mathbf{M}$ given by equations (2.17) and (2.18) to find the reflection coefficient R. Show that $T + R = 1$.

Problem 2.13
Verify that the transfer matrix $\mathbf{M}$ defined by the matrix elements in equations (2.17) and (2.19) satisfies the conservation of the probability of flux, equation (1.36), and the time-reversal symmetry formula, equation (1.50).

Problem 2.14
Use the transfer matrix $\mathbf{M}_{\text{step}}$ and the transfer matrix $\mathbf{M}_0$ for free propagation along the distance ℓ,

$$\mathbf{M}_0 = \begin{pmatrix} e^{ik'\ell} & 0 \\ 0 & e^{-ik'\ell} \end{pmatrix}, \tag{2.121}$$

to derive the transfer matrix $\mathbf{M}$ for the rectangular potential. Show that

$$\mathbf{M} = \mathbf{M}_{\text{step}}\mathbf{M}_0\mathbf{M}_{\text{step}}^{-1}. \tag{2.122}$$

Compare the result obtained with the formulas (2.17) and (2.18).

3 δ-Function Potential

In physical applications, it is often useful to consider a simplified form of the rectangular potential, namely, the δ-function potential,

$$V(x) = \frac{\hbar^2}{2m} \Lambda\, \delta(x). \tag{3.1}$$

The potential (3.1) can be obtained from the rectangular potential, defined by equation (2.2) in the limit of infinitesimally narrow barrier width,

$$2a \to 0, \tag{3.2}$$

and infinitesimally high barrier height,

$$V_0 = \frac{\hbar^2}{2m} \frac{\Lambda}{2a} \to \infty, \tag{3.3}$$

in such a way that the product $2a\, V_0 = \hbar^2 \Lambda/(2m)$ is constant. The potential (3.1) represents either a potential barrier ($\Lambda > 0$) or a potential well ($\Lambda < 0$) [7,15].

In this chapter, we will study first the transmission of a quantum particle through a single δ-function barrier. Then a system of N δ-function barriers will be examined with $N = 2, 3, \ldots$. For the special cases of $N = 1$ and 2 and $\Lambda < 0$, we will also analyze the condition of the existence of bound states and present the energies of the bound state.

3.1 Single δ-Function Potential

The calculation of the transfer matrix of a case of the single δ-function barrier is easy. Use formulas (2.17) for the transfer matrix of the rectangular barrier and perform the limiting procedure given by equations (3.2) and (3.3). To perform the limiting procedure from the formulas for the rectangular potential, we also need to express the wave vector $k' = \sqrt{2m(E - V_0)}/\hbar$, given by equation (2.5). Inserting V_0 from equation (3.3) and neglecting E with respect to V_0, we get

$$k' = i\sqrt{\Lambda/2a}. \tag{3.4}$$

Then we calculate the following two limits:

$$\lim_{a \to 0} \frac{\sin k'a}{k'} \sim a = 0 \tag{3.5}$$

and

$$\lim_{a \to 0} \frac{k'}{k} \sin 2k'a = -\frac{\Lambda}{k}. \tag{3.6}$$

These results should be inserted into the expressions (2.17) and (2.18) for the transfer matrix for rectangular potential barriers. We obtain

$$\mathbf{M}_\Lambda = \begin{pmatrix} 1 + \dfrac{\Lambda}{2ik} & +\dfrac{\Lambda}{2ik} \\[2mm] -\dfrac{\Lambda}{2ik} & 1 - \dfrac{\Lambda}{2ik} \end{pmatrix}. \tag{3.7}$$

The same expression for the transfer matrix can be obtained in the case of $\Lambda < 0$ (see problem 1.1).

3.1.1 Scattering Matrix

In the case of the δ-function potential, we can find an explicit form of the continuity equations, given by equations (1.8) and (1.9) in chapter 1. We start with Schrödinger's equation

$$-\frac{\hbar^2}{2m} \frac{\partial^2 \Psi(x)}{\partial x^2} + [V(x) - E]\Psi(x) = 0 \tag{3.8}$$

with potential $V(x)$ given by equation (3.1). First we integrate both sides of Schrödinger's equation (3.8) in the very narrow interval $-\varepsilon < x < \varepsilon$. Integration of the second derivative $\partial^2 \Psi(x)/\partial x^2$ is easy and gives us

$$\int_{-\varepsilon}^{+\varepsilon} \frac{\partial^2 \Psi(x)}{\partial x^2} \, dx = \frac{\partial \Psi(+\varepsilon)}{\partial x} - \frac{\partial \Psi(-\varepsilon)}{\partial x}. \tag{3.9}$$

To integrate the term $V(x)\Psi(x)$, we use the following property of the δ-function:

$$\int_{-\varepsilon}^{+\varepsilon} \delta(x) f(x) \, dx = f(0) \tag{3.10}$$

which holds for any function $f(x)$ that is continuous at $x = 0$.

Now we perform the limit of $\varepsilon \to 0$ and we obtain the following condition for the first derivative of the wave function at the point $x = 0$, which is the position of the δ-function potential:

$$\frac{\partial \Psi(0^+)}{\partial x} - \frac{\partial \Psi(0^-)}{\partial x} = \Lambda \Psi(0). \tag{3.11}$$

Equation (3.11) together with the condition of the continuity of the wave function,

$$\Psi(0^+) = \Psi(0^-), \tag{3.12}$$

provides us with all the necessary information for the construction of the scattering matrix **S**.

Consider now the wave function

$$\Psi_L(x) = A e^{ikx} + B e^{-ikx} \tag{3.13}$$

on the left side of the δ-function barrier, and the wave function

$$\Psi_R(x) = C e^{ikx} + D e^{-ikx} \tag{3.14}$$

on the right side. We substitute these wave functions into the continuity conditions (3.11) and (3.12) and we obtain the following equations for the coefficients A–D:

$$\begin{aligned} ik\,(C - D - A + B) &= \Lambda\,(A + B) \\ A + B &= C + D. \end{aligned} \tag{3.15}$$

To find the elements of the scattering matrix **S**, we need to express the coefficients of outgoing waves, B and C, in terms of the coefficients of incoming waves, A and D. From the first equation we have $C = A + B - D$. Inserting into the second equation gives us

$$B - D = \frac{\Lambda}{2ik}(A + B), \tag{3.16}$$

from which we express B in the form

$$B = \frac{\Lambda}{2ik - \Lambda} A + \frac{2ik}{2ik - \Lambda} D. \tag{3.17}$$

We insert this expression for B in the first equation (3.15) and express the coefficient C in terms of A and D. Finally, we obtain

$$\begin{pmatrix} B \\ C \end{pmatrix} = \mathbf{S} \begin{pmatrix} A \\ D \end{pmatrix}, \tag{3.18}$$

where the scattering matrix **S** is given by

$$\mathbf{S} = \begin{pmatrix} \dfrac{\Lambda}{2ik - \Lambda} & \dfrac{2ik}{2ik - \Lambda} \\[2ex] \dfrac{2ik}{2ik - \Lambda} & \dfrac{\Lambda}{2ik - \Lambda} \end{pmatrix}. \tag{3.19}$$

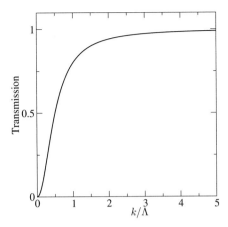

FIGURE 3.1. Transmission coefficient of a quantum particle through a single δ-function barrier as a function of wave vector k/Λ. Large values of k/Λ correspond to large energies of the incident particle, in accordance with equation (3.20).

3.1.2 Transmission Coefficient

For a free particle, the energy and the wave vector of the scattered particle are related by

$$E = \frac{\hbar^2 k^2}{2m}. \tag{3.20}$$

Using the explicit form of the transfer matrix $\mathbf{M}_\Lambda$ for the δ-function barrier given by equation (3.7), we find that the transmission coefficient T is given by

$$T = \frac{1}{|M_{22}|^2} = \frac{1}{1 + |M_{12}|^2} = \frac{4k^2}{4k^2 + \Lambda^2}. \tag{3.21}$$

Note that the transmission coefficient given by equation (3.21) does not depend on the sign of Λ. This means that, for a given energy of the incident particle, transmission is the same for the potential barrier and the potential well.

In figure 3.1 we plot the transmission coefficient T as a function of k/Λ. The transmission coefficient increases when the energy of the incoming particle increases, and is close to 1 when $k \gg \Lambda$.

3.1.3 Bound State

For negative Λ, we have an *attractive* potential in one dimension. In this case a bound state always exists. The wave function of the bound state decreases as $e^{-\kappa|x|}$ on both sides of the δ-function well. To find its energy, we put $k = i\kappa$ and apply the general condition $M_{22}(i\kappa) = 0$ given by equation (1.95) of section 1.4.3. From the explicit form of the transfer matrix (3.7), we obtain $\kappa = -\Lambda/2 = |\Lambda|/2$. The energy of the bound state of the single δ-function potential well, $E_{1b} = -\hbar^2\kappa^2/(2m)$, is then

$$E_{1b} = -\frac{\hbar^2}{2m}\frac{\Lambda^2}{4}, \tag{3.22}$$

and the wave function exponentially decreases on both sides of the impurity as

$$\Psi_b(x) \sim e^{-|\kappa x|} = e^{-|\Lambda x|/2}. \tag{3.23}$$

3.2 Two δ-Function Repulsive Potentials

Consider now two δ-function potentials separated by a distance of length ℓ. To determine the transmission and reflection amplitudes of this system, we calculate first the transfer matrix

$$\mathbf{M} = \mathbf{M}_{\Lambda_2} \begin{pmatrix} e^{ik\ell} & 0 \\ 0 & e^{-ik\ell} \end{pmatrix} \mathbf{M}_{\Lambda_1}. \tag{3.24}$$

Here, $\mathbf{M}_{\Lambda_1}$ ($\mathbf{M}_{\Lambda_2}$) is a transfer matrix of the first (second) δ-function barrier, respectively, given by equation (3.7), and the diagonal matrix

$$\begin{pmatrix} e^{ik\ell} & 0 \\ 0 & e^{-ik\ell} \end{pmatrix} \tag{3.25}$$

is the transfer matrix of the free space of length ℓ between the barriers.

From the explicit form of the transfer matrix $\mathbf{M}_\Lambda$, equation (3.7), we obtain the following expression for the transmission amplitude:

$$\frac{1}{t_{12}} = M_{22} = \left(1 - \frac{\Lambda_1}{2ik}\right)\left(1 - \frac{\Lambda_2}{2ik}\right)e^{-ik\ell} + \frac{\Lambda_1\Lambda_2}{4k^2}e^{+ik\ell}. \tag{3.26}$$

In what follows, we consider only the special case $\Lambda_1 = \Lambda_2 = \Lambda$. In this case,

$$\frac{1}{t_{12}} = M_{22} = \left(1 - \frac{\Lambda}{2ik}\right)^2 e^{-ik\ell} + \frac{\Lambda^2}{4k^2}e^{+ik\ell}, \tag{3.27}$$

and the transmission coefficient

$$T = |t_{12}|^2 \tag{3.28}$$

can be expressed in the form

$$T = \frac{1}{1 + \dfrac{\beta^2}{k^2\ell^2}\left[\cos k\ell + \dfrac{\beta}{2k\ell}\sin k\ell\right]^2} \tag{3.29}$$

(see problem 3.2 for details). The transmission coefficient given by equation (3.29) is expressed in terms of two *dimensionless* parameters. The first parameter,

$$\beta = \Lambda\ell, \tag{3.30}$$

defines the scattering properties of the system of two δ-function barriers separated by distance ℓ. Note that large values of β correspond either to the case of strong scattering potential Λ or to the case of large distance between two δ-function barriers.

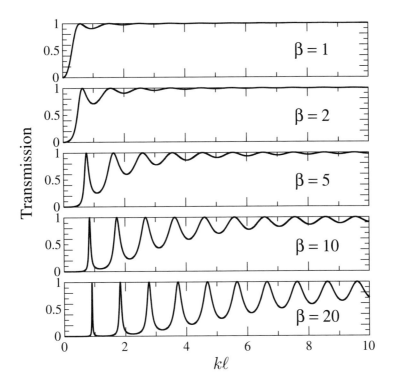

FIGURE 3.2. Transmission coefficient for double δ-function repulsive ($\Lambda > 0$) potentials for various strengths of the barrier $\beta = \Lambda \ell = 1, 2, 5, 10$, and 20.

The second dimensionless parameter in the expression for the transmission coefficient, equation (3.29), is $k\ell$,

$$k\ell = 2\pi \frac{\ell}{\lambda}, \tag{3.31}$$

which gives the ratio of the distance between the barriers, ℓ, to the wavelength λ of the scattered wave (electron), $\lambda = 2\pi/k$.

There are special values of $k\ell$ for which the transmission coefficient given by equation (3.29) is equal to 1. This is called resonant transmission. Indeed, $T = 1$, if

$$\cos k\ell + \frac{\beta}{2} \frac{\sin k\ell}{k\ell} = 0. \tag{3.32}$$

The resonant transmission exists for both the double δ-function repulsive ($\Lambda > 0$) and attractive ($\Lambda < 0$) potentials.

Equation (3.32) cannot be solved analytically. Only for the limit of $\beta \to \infty$ do we find that the resonant transmission appears when $k\ell = n\pi$. This is the case when the distance ℓ between two δ-function potentials is an integer multiple of one-half of the wavelength of the incident particle.

The transmission coefficient T as a function of parameter $k\ell$ is shown in figure 3.2 for various values of β. For small β, the transmission is close to 1 when $k\ell$ increases. This is easy to understand, since large values of k mean large kinetic energy of the incident particle, which simply does not "see" the weak barriers. This also agrees with equation (3.29), which gives T close to unity when $\beta/k\ell \ll 1$. As β increases, more and more transmission peaks with $T = 1$ appear. Peaks become narrower as β increases. In the limit $\beta \to \infty$, they become infinitesimally narrow and are located exactly at $k\ell = n\pi$ as expected from equation (3.32).

3.3 Bound States of Double δ-Function Attractive Potentials

The energy of the bound states of two δ-function attractive potentials can be again found with the use of the general condition $M_{22}(i\kappa) = 0$ derived in section 1.4.3 and given by equation (1.95). We insert $k = i\kappa$ ($\kappa > 0$) into the r.h.s. of equation (3.27) and obtain

$$M_{22}(k = i\kappa) = \left(1 + \frac{\beta}{2\kappa\ell}\right)^2 e^{\kappa\ell} - \frac{\beta^2}{4\kappa^2\ell^2} e^{-\kappa\ell} = 0. \tag{3.33}$$

This is the transcendental equation for κ,

$$e^{-2\kappa\ell} = \frac{(2\kappa\ell + \beta)^2}{\beta^2}. \tag{3.34}$$

After the substitution

$$\kappa\ell = u|\beta|/2 \tag{3.35}$$

($u > 0$), equation (3.34) simplifies to

$$e^{-|\beta|u/2} = |u - 1|. \tag{3.36}$$

The graphical solution of equation (3.36) is shown in figure 3.3. The energy of the bound state,

$$E_{2b} = -\frac{\hbar^2}{2m} \frac{\Lambda^2 u_b^2}{4}, \tag{3.37}$$

is plotted in figure 3.4. For small values of $|\beta|$, we have only one bound state. The second branch appears when $|\beta| = 2$. In the limit of $|\beta| \to \infty$, the energies of both bound states converge to the energy of the bound state of the single δ-function potential, E_{1b}, given by equation (3.22).

To understand the β dependence of the bound energy, first consider two δ-function attractive potentials located far from each other. In this case we obtain two bound states with the same energy, each localized around one impurity. The states are independent of each other because their wave functions decrease exponentially as a function of distance from the barrier, following the relation (3.23). As the two δ-function attractive potentials approach each other, we see from figure 3.4 that we have two branches for the bound state energy. The lower branch has lower energy than that of the single δ-function well. The wave function of the lower branch is symmetric, $\Psi_1(x) = \Psi_1(-x)$, as shown in figure 3.5, while the wave function of the upper branch is antisymmetric: $\Psi_2(x) = -\Psi_2(-x)$. This is a typical case of "bonding" and "antibonding" states. Each bound state of the δ-function well is split when the two δ-function wells are brought together (or, in the

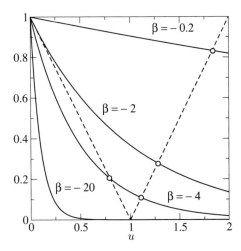

FIGURE 3.3. Graphical solution of equation (3.36) for obtaining the bound states of two δ-function attractive ($\Lambda < 0$) potentials. Dashed lines represent the r.h.s. of equation (3.36), which is either $u - 1$ for $u > 1$ or $1 - u$ for $u < 1$. Solid lines plot $e^{-|\beta|u/2}$ for four choices of β. Open circles denote solutions u_b. We see that for $|\beta| < 2$ there is only one solution $u_b > 1$. For $|\beta| > 2$ there are two solutions, $u_{b1} > 1$ and $u_{b2} < 1$. In the limit $|\beta| \to \infty$, both solutions converge to the same value, $u_{b1} \to 1^+$ and $u_{b2} \to 1^-$.

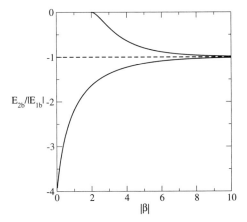

FIGURE 3.4. Energy of the bound states of two δ-function potential wells, $E_{2b}/|E_{1b}|$, as a function of $\beta = \Lambda\ell$ ($\Lambda < 0$). $E_{1b} = -(\hbar^2/2m)\Lambda^2/4$ is the energy of the bound state of a single δ-function potential given by equation (3.22). For $|\beta| < 2$, there is only one solution. Two branches exist for $|\beta| > 2$. Both converge to the single impurity bound energy when $|\beta| \to \infty$. As $\beta = \Lambda\ell$, the limit $|\beta| \to \infty$ means that the distance between two δ-function barriers increases to infinity.

same way, when the strength of the δ-function potential decreases) to form bonding (symmetric wave function with lower energy) and antibonding (antisymmetric wave function with higher energy than that of the isolated δ-function well) states. The energy difference between the bonding and antibonding states depends, as shown in figure 3.5, on the overlap between the wave functions of the isolated δ-function wells. Notice that,

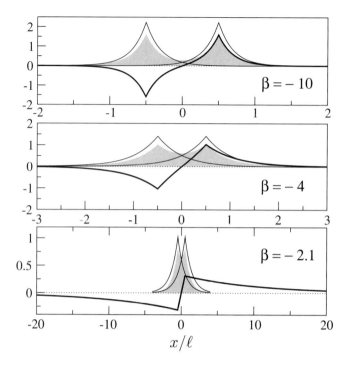

FIGURE 3.5. Wave functions of bound states of two δ-function attractive potentials for three values of β. Note that the horizontal axis is normalized to the distance between two δ-function barriers and that the antisymmetric solution becomes broader when β approaches -2. The symmetric solution is shown by the dashed area, and the antisymmetric solution is by the bold solid line.

as the distance between two δ-function wells decreases, the energy difference between the bonding and antibonding states increases. For the special case of two δ-function attractive potentials, when the distance between the two δ-function wells is equal to $2/\Lambda$ (i.e., $|\beta| = -\Lambda\ell = 2$) the antisymmetric wave function ceases to exist.

3.4 N Identical δ-Function Barriers

Now we proceed with the investigation of the more general case of transmission through N identical δ-function barriers. We are interested in the energy dependence of the transmission.

To determine the transmission coefficient, we must calculate the transfer matrix,

$$\mathbf{M}_\Lambda \mathbf{M}_0 \mathbf{M}_\Lambda \cdots \mathbf{M}_0 \mathbf{M}_\Lambda, \tag{3.38}$$

where $\mathbf{M}_\Lambda$ is the transfer matrix of a single δ-function barrier given by equation (3.7), and

$$\mathbf{M}_0 = \begin{pmatrix} e^{ik\ell} & 0 \\ 0 & \varepsilon^{-ik\ell} \end{pmatrix} \tag{3.39}$$

is the transfer matrix of the free space between the two barriers.

To simplify our analysis, we add an additional free space of length ℓ in front of the first δ-function barrier. We are allowed to do it, since, according to the results of section 1.4.1 [see equations (1.83) and (1.84)], the additional free space changes only the phase of the transmission and reflection amplitudes, but does not influence the value of the transmission and reflection coefficients. Then, the transfer matrix of the N barriers can be expressed as

$$\mathbf{M}_N = [\mathbf{M}_1]^N, \tag{3.40}$$

where

$$\mathbf{M}_1 = \mathbf{M}_0 \, \mathbf{M}_\Lambda. \tag{3.41}$$

After simple algebra we find that the matrix $\mathbf{M}_1$ is given by

$$\mathbf{M}_1 = \begin{pmatrix} e^{ik\ell}\left[1 + \dfrac{\beta}{2ik\ell}\right] & \dfrac{\beta}{2ik\ell}e^{ik\ell} \\[2ex] -\dfrac{\beta}{2ik\ell}e^{-ik\ell} & e^{-ik\ell}\left[1 - \dfrac{\beta}{2ik\ell}\right] \end{pmatrix}. \tag{3.42}$$

For one individual barrier, $N = 1$, we obtain from equation (3.42) that the transmission coefficient of a single barrier is

$$T_1 = \frac{1}{|M_{22}|^2} = \frac{1}{1 + \dfrac{\beta^2}{4k^2\ell^2}}. \tag{3.43}$$

We want to remind the readers that T_1 can also be obtained from the following formula, given by equation (1.102):

$$T_1 = \frac{1}{1 + |M_{12}|^2} = \frac{1}{1 + \left|\dfrac{r}{t}\right|^2} = \frac{1}{1 + \dfrac{\beta^2}{4k^2\ell^2}}. \tag{3.44}$$

These expressions are identical with formula (3.21) (remember that $\beta = \Lambda\ell$).

With the use of Chebyshev's identity [equation (1.106)], we can immediately write the explicit form of the transmission coefficient T_N for the system of N identical δ-function barriers,

$$T_N = |t_N|^2 = \frac{1}{1 + \left|\dfrac{r_1^2}{t_1^2}\right| \dfrac{\sin^2 Nq\ell}{\sin^2 q\ell}} = \frac{1}{1 + \dfrac{\beta^2}{4k^2\ell^2}\dfrac{\sin^2 Nq\ell}{\sin^2 q\ell}}, \tag{3.45}$$

where the parameter $q\ell$ is given by the trace of the transfer matrix $\mathbf{M}_1$. Using equations (1.101) and (3.42), we easily find that

$$2\cos q\ell = \mathrm{Tr}\,\mathbf{M}_1 = 2\cos k\ell + \beta\frac{\sin k\ell}{k\ell}. \tag{3.46}$$

The transmission coefficient through the system of N identical δ-function barriers is shown in figure 3.6 for $\beta = 3\pi$ and for different values of N. We see that there are intervals of $k\ell$ in which the transmission rapidly decreases when N increases. To explain these observations, we remind the reader of what we learned in section 1.4.2, namely, that the

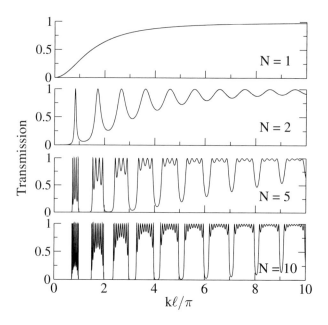

FIGURE 3.6. Transmission coefficient through N identical δ-function barriers as a function of $k\ell$. $N = 1$, 2, 5, and 10 (from top to bottom). $\beta = 3\pi$. A more detailed plot of the transmission coefficient for the first "transmission band" is shown in figure 3.8. One clearly sees the bands and the gaps for the $N = 10$ case.

transfer matrix $\mathbf{M}_N = [\mathbf{M}_1]^N$ has propagating solutions only when $|\text{Tr } \mathbf{M}_N| \leq 2$. In our case this condition is equivalent to the condition

$$|\text{Tr } \mathbf{M}_1| \leq 2 \tag{3.47}$$

(see problem 3.5 for details). We plot in figure 3.7 Tr $\mathbf{M}_1$ as a function of the wave vector of an incoming particle, $k\ell$. We see intervals of $k\ell$ where $|\text{Tr } \mathbf{M}_1| > 2$. These intervals are identical with those intervals in which the transmission coefficient rapidly decreases to zero when the number of barriers increases (figure 3.6).

A rapid decrease of the transmission when the Tr $\mathbf{M}_1 > 2$ can be understood also from the analytical expression for the transmission coefficient, equation (3.45). If Tr $|\mathbf{M}_1| > 2$, then we see from equation (3.46) that $q\ell$ is imaginary, so that the factor $\sin(Nq\ell)/\sin q\ell$ increases exponentially as $e^{(N-1)|q\ell|}$, when the number of δ-function barriers, N, increases. This, in turn, causes the transmission coefficient to decrease exponentially for large N.

On the other hand, the transmission is high in the interval where $|\text{Tr } \mathbf{M}_1| < 2$. In each such interval, $q\ell$ is real and varies from 0 to π. There are exactly $N - 1$ values of $q\ell$,

$$q\ell = \frac{n}{N}\pi, \quad n = 1, 2, \ldots, N - 1, \tag{3.48}$$

for which the transmission coefficient $T = 1$ (note that two other candidates for resonant transmission, namely, $q\ell = 0$ and π, do not give $T = 1$, since the ratio $\sin Nq\ell / \sin q\ell \neq 0$ for these values of $q\ell$). Indeed, we see $N - 1$ peaks in each transmission interval in figures 3.6 and 3.8.

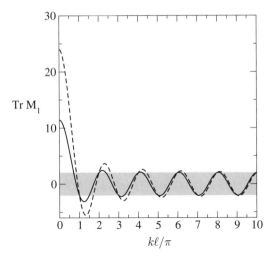

FIGURE 3.7. Tr $\mathbf{M}_1$ given by equation (3.46) as a function of $k\ell/\pi$ for $\beta = 3\pi$ (solid line) and $\beta_0 = 7\pi$ (dashed line). Only values of $k\ell$, that give $|\text{Tr } \mathbf{M}_1| < 2$ (dashed area) will allow transmission in the limit of large N.

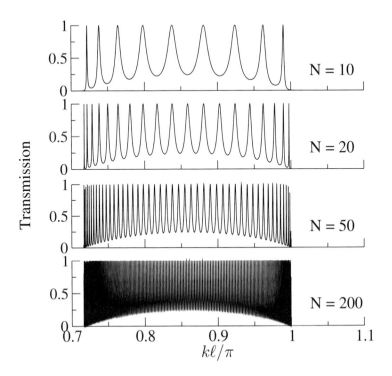

FIGURE 3.8. Detailed plot of the transmission coefficient through N δ-function barriers ($\beta = 3\pi$). $N = 10$, 20, 50, and 200. Notice that for each N there are $N-1$ transmission peaks which give $T = 1$, and $N-2$ transmission minima.

It looks reasonable to assume that each interval of the allowed transmission converges to the transmission bands with $T \equiv 1$ in the limit of $N \to \infty$. However, this is not the case. As we see in figure 3.8, the transmission coefficient T rapidly oscillates for large N as a function of $k\ell$. These oscillations do not disappear in the limit $N \to \infty$. This is clearly seen from the analytical formula for transmission, given by equation (3.45) We see that for values

$$q\ell = \frac{\pi}{N}\left(n + \frac{1}{2}\right) \tag{3.49}$$

the denominator in equation (3.45) is $1 + |r_1/t_1|^2 / \sin^2 q\ell > 1 + |r_1/t_1|^2$. This means that for these particular values of $q\ell$ the transmission coefficient through N δ-function barriers is even *smaller* than the transmission coefficient for a single δ-function barrier. We conclude that the transmission coefficient, given by equation (3.45), oscillates between the maximum value 1 when $q\ell$ satisfies the condition (3.48) and the minimum value $1/(1 + |r/t|^2 / \sin^2 q\ell)$ when $q\ell$ is given by equation (3.49). Since $|r/t|^2 = \beta^2/(4k^2\ell^2)$, it is clear that these oscillations are less pronounced when $\beta \ll k\ell$. Nevertheless, to obtain transmission bands, we have to consider an *infinite* lattice of δ-function barriers. We will discuss this problem in chapter 4.

3.5 Supplementary Notes

3.5.1 Bonding and Antibonding States

Consider a system of two identical potential wells located far away from each other at $x_1 = -a$ and $x_2 = +a$. If the wave functions of the particle in a given potential well, $\Phi_1(x) = \Phi_0(x - a)$ and $\Phi_2(x) = \Phi_0(x + a)$, decrease sufficiently fast as a function of the distance form the potential well, so that $\int dx\ \Phi_0^*(x - a)\Phi_0(x + a) = 0$, then the system is degenerate, with eigenenergies $E_1 \equiv E_2 = E_0$, where $E_0 = E_{1b}$ is the bound state of the single potential barrier, given by equation (3.22). The corresponding eigenfunctions are $\Phi_1 = \Phi_0(x - a)$ and $\Phi_2 = \Phi_0(x + a)$. In this representation the Hamiltonian matrix of such a system can be written as

$$\mathcal{H}_0 = \begin{pmatrix} E_0 & 0 \\ 0 & E_0 \end{pmatrix}, \tag{3.50}$$

and the eigenvectors of the degenerate system are simply

$$u_1 = \begin{pmatrix} 1 \\ 0 \end{pmatrix} \quad \text{and} \quad u_2 = \begin{pmatrix} 0 \\ 1 \end{pmatrix}, \tag{3.51}$$

which means that the particle is either in the first well or in the second one.

Assume now that we shift the potential wells closer to each other, so that the overlap of the wave functions,

$$V = \int dx\ \Phi_1(x)^* \Phi_2(x), \tag{3.52}$$

is nonzero. This enables the particle to tunnel from one potential well into the second one. In the limit $V \ll 1$, we approximate the Hamiltonian matrix by

$$\mathcal{H} = \begin{pmatrix} E_0 & V \\ V & E_0 \end{pmatrix}. \tag{3.53}$$

From equation (A.24) we see that the eigenvalues are given by the solution of the equation $\det (\mathcal{H} - E) = 0$, which in our case has the form

$$(E_0 - E)^2 - V^2 = 0. \tag{3.54}$$

It is easy to show that the eigenvalues are of the form

$$E_b = E_0 - V \tag{3.55}$$

and

$$E_a = E_0 + V. \tag{3.56}$$

We call the state with energy E_b the *bonding* state, and the state with energy E_a the *antibonding* state. The eigenvectors of the bonding and antibonding states are

$$\Psi_b = \frac{1}{\sqrt{2}} \left[\Phi_1(x) + \Phi_2(x) \right] = \frac{1}{\sqrt{2}} \left[\Phi_0(x-a) + \Phi_0(x+a) \right] \tag{3.57}$$

and

$$\Psi_a = \frac{1}{\sqrt{2}} \left[\Phi_1(x) - \Phi_2(x) \right] = \frac{1}{\sqrt{2}} \left[\Phi_0(x-a) - \Phi_0(x+a) \right]. \tag{3.58}$$

We see that the bonding state, which possesses the lower energy E_b, is *symmetric*, and the antibonding state with higher energy E_a is *antisymmetric*. Note that, the symmetry of the wave functions Ψ_a and Ψ_b is a consequence of the symmetry of the system with respect to space inversion $x \to -x$. Also, note that a nonzero overlap V, no matter how small it is, always causes the splitting of a degenerate spectrum into two nondegenerate energy levels.

3.6 Problems

Problems with Solutions

Problem 3.1
Derive the transfer matrix for the single δ-function attractive potential $(\Lambda < 0)$.

Solution. The derivation proceeds in the same way as for the case $\Lambda > 0$, but the wave vector k' is now given by the relation

$$k' = \sqrt{\frac{|\Lambda|}{2a}}. \tag{3.59}$$

Note that, k' is real, in contrast to the case $\Lambda > 0$ when k' is imaginary [equation (3.4)]. We can perform all the limiting procedures described in section 3.1. For instance, $\sin 2k'a \approx \sqrt{2|\Lambda|a}$ and the matrix element M_{11} in equation (2.17) reduces to $M_{11} = 1 + i|\Lambda|/2k$. Since Λ is negative, we have $\Lambda = -|\Lambda|$ and recover expression (3.7).

Problem 3.2
Starting from the expression for the transmission amplitude, equation (3.26), derive the transmission coefficient T given by equation (3.29) for the case $\Lambda_1 = \Lambda_2 = \Lambda$.

Solution. To simplify the notation, we define $a = \Lambda/(2k)$. Then equation (3.26) reduces to

$$t_{12}^{-1} = (1 + ia)^2 e^{\,ik\ell} + a^2 e^{ik\ell} \tag{3.60}$$

and

$$T^{-1} = |t_{12}|^{-2} = \left[(1 + ia)^2 e^{-ik\ell} + a^2 e^{+ik\ell}\right]\left[(1 - ia)^2 e^{+ik\ell} + a^2 e^{-ik\ell}\right], \tag{3.61}$$

which gives

$$T^{-1} = 1 + 2a^2 + 2a^4 + 2a^2\left[(1 - a^2)\cos 2k\ell + 2a \sin 2k\ell\right]. \tag{3.62}$$

After rearranging the terms on the r.h.s. of equation (3.62), we obtain

$$T^{-1} = 1 + 2a^2\left[(1 + \cos 2k\ell) + a^2(1 - \cos 2k\ell) + 2a \sin 2k\ell\right]. \tag{3.63}$$

By using the relations $1 + \cos 2x = 2\cos^2 x$, $1 - \cos 2x = 2\sin^2 x$, and $\sin 2x = 2 \sin x \cos x$, we obtain

$$T^{-1} = 1 + 4a^2\left[\cos^2 k\ell + a^2 \sin^2 k\ell + 2a \sin k\ell \cos k\ell\right]. \tag{3.64}$$

Inserting $a = \Lambda/2k = \beta/2k\ell$, we obtain equation (3.29).

Problem 3.3
Explain why the single rectangular potential studied in chapter 2 gives a resonance transmission $T = 1$, but the transmission is always smaller than 1 for a single δ-function potential.

Solution. Note that, for the case of the rectangular potential, the wave can be reflected from the two boundaries—namely, the left and right boundaries of the rectangular potential. Interference of the reflected waves results in $T = 1$ for $k\ell = n\pi$ [equation (2.28)]. This effect is missing for the δ-function potential barrier.

Problem 3.4
Note that the system of two δ-function attracting potentials with $\beta = -2$ differs qualitatively from any other value of β (see figure 3.9). Use the analytical formula for the transmission coefficient, given by equation (3.29) and explain why $T \to 1$ when $\beta = -2$ and $k\ell \to 0$.

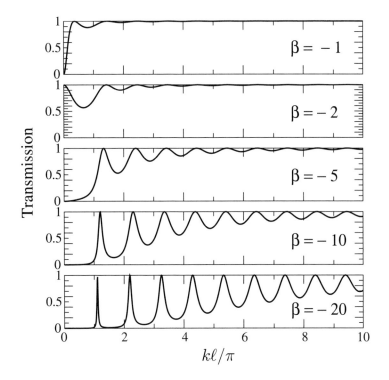

FIGURE 3.9. Transmission coefficient for two δ-function potential wells for various strengths of *negative* $\beta = -1, -2, -5, -10,$ and -20. Note that, $T \to 1$ when $\beta = -2$ and $k\ell \to 0$, and see problem 3.4 for explanation. Note also that the resonance transmission peaks $T = 1$ are located above the values $k\ell = n\pi$, in contrast to the case of positive β.

Solution. By inserting $\beta = -2$ in the denominator of the result equation (3.29) we obtain the result that

$$T^{-1} = 1 + \frac{4}{k^2\ell^2}\left[\cos k\ell - \frac{1}{k\ell}\sin k\ell\right]^2. \qquad (3.65)$$

In the limit $k\ell \to 0$, we Taylor-expand both $\cos k\ell$ and $\sin k\ell$, $\cos x = 1 - x^2/2 + \mathcal{O}(x^4)$ and $(\sin x)/x = 1 - x^2/6 + \mathcal{O}(x^4)$, and obtain

$$T^{-1} = 1 + \frac{4}{k^2\ell^2}\left[1 - \frac{(k\ell)^2}{2} - \frac{1}{k\ell}\left(k\ell - \frac{(k\ell)^3}{6}\right)\right]^2, \qquad (3.66)$$

so that $T^{-1} = 1 + 4(k\ell)^2/9$ converges to 1 when $k\ell \to 0$. Consequently,

$$\lim_{k\ell \to 0} T = 1. \qquad (3.67)$$

Problem 3.5
Show that the two conditions $|\text{Tr } \mathbf{M}_1| < 2$ and $|\text{Tr } \mathbf{M}_N| < 2$, where $\mathbf{M}_N = [\mathbf{M}_1]^N$, are satisfied simultaneously.

Solution. When $|\text{Tr } \mathbf{M}_1| < 2$, then the matrix $\mathbf{M}_1$ has the eigenvalues $\lambda_{1,2} = e^{\pm iq\ell}$ and $q\ell$ is real. Since the eigenvalues of the matrix $\mathbf{M}_N$ are given by $\lambda_{1,2}^N = e^{\pm iNq\ell}$, we immediately

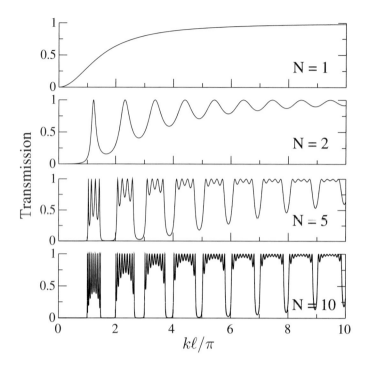

FIGURE 3.10. Transmission through N identical δ-function attractive potentials ($\beta = -3\pi$). Note the positions of the transmission peaks. Compare them with the transmission peaks for $\beta = 3\pi$ (figure 3.6).

see that $|\mathrm{Tr}\,\mathbf{M}_N| = |2\cos Nq\ell| < 2$. Similarly, if $q\ell$ is imaginary so that $|\lambda_1| > 1$, then the absolute values of both traces are larger than 2.

Problems without Solutions

Problem 3.6
Prove that the scattering matrix $\mathbf{S}$ given by equation (3.19) is unitary (i.e., $\mathbf{S}^\dagger\mathbf{S} = 1$). Show also that the $\mathbf{S}$ matrix satisfies the condition $\mathbf{S}^* = \mathbf{S}^{-1}$.

Problem 3.7
Derive an explicit form of the transfer matrix $\mathbf{M}_\Lambda$ from the elements of the scattering matrix, given by equation (3.19).

Problem 3.8
Find the transmission coefficient through two δ-function potential barriers, equation (3.29), in the limit of $\beta \to \infty$. Show that condition for transmission resonance, $T = 1$, is

$$k\ell = n\pi, \qquad n = \pm 1, \pm 2, \ldots, \tag{3.68}$$

and does not depend on the sign of β.

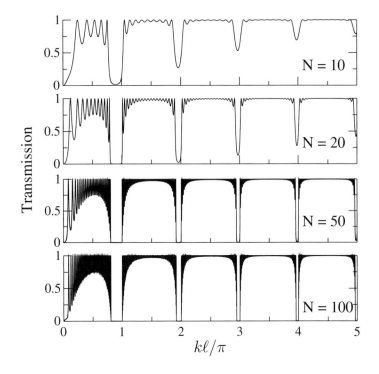

FIGURE 3.11. Transmission through N identical δ-function attractive potentials $(\beta = -\pi/2)$. Note the positions of the transmission peaks and compare them with those for $\beta = -3\pi$ (figure 3.10).

Problem 3.9
Apply the general formula for the transmission coefficient through N identical δ-function barriers, given by equations (3.45) and (3.46), to the special case $N = 2$, and derive an analytical expression for the transmission coefficient through two δ-function barriers. Compare your result with equation (3.29).

Problem 3.10
Study the propagation of a quantum particle through a system of two δ-function potential wells. Plot the transmission coefficient as a function of $k\ell$ for negative $\beta = -1, -2, -5, -10$, and -20. Compare the calculated results (shown in figure 3.9) with those for the positive values of β (figure 3.2). Plot Tr $\mathbf{M}_1$ as a function of the incoming wave vector $k\ell$ to explain the position of the transmission gaps.

Problem 3.11
Calculate the transmission coefficient for a system of N identical δ-function potential wells with a negative β. Use $\beta = -\pi/2$ and -3π. The results are shown in figures 3.10 and 3.11.

4 Kronig-Penney Model

In section 3.4 we studied the transmission of a quantum particle (electron) through N identical δ-function repulsive or attractive potentials. We calculated how the transmission coefficient depends on the parameter $k\ell$, where ℓ is the distance between two neighboring δ-function potential barriers and k is the wave vector of the incident particle. We found intervals of $k\ell$ in which the transmission coefficient decreases exponentially as N increases and becomes infinitesimally small in the limit of $N \to \infty$. These intervals were separated by other intervals in which the transmission coefficient, as a function of $k\ell$, oscillates and is close to 1 if the δ-function potential is weak.

In this chapter we introduce an *infinite periodic* system of δ-function potential barriers. This model is called the *Kronig-Penney* model [23, 30].

The Kronig-Penney model represents the simplest model of a periodic system. The analysis of the transmission properties of the Kronig-Penney model enables us to understand the main properties of electronic transport in solids. Solids generally have a crystalline structure, that is, the ions are arranged in a way that exhibits periodicity in space. So when a free particle propagates in such a periodic potential, its motion will be affected by the spatial periodicity.

We will calculate the energy spectrum of the Kronig-Penney model, which consists of bands of allowed energies, separated by gaps. No transmission is allowed in gaps. We derive the density of propagating states inside each transmission band and discuss the properties of bound states.

4.1 The Periodic Model

Schrödinger's equation for a quantum particle in the Kronig-Penney model reads

$$-\frac{\hbar^2}{2m}\frac{\partial^2\Psi(x)}{\partial x^2} + \sum_{n=-\infty}^{+\infty}\frac{\hbar^2}{2m}\frac{\beta_n}{\ell}\delta(x-n\ell)\Psi(x) = E\Psi(x). \tag{4.1}$$

In equation (4.1), β_n is the strength of the nth δ-function potential. In a periodic model, we consider $\beta_n = \beta$ for each site n. The parameter ℓ is the spacing between two neighboring δ-function potentials. For $\beta > 0$, we have a periodic lattice of δ-function repulsive potentials, while the choice of $\beta < 0$ corresponds to an infinite periodic system of δ-function attractive potentials.

For the case of zero potential ($\beta_n = 0$ for all n), equation (4.1) gives the propagation of a free particle. In this case the energy of the particle is

$$E = \frac{\hbar^2 k^2}{2m}, \tag{4.2}$$

and the corresponding wave function $\Psi(x) = e^{ikx}$. The wave function at position $x+a$ is $\Psi(x+a) = e^{ik(x+a)} = e^{ika}\Psi(x)$ for any value of a. We therefore have that

$$|\Psi(x+a)|^2 = |\Psi(x)|^2, \tag{4.3}$$

i.e., the probabilities of finding the particle at positions x and $x+a$ are the same. In the *periodic* Kronig-Penney model, with $\beta_n = \beta$ for all n, the translational symmetry of the system is broken. The relation (4.3) holds only for $a = n\ell$, where n is an integer number. We will show that the wave function has the form $\Psi(x) = e^{iqx}u_q(x)$, where q is the wave vector of the particle propagating in the periodic Kronig-Penney model, and $u_q(x) = u_q(x+\ell)$ is a periodic function with the spatial period of the Kronig-Penney model. This statement, known as *Bloch's theorem* [1,6,7,10,17,23,30,39], was first derived by F. Bloch. It is also known in the mathematical literature as *Floquet's theorem*.

Since there is no potential in the region between two neighboring barriers, $n\ell < x < (n+1)\ell$, the wave function of an electron in this region can be written as a superposition of plane waves of a free particle,

$$\Psi_n(x) = A_n e^{ikx} + B_n e^{-ikx}. \tag{4.4}$$

The energy of a free particle with wave vector k is given by equation (4.2). It is convenient to define the wave functions in a discrete set of spatial points, $x_n = n\ell^-$, located infinitesimally close to the nth barrier,

$$\Psi_n = \Psi(x = n\ell^-). \tag{4.5}$$

Then we can express the wave function Ψ_{n+1} in terms of the wave function Ψ_n as

$$\begin{pmatrix}\Psi_{n+1}^+ \\ \Psi_{n+1}^-\end{pmatrix} = \mathbf{M}_n \begin{pmatrix}\Psi_n^+ \\ \Psi_n^-\end{pmatrix}, \tag{4.6}$$

where Ψ_n^+ (Ψ_n^-) is the right- (left-)moving wave on the nth site, respectively. The transfer matrix $\mathbf{M}_n$ is formally identical with that given by equation (3.42) but with β substituted by β_n:

$$\mathbf{M}_n = \begin{pmatrix} e^{ik\ell}\left[1 + \dfrac{\beta_n}{2ik\ell}\right] & \dfrac{\beta_n}{2ik\ell}e^{ik\ell} \\ -\dfrac{\beta_n}{2ik\ell}e^{-ik\ell} & e^{-ik\ell}\left[1 - \dfrac{\beta_n}{2ik\ell}\right] \end{pmatrix}. \tag{4.7}$$

For the periodic case, we consider all the δ-function barriers equal to each other, $\beta_n = \beta$, and put $\mathbf{M}_n = \mathbf{M}_\beta$. It is easy to verify that $\det \mathbf{M}_\beta = 1$ and

$$\mathrm{Tr}\,\mathbf{M}_\beta = \beta\,\frac{\sin k\ell}{k\ell} + 2\cos k\ell. \tag{4.8}$$

Then, two eigenvalues of $\mathbf{M}_\beta$ have the form

$$\lambda_{1,2} = e^{\pm iq\ell}. \tag{4.9}$$

Since $\mathrm{Tr}\,\mathbf{M}_\beta = \lambda_1 + \lambda_2 = e^{iq\ell} + e^{-iq\ell} = 2\cos q\ell$, we obtain from equation (4.8)

$$2\cos q\ell = \beta\,\frac{\sin k\ell}{k\ell} + 2\cos k\ell. \tag{4.10}$$

The left side of equation (4.10) must always be bounded by 2 in order to have a propagating solution (real q). This provides us with the restrictions on the possible ranges of energy $E = \hbar^2 k^2/(2m)$ of the particle. If $\beta = 0$, we obtain that $q = k$ and the particle is free to move.

4.2 Allowed Energy Bands

Figure 4.1 shows a plot of the function

$$f(k\ell) = \cos k\ell + \frac{\beta}{2}\frac{\sin k\ell}{k\ell} \tag{4.11}$$

as a function of $k\ell$. The horizontal dotted lines represent the bands of $\cos q\ell$ and the forbidden regions of $k\ell$, for which the function $f(k\ell)$ lies outside the ±1 limits. We therefore have allowed energy bands separated by regions that are forbidden, called gaps. There are no allowed states with energy $E(k)$ if $k\ell$ lies in a gap.

The allowed energies correspond to values of $k\ell$ for which $q\ell$ is real in equation (4.10). This means that $|\cos q\ell| < 1$. Using equation (4.10), we can write the condition for allowed energies as follows:

$$-1 \le \cos k\ell + \frac{\beta}{2}\frac{\sin k\ell}{k\ell} \le 1. \tag{4.12}$$

Now we will find expressions for the upper and lower edges of the transmission bands. First, consider the case of *positive* β. From figure 4.1, we see that the lower and upper band edges, q_-^m and q_+^m, of the mth transmission band can be found from the

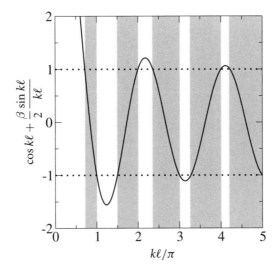

FIGURE 4.1. Plot of $f(k\ell) = \cos k\ell + (\beta/2)(\sin k\ell)/k\ell$ as a function of $k\ell/\pi$ for $\beta = 3\pi$. Dotted horizontal lines indicate the bounds ± 1 of $f(k\ell)$. Only energy states for which $|\cos q\ell| \leq 1$ are allowed. No transmission is possible for a value of k for which $|\cos q\ell| > 1$. The intervals of allowed energies are highlighted by shaded areas.

conditions

$$\cos q_+^m \ell = -1 \quad \text{and} \quad \cos q_-^m \ell = 1, \quad m = 1, 3, 5, \ldots,$$

$$\cos q_-^m \ell = -1 \quad \text{and} \quad \cos q_+^m \ell = 1, \quad m = 2, 4, 6, \ldots. \tag{4.13}$$

Note that for

$$k\ell = m\pi, \quad m = 1, 2, \ldots, \tag{4.14}$$

we have $\sin m\pi = 0$, and, consequently, also $q\ell \equiv k\ell = m\pi$. Since $\cos m\pi = (-1)^m$, we identify the value $q\ell = m\pi$ with the upper edge of the mth band:

$$q_+^m \ell = m\pi. \tag{4.15}$$

Note that equation (4.15) also gives the onset of the energy gaps, the regions where the wave particle cannot propagate since the wave vector is imaginary. The existence of energy gaps can be understood qualitatively as follows. In a first approximation, the particles are free to move, except that there is destructive interference when the waves are reflected from successive δ-function barriers and obtain a phase difference that is equal to an integral number of 2π. This is the condition satisfying equation (4.15). This condition is also called *Bragg's condition*.

To find the lower edge of the mth band, we solve the equation

$$\cos k\ell + \frac{\beta}{2} \frac{\sin k\ell}{k\ell} = (-1)^{m+1}. \tag{4.16}$$

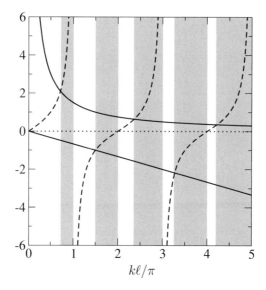

FIGURE 4.2. Graphical solution of equations (4.17) for the case $\beta = 3\pi$. Dashed lines show $\tan k\ell/2$; solid lines are $\beta/(2k\ell)$ and $-2k\ell/\beta$. Upper band edges are given by $k\ell = m\pi$, $m = 1, 2, \ldots$. Lower bands are determined by the cross sections of the solid and dashed lines. Bands of allowed energies are displayed by shaded areas.

Using the relations $1 + \cos x = 2\cos^2(x/2)$, $1 - \cos x = 2\sin^2(x/2)$, and $\sin x = 2\sin(x/2)\cos(x/2)$, we obtain from equation (4.16) that the lower band edge is given by the solution of the following equations:

$$\tan \frac{k\ell}{2} = +\frac{\beta}{2k\ell}, \quad m \text{ odd},$$

$$\tan \frac{k\ell}{2} = -\frac{2k\ell}{\beta}, \quad m \text{ even}.$$

(4.17)

The graphical solution of equations (4.17) is given in figure 4.2.

Figure 4.3 is a plot of $k^2\ell^2 = (2m/\hbar^2)E\ell^2$ versus $q\ell$. For completeness, we show in figure 4.4 also the $q\ell$ dependence of the vector $k\ell$. It is evident that, in the limit of $\beta = 0$, we obtain $q\ell \equiv k\ell$, so that all values of k are allowed.

We see in figure 4.3 that the energy spectrum consists of continuous bands separated by forbidden gaps. The band structure of the energy of electrons in a periodic structure is a direct consequence of the wave nature of quantum particles. This very simple picture of band structure accounts qualitatively for the distinction between conductors and insulators discussed in the next chapter.

The case of negative β must be treated separately. The main difference between an infinite periodic array of δ-function attractive ($\beta < 0$) and repulsive ($\beta > 0$) potentials lies in the form of the energy spectrum. When performing the limit $k\ell \to 0$, we find that

$$\text{Tr } \mathbf{M}_\beta = 2\cos q\ell = 2 + \beta.$$

(4.18)

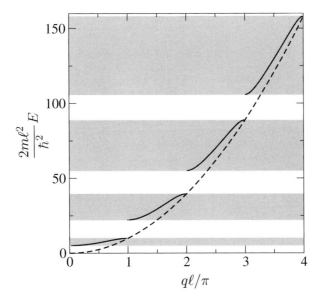

FIGURE 4.3. Energy of the electron (measured in units $\hbar^2/(2m\ell^2)$ as a function of $q\ell$ for $\beta = 3\pi$. The dashed line is the free-electron solution $E = \hbar^2 k^2/(2m) = \hbar^2 q^2/(2m)$, which corresponds to the case $\beta = 0$. The shaded areas give the allowed energy bands.

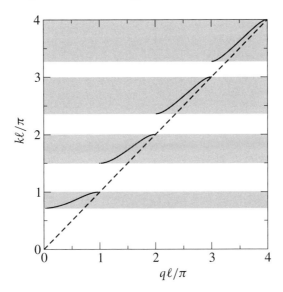

FIGURE 4.4. Relation between wave vectors k and q for the Kronig-Penney model with $\beta = 3\pi/2$.

We immediately see that in this limit $\text{Tr } \mathbf{M}_\beta > 2$ for all *positive* values of β, but $|\text{Tr } \mathbf{M}_\beta| < 2$ for

$$-4 < \beta < 0. \tag{4.19}$$

For these values of β, there is no gap at the beginning of the spectrum of the Kronig-Penney model, as is shown in figure 4.5. The lower and upper band edges of

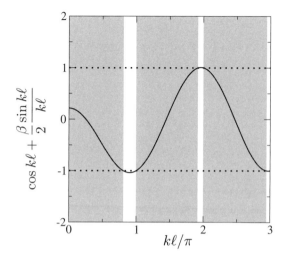

FIGURE 4.5. Plot of $f(k\ell) = \cos k\ell + (\beta/2)(\sin k\ell)/k\ell$ as a function of $k\ell/\pi$ for negative $\beta = -\pi/2$. Dotted horizontal lines indicate the bounds ± 1 of $f(k\ell)$. Only energy states for which $|\cos q\ell| \leq 1$ are allowed. No transmission is possible for values of k for which $|\cos q\ell| > 1$. Intervals of allowed energies are highlighted by shading. Note that the first band of allowed energies starts at $k\ell = 0$.

the mth band can be determined in the same way as for $\beta > 0$. However, the values of the wave vector

$$k\ell = (m-1)\pi, \quad m = 1, 2, \ldots, \tag{4.20}$$

determine the position of the *lower* band edge of the mth band.

The *upper* band edges can be calculated from the equation

$$\cos k\ell + \frac{\beta}{2}\frac{\sin k\ell}{k\ell} = (-1)^m \tag{4.21}$$

[compare this equation with equation (4.16) for the lower band edges of the model with β positive]. This leads to the following equations:

$$\tan\frac{k\ell}{2} = +\frac{2k\ell}{|\beta|}, \quad m = 1, 3, \ldots,$$

$$\tag{4.22}$$

$$\tan\frac{k\ell}{2} = -\frac{|\beta|}{2k\ell}, \quad m = 2, 4, \ldots.$$

A graphical solution for $\beta = \pi/2$ is shown in figure 4.6.

For $\beta < -4$, the lowest energy band disappears. The first transmission band starts for $k\ell = \pi$. There are no longer any allowed energies for $k\ell < \pi$. We can use the solutions of equations (4.20) and (4.22) for the estimation of the position of the mth band, but must substitute $m \to m+1$ since the lowest transmission band no longer exists. We will demonstrate the properties of the spectrum of the Kronig-Penney model in the next section.

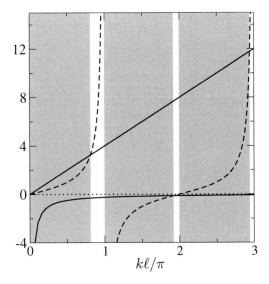

FIGURE 4.6. Graphical solution of equations (4.22) for $\beta = -\pi/2$. Dashed line is $\tan k\ell/2$. Note that there is no gap at the beginning of the energy spectrum.

4.3 The Density of States

The density of states of the one-dimensional quantum system is defined in section 4.7.1 by the following relation:

$$n(E) = \frac{1}{2\pi} \frac{1}{\left|\dfrac{\partial E}{\partial q}\right|}. \tag{4.23}$$

To calculate the density of states of the Kronig-Penney model, we start with equation (4.10). We have to express the energy E in terms of the wave vector q of the electron,

$$n(E) = \frac{1}{2\pi} \left|\frac{\partial q}{\partial E}\right| = \left|\frac{1}{2\pi} \frac{\partial k}{\partial E} \frac{\partial q}{\partial k}\right| = \left|\frac{1}{\pi} \frac{1}{\partial E/\partial k} \frac{\partial q\ell}{\partial \cos q\ell} \frac{\partial \cos q\ell}{\partial k\ell}\right| \tag{4.24}$$

$$= \frac{\sqrt{m}}{\sqrt{2}\pi\hbar\sqrt{E}} \left|\frac{1}{\sin q\ell} \frac{\partial \cos q\ell}{\partial k\ell}\right|,$$

which gives an explicit expression for the density of states,

$$n(E) = \frac{\sqrt{m}}{\sqrt{2}\pi\hbar\sqrt{E}} \left|\frac{\dfrac{\beta}{2}\left(\dfrac{\sin k\ell}{(k\ell)^2} - \dfrac{\cos k\ell}{k\ell}\right) + \sin k\ell}{\left[1 - \left(\dfrac{\beta}{2}\dfrac{\sin k\ell}{k\ell} + \cos k\ell\right)^2\right]^{1/2}}\right|. \tag{4.25}$$

The density of states $n(E)$ is shown in figure 4.7 for two positive values of β and in figure 4.8 for two negative values of β. These plots exhibit all the properties of the energy

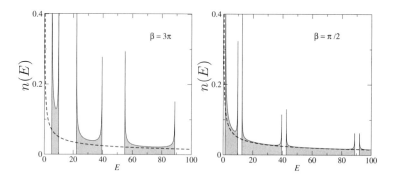

FIGURE 4.7. Density of states of the Kronig-Penney model for $\beta = 3\pi$ (left) and $\pi/2$ (right). The dashed line is the density of states for the free-electron case ($\beta = 0$).

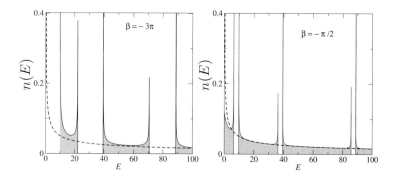

FIGURE 4.8. Density of states for a *negative* $\beta = -3\pi$ and for $\beta = -\pi/2$. The dashed line is the density of states for the free-electron case ($\beta = 0$). Note that, values of $q\ell = m\pi$, $m = 1, 2, \ldots$, give the *lower* edge of the transmission bands. Note also that for $0 > \beta > -4$ the first transmission band starts at $E = 0$.

spectrum discussed in the previous chapter. In particular, we see that the energy gaps are broader when the absolute value of $|\beta|$ increases. In the limit of $\beta \to \infty$, transmission bands reduce to isolated energy levels $E(k)$ with $k\ell = m\pi$. These are the energy levels of a particle moving inside an infinite potential well of width ℓ. In the opposite limit of $\beta \to 0$, the gaps are very narrow and the density of states converges to the density of states of the free electron, derived in section 4.7.1 and given by equation (4.94).

In each transmission band we see that the density of states diverges when the energy approaches band edges. This is a typical feature of the density of states of a quantum particle propagating in a one-dimensional system; the divergences are called van Hove singularities. A general formula for a van Hove singularity in a one-dimensional system is

$$n(E) \sim \frac{1}{\sqrt{|E - E_{\text{edge}}|}}. \tag{4.26}$$

This singularity follows from the quadratic dependence of the energy E on the wave vector q close to the band edge.

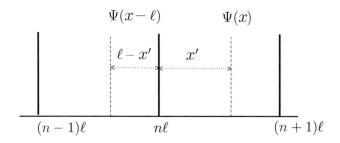

FIGURE 4.9. Definition of x'.

4.4 Wave Function

We see that an electron with energy E inside the band propagates with a wave vector q given by equation (4.10). To find an explicit form of the wave function, consider the more general transfer matrix $\mathbf{M}_\beta(x')$ that relates two wave functions at distance ℓ. With the notation $x = n\ell + x'$, $0 \le x' < \ell$ (figure 4.9) we can write

$$\Psi(x) = \mathbf{M}_\beta(x')\Psi(x - \ell). \tag{4.27}$$

Using the multiplication properties of transfer matrices, we find that the transfer matrix $\mathbf{M}_\beta(x')$ is given by

$$\mathbf{M}_\beta(x') = \begin{pmatrix} e^{ikx'} & 0 \\ 0 & e^{-ikx'} \end{pmatrix} \mathbf{M}_\Lambda \begin{pmatrix} e^{ik(\ell-x')} & 0 \\ 0 & e^{-ik(\ell-x')} \end{pmatrix}. \tag{4.28}$$

In equation (4.28) we used the transfer matrix of a single δ-function potential, $\mathbf{M}_\Lambda$, given by equation (3.7). After matrix multiplication, we obtain

$$\mathbf{M}_\beta(x') = \begin{pmatrix} e^{ik\ell}\left(1 + \dfrac{\beta}{2ik\ell}\right) & e^{ik(2x'-\ell)}\dfrac{\beta}{2ik\ell} \\ -e^{-ik(2x'-\ell)}\dfrac{\beta}{2ik\ell} & e^{-ik\ell}\left(1 - \dfrac{\beta}{2ik\ell}\right) \end{pmatrix}. \tag{4.29}$$

Note that, for $x' \to \ell$, the matrix $\mathbf{M}_\beta(x')$ converges to the transfer matrix $\mathbf{M}_\beta$ given by equation (4.7).

It is easy to verify (problem 4.13) that the eigenvalues of $\mathbf{M}_\beta(x')$ are $e^{\pm iq\ell}$, and do not depend on x'. The corresponding eigenvectors are

$$v_q(x') = \frac{1}{\sqrt{1 + |a(x')|^2}} \begin{pmatrix} 1 \\ a(x') \end{pmatrix} \tag{4.30}$$

and

$$v_{-q}(x') = \frac{1}{\sqrt{1+|a(x')|^2}} \begin{pmatrix} a^*(x') \\ 1 \end{pmatrix}, \tag{4.31}$$

where

$$a(x') = -\frac{2ik\ell}{\beta} e^{ik(\ell-2x')} \left[e^{ik\ell} \left(1 + \frac{\beta}{2ik\ell} \right) - e^{iq\ell} \right]. \tag{4.32}$$

Thus, if $\Psi(x)$ is an eigenvector of matrix $\mathbf{M}_\beta(x')$, given by equation (4.29), then it satisfies the relation

$$\Psi(x) = \Psi(n\ell + x') = e^{iq\ell} \, \Psi(n\ell + x' - \ell) = \cdots = e^{iqn\ell} v_q(x'). \tag{4.33}$$

The last expression can be written as

$$\Psi(x) = e^{iqx} u_q(x), \tag{4.34}$$

where

$$u_q(x) = e^{-iqx'} v_q(x'), \qquad 0 < x' \le \ell. \tag{4.35}$$

Clearly, $u_q(x + \ell) = u_q(x)$. We obtain the result that the wave function $\Psi(x)$, given by equation (4.34), has the form of a Bloch function.

4.5 Single Impurity

Consider now a periodic Kronig-Penney model with a single impurity represented by the δ-function potential

$$V_0 = \frac{\hbar^2}{2m} \frac{\beta_0}{\ell} \delta(x) \tag{4.36}$$

located at site $n = 0$. Using the transfer matrix technique developed in chapter 1, we calculate the energy of the bound state of the quantum particle localized at the impurity site. We find that bound states exist only when $\beta_0 < \beta$ (if β is positive) or $\beta_0 > \beta$ (if β is negative). When these conditions are satisfied, then the impurity creates exactly one bound state in each gap.

If the energy of the electron lies inside the band, then the electron is scattered on the impurity and the transfer matrix method allows us to calculate the transmission coefficient.

The impurity is characterized by the transfer matrix

$$\mathbf{M}_{\beta_0} = \begin{pmatrix} e^{ik\ell} \left[1 + \dfrac{\beta_0}{2ik\ell} \right] & \dfrac{\beta_0}{2ik\ell} e^{ik\ell} \\[2ex] -\dfrac{\beta_0}{2ik\ell} e^{-ik\ell} & e^{-ik\ell} \left[1 - \dfrac{\beta_0}{2ik\ell} \right] \end{pmatrix}. \tag{4.37}$$

The transfer matrix technique, discussed in chapter 1, can be used only in the representation in which the transfer matrix $\mathbf{M}_\beta$, which determines the propagation of the electron on both sides of the impurity, is diagonal,

$$\mathbf{Q}^{-1}\mathbf{M}_\beta\mathbf{Q} = \begin{pmatrix} e^{iq\ell} & 0 \\ 0 & e^{-iq\ell} \end{pmatrix}. \tag{4.38}$$

Matrices $\mathbf{Q}$ and $\mathbf{Q}^{-1}$ are given in problem 4.1. The first step of our analysis is therefore the similarity transformation

$$\mathbf{M}_{\beta_0} \rightarrow \mathbf{M}'_{\beta_0} = \mathbf{Q}^{-1}\mathbf{M}_{\beta_0}\mathbf{Q}. \tag{4.39}$$

Inserting expressions (4.37) and (4.70) into equation (4.39), we find, after rather complicated algebra, the matrix elements of $\mathbf{M}_{\beta_0}$. We do not perform the calculations here (details are given in problem 4.2), only present the final result for the element $[M'_{\beta_0}]_{22}(q)$,

$$\left[M'_{\beta_0}\right]_{22}(q) = e^{-iq\ell}\left[1 + i\frac{\beta_0 - \beta}{2k\ell}\frac{\sin k\ell}{\sin q\ell}\right], \tag{4.40}$$

which enables us to calculate both the energy of the bound state and the transmission.

4.5.1 Bound State

Now we use our general result, derived in section 1.4.3, that the energy of the bound state is determined by the solution of the equation

$$\left[M'_{\beta_0}\right]_{22}(q = i\kappa) = 0, \tag{4.41}$$

Inserting $q = i\kappa$ into equation (4.40) we obtain

$$\sinh\kappa\ell = (\beta - \beta_0)\frac{\sin k\ell}{2k\ell}. \tag{4.42}$$

This equation, together with the equation

$$\cosh\kappa\ell = \cos k\ell + \beta\frac{\sin k\ell}{2k\ell}, \tag{4.43}$$

which follows from equation (4.10) for $q = i\kappa$, determines two parameters: κ_b, which characterizes the exponential decrease of the wave function on both sides of the impurity, and the wave vector $k = k_b$, which gives the energy of the bound state,

$$E_b = \frac{\hbar^2 k_b^2}{2m}. \tag{4.44}$$

With the use of the relation $\cosh^2 x - \sinh^2 x = 1$, we obtain from equations (4.42) and (4.43) the nonlinear equation for k_b,

$$\left(\cos k_b\ell + \frac{\beta}{2}\frac{\sin k_b\ell}{k_b\ell}\right)^2 = (\beta - \beta_0)^2\frac{\sin^2 k_b\ell^2}{4k_b^2\ell^2} + 1. \tag{4.45}$$

We solve equations (4.45) numerically. Figure 4.10 shows the positions of the energies of bound states for $\beta = 3\pi$.

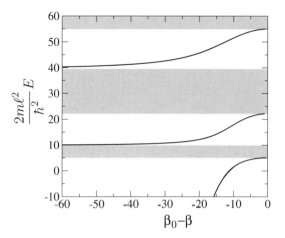

FIGURE 4.10. Energy of the bound states localized around a single δ-function impurity with β_0 for the Kronig-Penney model with $\beta = 3\pi$. For each energy band, we have exactly one impurity state in the gap below the transmission band. Note that there are also bound states with *negative* energy in the lowest gap.

Note that, for positive β, we have a bound state only when $\beta_0 < \beta$. This can be understood from equations (4.42) and (4.43). Consider first the case when $\sin k\ell > 0$. Then, from equation (4.43), we have that $\cosh \kappa\ell > 1$. Since $\kappa > 0$, we have from equation (4.42) that $\beta - \beta_0 > 0$. In the case $\sin k\ell < 0$, we have from equation (4.43) that $\cosh \kappa < -1$. This is possible only when $\kappa = \kappa' + i\pi$ and κ' is real and positive [we remind the reader that $\cosh(x + i\pi) = -\cosh x$ and $\sinh(x + i\pi) = -\sinh x$][1]. Inserting κ into equation (4.42) we again find that $\beta_0 < \beta$. Thus, indeed, the bound state exists only when $\beta_0 < \beta$. In the same way we can show that for $\beta < 0$ the bound state exists only when $\beta_0 > \beta$.

4.5.2 Transmission through a Single Impurity

We can use the explicit expression for the transfer matrix element $\left[M_{\beta_0}\right]_{22}$, given by equation (4.40) for the calculation of the transmission coefficient

$$T = |t|^2 = \frac{1}{\left|\left[M'_{\beta_0}\right]_{22}\right|^2}.$$

(4.46)

Inserting $[M'_{\beta_0}]_{22}$ from equation (4.40) we easily calculate

$$T = \frac{1}{1 + \left(\dfrac{\beta - \beta_0}{2k\ell} \dfrac{\sin k\ell}{\sin q\ell}\right)^2}.$$

(4.47)

Figure 4.11 plots the transmission coefficient as a function of $k\ell$ for $\beta = 3\pi$ and a few values of β_0. $k\ell$ lies inside the first transmission band. As expected, transmission

---

[1] This solution means that $q = i\kappa' + \pi$ so that $e^{iq\ell} = -e^{-\kappa'}$.

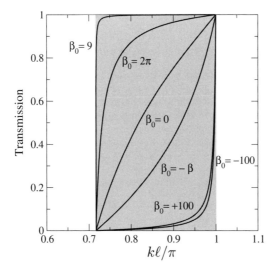

FIGURE 4.11. Transmission coefficient T of the electron through a single impurity as a function of $k\ell$ for the Kronig-Penney model with $\beta = 3\pi$. The impurity is defined by equation (4.36). Note that $T \rightarrow 1$ when $k\ell \rightarrow \pi$. Other energy bands exhibit qualitatively similar $k\ell$ dependence of the transmission coefficient.

is very small when $\beta_0 \rightarrow \infty$. Note that $\sin q\ell \rightarrow 0$ but $\sin k\ell \neq 0$ at the lower band edge. Therefore, the transmission coefficient $T \rightarrow 0$ for any $\beta_0 \neq \beta$ when the energy E approaches the lower band edge. However, the transmission is always equal to 1 at the upper band edge, when $k\ell = \pi$. This is easy to understand, since the wave function of the electron is exactly zero at the position of the impurity in this particular case (see also problems 4.5 and 4.7).

4.6 N δ-Function Barriers versus Infinite Kronig-Penney Model

To end this chapter, we want to explain a crucial difference between the system of N δ-function potential barriers discussed in chapter 3 and the *infinite* array of δ-function potential barriers discussed in this chapter.

In the case of N barriers, we calculated in chapter 3 the transmission coefficients between the incoming wave and the outgoing wave. Both waves propagate in the free space *outside* the region of barriers. Their wave vector is k, the energy $E = \hbar^2 k^2/(2m)$, and the x dependence is e^{ikx}. We found in chapter 3 that the transmission coefficient between these states oscillates as a function of the wave vector $k\ell$.

Contrary to this situation, in the case of an *infinite* array of δ-function barriers, there are only propagating solutions of the form e^{iqx} for each *real* value of the wave vector q, given by equation (4.10). We *do not* calculate the transmission, but energies, which allow propagation of the particle in an *infinite* array. Although the positions of transmission bands coincide with intervals of nonzero transmission through N δ-function barriers, the physics is different. In no case can we claim that an infinite crystal can be obtained in the limit $N \to \infty$ of δ-function barriers, as discussed in chapter 3. To obtain a crystal, we not only need an infinite number of barriers, we also have to remove two (semi-infinite) leads, which are always considered in chapter 3.

4.7 Supplementary Notes

4.7.1 Density of States

The density of states for electrons [1,11,17,23] in a band yields the number of states in a certain energy range. This function is very important in electronic processes, particularly in transport phenomena, as well as in lattice vibrations. The density of states $n(E)$ is defined such that $n(E)dE$ gives the number of states per unit volume in the energy range $(E, \ E+dE)$. In order to calculate $n(E)$ we need to know the relation between the energy E and the wave vector k. For free electrons we have $E = \hbar^2 k^2/(2m)$ while the momentum $p = \hbar k$ and the wave function is given by

$$\Psi(x) = Ae^{ikx}. \tag{4.48}$$

We shall now consider the effect of boundary conditions on the solution (4.48). There are two types of boundary conditions usually used, is periodic and fixed boundary conditions. If we work in one dimension and the length of the system is L, then periodic boundary conditions mean that

$$\Psi(x) = \Psi(x+L), \tag{4.49}$$

where $\Psi(x)$ is given by equation (4.48). If we substitute equation (4.48) into equation (4.49), we find that

$$e^{ikL} = 1. \tag{4.50}$$

This equation imposes a condition on the allowed values of the wave vector k. Only those values that satisfy equation (4.50) are allowed to be used. Noting that $e^{i2\pi m} = 1$ for any integer m, we conclude from equation (4.50) that the allowed values of k are

$$k = m\frac{2\pi}{L}, \tag{4.51}$$

where $m = 0, \pm 1, \pm 2, \ldots$. If these values are plotted along the k axis, they form a one-dimensional mesh of regularly spaced points. The spacing between these points is $2\pi/L$. Each k value given by equation (4.51) represents one eigenstate of our system. Suppose we choose an arbitrary interval dk in k-space and look for the number of eigenstates whose k's lie in this interval. We assume that L is large, so that the points are quasi-continuous. Since the spacing between the points on the k axis is $2\pi/L$, the number of eigenstates in k space is $(L/2\pi)dk$. We define

$$g(k) = 2\frac{L}{2\pi} \tag{4.52}$$

as the density of states in k space. The additional factor of 2 reflects the fact that each quantum state can be occupied by two electrons with opposite spins.

The vector k and the energy E are interrelated via the dispersion relation $E = \hbar^2 k^2/(2m)$. We can find the number of eigenstates, $n'(E)$, in the energy range dE lying in $(E, E + dE)$. Then the density of states $n'(E)$ is defined such that

$$n'(E)dE = g(k)dk. \tag{4.53}$$

We then obtain

$$n'(E) = g(k)\frac{1}{\left|\dfrac{\partial E}{\partial k}\right|} = 2\frac{L}{2\pi}\frac{1}{\left|\dfrac{\partial E}{\partial k}\right|}. \tag{4.54}$$

The density of states per unit length is

$$n(E) = n'(E)/L = \frac{1}{\pi}\frac{1}{\left|\dfrac{\partial E}{\partial k}\right|}. \tag{4.55}$$

We can easily extend the above results to the three-dimensional case. The solution analogous to equation (4.48) is now

$$\Psi(x, y, z) = A_{k_x, k_y, k_z}\, e^{ik_x x} e^{ik_y y} e^{ik_z z}. \tag{4.56}$$

If we impose periodic boundary conditions we obtain

$$(k_x, k_y, k_z) = \left(n_x \frac{2\pi}{L_x}, n_y \frac{2\pi}{L_y}, n_z \frac{2\pi}{L_z}\right), \tag{4.57}$$

where n_x, n_y, and n_z are integers. For simplicity we will assume that $L_x = L_y = L_z = L$. If we plot these values in $\vec{k}$-space, we obtain a three-dimensional cubic mesh. The volume assigned to each point in this $\vec{k}$-space is $(2\pi/L)^2$. So the density of states in the $\vec{k}$-space is given by

$$g(k) = 2\left(\frac{L}{2\pi}\right)^3. \tag{4.58}$$

We now wish to find the number of eigenstates, $R(k)$, inside a sphere of radius $k = \sqrt{k_x^2 + k_y^2 + k_z^2}$. The volume of this sphere is $(4\pi/3)k^2$ and, since the volume per each state is $(2\pi/L)^3$, it follows that $R(k)$ is

$$R(k) = 2\left(\frac{L}{2\pi}\right)^3 \frac{4\pi}{3}k^3. \tag{4.59}$$

By substituting $k^2 = 2mE/\hbar^2$ into equation (4.59) we obtain

$$R(E) = 2\left(\frac{L}{2\pi}\right)^3 \frac{4\pi}{3}\left(\frac{2mE}{\hbar^2}\right)^{3/2}. \tag{4.60}$$

$R(E)$ is the number of eigenstates with energy less than E. To obtain the density of states $n'(E)$ for the three-dimensional case, we need to take the derivative of $R(E)$

with respect to E, i.e.,

$$n'(E) = \frac{\partial R(E)}{\partial E} = \left(\frac{L}{\pi}\right)^3 \frac{4\pi}{3} \left(\frac{2m}{\hbar^2}\right)^{3/2} \frac{3}{2}\sqrt{E}. \tag{4.61}$$

The density of states per unit volume is then

$$n(E) = \frac{n'(E)}{L^3}. \tag{4.62}$$

Notice that, for the calculation of the density of states $n(E)$, it is always easier to calculate $R(k)$ first as given for the three-dimensional case, then substitute the wave vector k with respect to energy E to obtain $R(E)$, and then take the derivative.

We would like to define the velocity and the effective mass m^* of a particle that is moving through a periodic potential. In the case of a free particle, the velocity is given by $p/m = \hbar k/m$, which can be written as

$$\vec{v} = \frac{1}{\hbar}\vec{\nabla}_k E(\vec{k}), \tag{4.63}$$

and in one dimension

$$v = \frac{1}{\hbar}\frac{\partial E}{\partial k}. \tag{4.64}$$

This means that the velocity is proportional to the slope of the energy curve $E = E(k)$.

The effective mass m^* of a particle moving through a periodic potential is given by

$$m^* = \frac{\hbar^2}{\dfrac{\partial^2 E}{\partial k^2}}. \tag{4.65}$$

If there is no periodic potential ($\beta = 0$), $E = \hbar^2 k^2/(2m)$ and by using equation (4.65) we obtain that $m^* = m$. If the relation between the energy E and the wave vector k is not quadratic but more complicated, we can have positive or negative effective mass m^*. We will show a few examples of systems with negative mass later.

4.7.2 Periodic Arrangement of Potential Barriers

As discussed in chapter 3, the δ-function potential represents the limit of an infinitesimally narrow rectangular barrier. Using the transfer matrix for the rectangular potential and the techniques developed earlier in this chapter, we will calculate the energy spectrum of the propagating states for a periodic arrangement of rectangular barriers [30] shown in figure 4.12. The potential barriers are characterized by the height of the potential V_0. The width of the barriers is $2a$ and the distance between barriers $2b$.

The spatial period of the model is now

$$\ell = 2a + 2b. \tag{4.66}$$

$$\ell = 2a + 2b$$

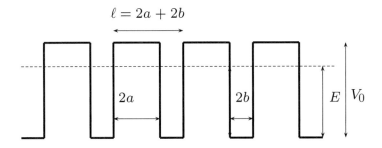

FIGURE 4.12. Periodic array of potential barriers.

Since the model is periodic, it is sufficient to derive the transfer matrix for only one spatial period. It is given by the matrix multiplication

$$\mathbf{M} = \mathbf{M}_0 \mathbf{M}_{\text{box}}, \tag{4.67}$$

where

$$\mathbf{M}_0 = \begin{pmatrix} e^{i2kb} & 0 \\ 0 & e^{-i2kb} \end{pmatrix} \tag{4.68}$$

is the transfer matrix for the free space between two barriers and $\mathbf{M}_{\text{box}}$ is the transfer matrix of a rectangular potential, defined by equations (2.17) and (2.18).

The wave vector $q\ell$ of a propagating state is given as the *real* solution of the equation

$$\cos q\ell = \cos 2k'a \, \cos 2kb - \frac{1}{2}\left(\frac{k}{k'} + \frac{k'}{k}\right)\sin 2k'a \, \sin 2kb, \tag{4.69}$$

where $k = \sqrt{2mE}/\hbar$ and $k' = \sqrt{2m(E - V_0)}/\hbar$. Note that equation (4.69) can be used for both potential barriers ($V_0 > 0$) and potential wells ($V_0 < 0$).

The periodic potential, shown in figure 4.12, will be discussed later in chapter 13.

4.8 Problems

Problems with Solutions

Problem 4.1
Define the matrix

$$\mathbf{Q} = \begin{pmatrix} 1 & a^* \\ a & 1 \end{pmatrix} \tag{4.70}$$

where

$$a = a_+(x' = \ell) = -\frac{2ik\ell}{\beta}e^{-ik\ell}\left[e^{ik\ell}\left(1 + \frac{\beta}{2ik\ell}\right) - e^{\pm iq\ell}\right], \tag{4.71}$$

and the function $a(x')$ is given by equation (4.32). Show that

$$\mathbf{Q}^{-1}\mathbf{M}_\beta\mathbf{Q} = \mathbf{Q}^{-1} \begin{pmatrix} e^{ik\ell}\left[1 + \dfrac{\beta}{2ik\ell}\right] & \dfrac{\beta}{2ik\ell}e^{ik\ell} \\ -\dfrac{\beta}{2ik\ell}e^{-ik\ell} & e^{-ik\ell}\left[1 - \dfrac{\beta}{2ik\ell}\right] \end{pmatrix} \mathbf{Q}$$

$$= \begin{pmatrix} e^{iq\ell} & 0 \\ 0 & e^{-iq\ell} \end{pmatrix}. \tag{4.72}$$

From equation (A.9) we have that

$$\mathbf{Q}^{-1} = \frac{1}{\det\mathbf{Q}} \begin{pmatrix} 1 & -a^* \\ -a & 1 \end{pmatrix}, \quad \det\mathbf{Q} = 1 - |a|^2. \tag{4.73}$$

Solution. We use an explicit form of the eigenvectors of the transfer matrix, given by equations (4.30)–(4.32). We have

$$\mathbf{M}_\beta \begin{pmatrix} 1 \\ a \end{pmatrix} = e^{+iq\ell} \begin{pmatrix} 1 \\ a \end{pmatrix} \tag{4.74}$$

and

$$\mathbf{M}_\beta \begin{pmatrix} a^* \\ 1 \end{pmatrix} = e^{-iq\ell} \begin{pmatrix} a^* \\ 1 \end{pmatrix}, \tag{4.75}$$

where $q\ell$ is given by equation (4.10). Using equation (A.35) we obtain

$$\mathbf{M}_\beta\mathbf{Q} = \mathbf{Q} \begin{pmatrix} e^{iq\ell} & 0 \\ 0 & e^{-iq\ell} \end{pmatrix}. \tag{4.76}$$

This immediately gives required result, equation (4.72).

Problem 4.2
Calculate the matrix

$$\mathbf{M}'_{\beta_0} = \mathbf{Q}^{-1}\mathbf{M}_{\beta_0}\mathbf{Q}, \tag{4.77}$$

where matrices $\mathbf{Q}$ and $\mathbf{Q}^{-1}$ are defined by equations (4.70) and (4.73), respectively, and $\mathbf{M}_{\beta_0}$ is the transfer matrix with the δ-function impurity of strength β_0 given by equation (4.37),

$$\mathbf{M}_{\beta_0} = \begin{pmatrix} e^{ik\ell}\left[1 + \dfrac{\beta_0}{2ik\ell}\right] & \dfrac{\beta_0}{2ik\ell}e^{ik\ell} \\ -\dfrac{\beta_0}{2ik\ell}e^{-ik\ell} & e^{-ik\ell}\left[1 - \dfrac{\beta_0}{2ik\ell}\right] \end{pmatrix}. \tag{4.78}$$

Solution. We calculate

$$\mathbf{M}'_{\beta_0} = \mathbf{Q}^{-1}\mathbf{M}_{\beta_0}\mathbf{Q} = \mathbf{Q}^{-1}\mathbf{M}_{\beta}\mathbf{Q} + \mathbf{Q}^{-1}\left[\mathbf{M}_{\beta_0} - \mathbf{M}_{\beta}\right]\mathbf{Q}. \tag{4.79}$$

We use the result of problem 4.1 and the explicit form of the matrices $\mathbf{M}_{\beta}$ and $\mathbf{M}_{\beta_0}$. We obtain

$$\mathbf{M}'_{\beta_0} = \begin{pmatrix} e^{iq\ell} & 0 \\ 0 & e^{-iq\ell} \end{pmatrix} + \frac{\beta_0 - \beta}{2ik\ell}\mathbf{X} \tag{4.80}$$

where

$$\begin{aligned} \mathbf{X} &= \mathbf{Q}^{-1}\begin{pmatrix} e^{ik\ell} & e^{ik\ell} \\ -e^{-ik\ell} & -e^{-ik\ell} \end{pmatrix}\mathbf{Q} \\ &= \frac{1}{\det \mathbf{Q}}\begin{pmatrix} e^{ik\ell}(1+a) + e^{-ik\ell}(1+a)a^* & e^{ik\ell}(1+a^*) + e^{-ik\ell}(1+a^*)a^* \\ -e^{ik\ell}(1+a)a - e^{-ik\ell}(1+a^*)a^* & -e^{ik\ell}(1+a^*)a - e^{-ik\ell}(1+a^*) \end{pmatrix}. \end{aligned} \tag{4.81}$$

Inserting a from equation (4.71), we can calculate all matrix elements of $\mathbf{X}$. The algebra is too difficult to be presented here. Therefore, we present only the final results,

$$X_{22} = -\frac{\sin k\ell}{\sin q\ell}e^{-iq\ell} \quad \text{and} \quad X_{12} = \frac{\sin k\ell}{\sin q\ell}e^{ik\ell}. \tag{4.82}$$

Inserting $\mathbf{X}$ into equation (4.80) we finally obtain

$$\mathbf{M}'_{\beta_0=0} = \begin{pmatrix} e^{iq\ell}\left[1 + \dfrac{\tilde{\beta}}{2ik\ell}\right] & \dfrac{\tilde{\beta}}{2ik\ell}e^{ik\ell} \\ -\dfrac{\tilde{\beta}}{2ik\ell}e^{-ik\ell} & e^{-iq\ell}\left[1 - \dfrac{\tilde{\beta}}{2ik\ell}\right] \end{pmatrix}, \tag{4.83}$$

where

$$\tilde{\beta} = (\beta_0 - \beta)\frac{\sin k\ell}{\sin q\ell}. \tag{4.84}$$

Problem 4.3
In problem 4.2 consider the special case $\beta = 0$.

Solution. In the case of $\beta = 0$, we have $a \equiv 0$ and the matrix $\mathbf{M}'_{\beta_0} \equiv \mathbf{M}_{\beta_0}$ is the transfer matrix for the δ-function potential, discussed in chapter 3.

Problem 4.4
Calculate the transmission through N identical impurities located at $x = n\ell$, $n = 1, 2, \ldots, N$. Consider the special case when the strength of the impurities $\beta_0 = 0$.

Solution. Consider first the special case of a single impurity located at $x = 0$ with $\beta_0 = 0$. The transfer matrix of the impurity reads

$$
\mathbf{M}_{\beta_0=0} = \begin{pmatrix} e^{ik\ell} & 0 \\ 0 & e^{-ik\ell} \end{pmatrix} \tag{4.85}
$$

so that it seems there is no scattering. However, the electron propagates as $e^{\pm iqx}$ in the regions $x > 0$ and $x < 0$. Such a wave propagates only in a periodic array of δ-function potentials and is scattered at $x = 0$. Note that $\mathbf{M}'_{\beta_0=0}$, given by equations (4.83) and (4.84), is not a diagonal matrix.

The transmission is given by equation (4.47) with $\beta_0 = 0$,

$$
T_1(\beta_0 = 0) = \frac{1}{1 + \left(\dfrac{\beta}{2k\ell} \dfrac{\sin k\ell}{\sin q\ell} \right)^2}. \tag{4.86}
$$

In the case of $N > 1$ impurities, we use Chebyshev's formula, given by equation (1.106). Note, however, that the particle propagates in the leads with the wave vector q and that $\mathrm{Tr}\, \mathbf{M}_{\beta_0=0} = 2\cos k\ell$. With the use of the transmission through a single impurity given by equation (4.86), we obtain

$$
T_N(\beta_0 = 0) = \frac{1}{1 + \left(\dfrac{\beta}{2k\ell} \dfrac{\sin Nk\ell}{\sin q\ell} \right)^2}. \tag{4.87}
$$

Figure 4.13 plots the transmission through $N = 20$ impurities for the energy from the first band. The transmission coefficient oscillates as a function of $k\ell$.

Problem 4.5
Consider the transmission of an electron through a single impurity. Prove that the transmission coefficient T, given by equation (4.47) equals 1 at the upper band edge (we consider the Kronig-Penney model with $\beta > 0$).

Solution. In the case of positive β, we have $k\ell = n\pi$ at the upper band edge. Consider $k\ell = n\pi - x$, $x \ll 1$, and calculate

$$
\sin^2 q\ell = 1 - \cos^2 q\ell = 1 - \cos^2 k\ell - \frac{\beta^2}{4k^2\ell^2} \sin^2 k\ell - 2\frac{\beta}{2k\ell} \sin k\ell \cos k\ell. \tag{4.88}
$$

Now we insert $\sin k\ell = \sin(n\pi - x) = (-1)^{n+1} \sin x \approx (-1)^{n+1} x + \cdots$ and $\cos k\ell = \cos(n\pi - x) = (-1)^n \cos x \approx (-1)^n (1 - x^2/2 + \cdots)$ and obtain

$$
\sin^2 q\ell \approx \frac{\beta}{k\ell} x + \cdots \tag{4.89}
$$

and

$$
\sin^2 k\ell = \sin^2(n\pi - x) = \sin^2 x \approx x^2 + \cdots . \tag{4.90}
$$

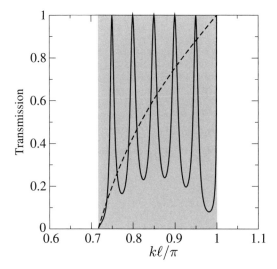

FIGURE 4.13. Transmission coefficient T_N given by equation (4.87) of $N = 20$ impurities with $\beta_0 = 0$ as a function of $k\ell$, for the Kronig-Penney model with $\beta = 3\pi$. The dashed line is the transmission T_1, given by equation (4.86).

Since $x \ll 1$, we neglected all terms proportional to higher orders in x. Using these expressions, we obtain that the ratio

$$\frac{\sin^2 k\ell}{\sin q\ell} \approx x \to 0 \quad \text{when} \quad x \to 0, \tag{4.91}$$

so that indeed $T \to 1$ when $k\ell \to n\pi^-$.

Problems without Solutions

Problem 4.6
Consider the limit $\beta \to 0$, which corresponds to the case of free-particle propagation. Use the graphical solution of equations (4.17) shown in figure 4.2 to prove that the band gaps become infinitesimally narrow for $\beta \to 0$ and eventually disappear for $\beta = 0$.

Problem 4.7
Use again the graphical solution of equations (4.17) and show that the band of allowed energies becomes narrower when β increases. Discuss the limit of $\beta \to \infty$. Show that in this limit there are only propagating states with

$$k\ell = q\ell = n\pi. \tag{4.92}$$

Their energy is

$$E_n = \frac{\hbar^2 \pi^2}{2m\ell^2} n^2, \quad n = 1, 2, \dots. \tag{4.93}$$

Note that the condition (4.92) has a simple physical explanation. If the δ-function potential barriers are infinitesimally strong, then transmission through the system is possible only if the distance ℓ between two neighboring barriers equals an integer multiple of half of the particle wavelength, $\lambda = 2\pi/k$. In this particular case, the electron wave function is zero at the position of the δ-function potential, so that the propagation does not depend on β. See also problem 3.8.

Problem 4.8

Calculate the density of states for the free particle with the energy $E = \hbar^2 k^2/(2m)$ propagating in a one-dimensional system. Prove that the density of states is given by

$$n(E) = \frac{\sqrt{m}}{\pi\hbar\sqrt{2E}}, \tag{4.94}$$

which diverges as $E^{-1/2}$ as $E \to 0$.

This singular behavior is typical for one-dimensional systems. In dimensions 2 and 3, the wave vector k possesses two (three) components. Since the energy E does not depend on the orientation of the wave vector, we have to sum over all possible orientations of the wave vector.

Problem 4.9

Show that the density of states $n(E)$ for the free-electron case $E = \hbar^2 k^2/(2m)$ is given by [11]

$$n(E) \begin{cases} = \dfrac{m^{3/2}}{\sqrt{2\pi^2\hbar^3}}\sqrt{E} & (3d), \\[2ex] = \dfrac{m}{2\pi\hbar^2} & (2d), \\[2ex] = \dfrac{\sqrt{m}}{\pi\hbar\sqrt{2}}\dfrac{1}{\sqrt{E}} & (1d). \end{cases} \tag{4.95}$$

Problem 4.10

Show that the density of states for classical waves, where $\omega = ck$, is given by

$$n(\omega) \begin{cases} = \dfrac{\omega^2}{2\pi^2 c^3} & (3d), \\[2ex] = \dfrac{\omega}{2\pi c^2} & (2d), \\[2ex] = \dfrac{1}{\pi c} & (1d). \end{cases} \tag{4.96}$$

Problem 4.11

Show that the density of states of the Kronig-Penney model equation (4.25) reduces in the limit of $\beta \to 0$ to the free-electron density, given by equation (4.94).

Problem 4.12

Verify that the density of states $n(E)$ given by equation (4.25) indeed diverges in the vicinity of each band edge, in agreement with the relation (4.26).

Problem 4.13
Show that the transfer matrix

$$\mathbf{M}_\beta(x') = \begin{pmatrix} e^{ikx'} & 0 \\ 0 & e^{-ikx'} \end{pmatrix} \mathbf{M}_\Lambda \begin{pmatrix} e^{ik(\ell-x')} & 0 \\ 0 & e^{-ik(\ell-x')} \end{pmatrix}, \tag{4.97}$$

where $\mathbf{M}_\Lambda$ is the transfer matrix of a single δ-function impurity, given by equation (3.7), has eigenvalues $e^{iq\ell}$ and $e^{-iq\ell}$, given by equation (4.10). Note that the transfer matrix (4.97) equals the transfer matrix (4.7) for the special case $x' = \ell$.

Problem 4.14
Analyze the problem of a single impurity in the Kronig-Penney model with a negative β.

Problem 4.15
For the case of a single impurity, given by equation (4.36), calculate the value of β_0 for which the energy of the bound state in the lowest gap is zero, $E_b = 0$.

Problem 4.16
Use equation (4.10) to prove that the periodic Kronig-Penney model with negative $\beta > -4$ allows propagating states with real $q\ell$ also for energy $E < 0$.

5 Tight Binding Model

In this chapter, we introduce the most important ideas of electron propagation in periodic lattices, such as energy bands and gaps, the density of states, effective mass and group velocity of the electron, and the Fermi energy. We also derive the transfer matrix that enables us to find the energy of a bound state and to calculate the transmission of an electron through a system of N particles.

We begin by introducing and examining a very simple model, the so-called *tight binding model* [11, 23, 75, 103, 115], defined by Schrödinger's equation,

$$i\hbar \frac{\partial c_n}{\partial t} = \varepsilon_n c_n + V_n c_{n+1} + V_n^* c_{n-1}. \tag{5.1}$$

The tight binding model given by equation (5.1) is a very good approximate model for electrons propagating in a one-dimensional solid. The tight binding model can be obtained as follows: We assume that electrons are well localized around the ions, as shown in figure 5.1. Then the wave function $\Psi(x)$ of the electron can be written in the form

$$\Psi(x) = \sum_n c_n \Phi_n(x), \tag{5.2}$$

where $\Phi_n(x)$ is the eigenfunction of the isolated atom n, and c_n is the amplitude of the probability that the electronic wave function is localized at site n. We multiply Schrödinger's equation

$$\mathcal{H}\Psi(x) = i\hbar \frac{\partial \Psi(x)}{\partial t}, \tag{5.3}$$

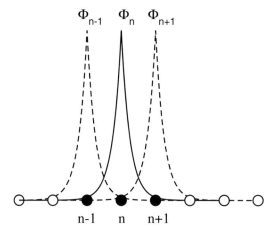

FIGURE 5.1. One-dimensional crystalline solid. Atoms are positioned in lattice sites $x = na$. The overlap of the wave functions of two nearest-neighboring electrons defines the hopping term V in the tight binding model given by equation (5.1).

where $\mathcal{H}$ is the Hamiltonian of the electron in a solid, by Φ_n^* and integrate over x. The site's energies are given by the integral

$$\varepsilon_n = \langle \Phi_n^* | \mathcal{H} | \Phi_n \rangle = \int dx \, \Phi_n^*(x) \mathcal{H} \Phi_n(x). \tag{5.4}$$

The corresponding hopping matrix element $V = \langle \Phi_n^* | H | \Phi_{n+1} \rangle$ is given by the overlap integral between the orbitals at sites n and $n+1$. Since we assumed that the wave function $\Phi_n(x)$ is localized around the ion n, we consider all terms $\langle | \Phi_n(x) | \mathcal{H} | \Phi_m(x) \rangle$ with $|n - m| > 1$ to be small, and neglect them. In other words, if the electronic wave function is well localized around each atom, then the overlap of next-nearest-neighbor wave functions Φ_n and Φ_{n+2} is small and can be neglected. We obtain the tight binding model, given by equation (5.1). In what follows we assume that V_n does not depend on the lattice size n, and is real, $V_n = V = V^*$.

Equation (5.1) describes the propagation of the particle on a *discrete* lattice. In this case, the wave function of the electron, c_n, is defined only on a discrete set of sites, $x_n = na$, where a is the lattice period. The electron propagates by hopping from one lattice site to the next-nearest-neighbor one.

It is worth noting that the tight binding model can be introduced for various physical problems. For instance, we will show later that the system of coupled oscillators can also be described by a tight binding model. This mathematical equivalence will be very useful, since it allows us to obtain a physical picture of wave propagation in a quantum-mechanical system, in analogy with the wave propagation in a classical system of coupled oscillators.

In what follows, we will consider a stationary Schrödinger equation. We assume

$$c_n(t) = c_n \, e^{i\mathcal{E}t}. \tag{5.5}$$

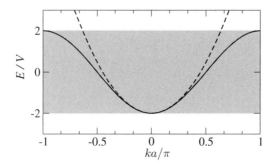

FIGURE 5.2. Allowed energies of the periodic tight binding model with $\varepsilon_n \equiv 0$. Allowed energies $\mathcal{E}(k)/V = -2\cos ka$ lie in the interval $-2 \leq \mathcal{E}/V \leq 2$. The dashed line shows the free-electron dispersion relation $\mathcal{E}/V = -2 + k^2 a^2$ given by equation (5.10).

Inserting (5.5) into (5.1), we obtain

$$-\mathcal{E} c_n = \varepsilon_n c_n + V c_{n+1} + V c_{n-1}. \tag{5.6}$$

5.1 Periodic Model

5.1.1 Dispersion Relation

In the periodic case [11,23], where $\varepsilon_n = 0$ for all sites n, we expect the solution of equation (5.6) to be expressed as follows:

$$c_n = A e^{\pm ikan}. \tag{5.7}$$

Here, k is the wave vector. Inserting expression (5.7) into equation (5.6), we obtain the dispersion relation for the energy of electrons as

$$\mathcal{E}(k) = -2V\cos ka \tag{5.8}$$

for both right- and left-moving electrons.

The dispersion relation (5.8) is shown in figure 5.2 and is compared with the energy of a free particle. We see that, in contrast to continuum models, the energy of the electron in the tight binding model is restricted to the interval

$$-2V \leq \mathcal{E} \leq 2V, \tag{5.9}$$

called the energy band, or simply the band. The relation (5.8) is called the *band structure*. There are no propagating states with energy $|\mathcal{E}| > 2V$, so that the width of the band is $4V$. The higher V, the wider the band.

In the limit of $ka \ll 1$, we can expand $\cos ka$ in a Taylor series. The energy $\mathcal{E}$ then reads

$$\mathcal{E} \approx -2V + Vk^2 a^2, \quad ka \ll 1, \tag{5.10}$$

or equivalently

$$\mathcal{E} + 2V = \frac{\hbar^2 k^2}{2m^*}, \tag{5.11}$$

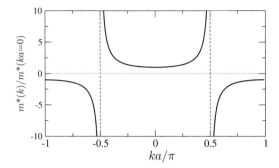

FIGURE 5.3. Effective mass m^*, given by equation (5.12), as a function of the wave vector ka in the tight binding periodic model.

where we have introduced the *effective mass* m^*,

$$m^* = \frac{\hbar^2}{2a^2 V}. \tag{5.12}$$

It is seen that the effective mass m^* is inversely proportional to the overlap integral V. This is intuitively reasonable, since the greater the overlap of the atomic wave functions, the easier it is for the electron to tunnel from one atomic site to another, and, hence, the smaller is the inertia (or mass) of the electron. Conversely, a small overlap V leads to a large effective mass.

5.1.2 Effective Mass

In general, the effective mass [23] can be defined by the equation

$$\frac{1}{m^*} = \frac{1}{\hbar^2} \frac{\partial^2 \mathcal{E}(k)}{\partial k^2}. \tag{5.13}$$

In figure 5.3, we plot the effective mass m^* *versus* the wave vector ka for the one-dimensional tight binding model. Near the bottom of the band, m^* has a constant value, which is positive, because the quadratic relation (5.11) is satisfied near the bottom of the band. But, as ka increases, m^* is no longer constant, becoming a function of ka, because the quadratic relation (5.11) is no longer valid. At the point $k_c = \pi/(2a)$, the effective mass diverges. Beyond the point $k_c = \pi/(2a)$, the effective mass m^* becomes negative. This region corresponds to the top of the energy band. If we define $k' = (\pi/a) - k$ then we see that the energy $\mathcal{E}(k')$ near the maximum point can be expanded as

$$\mathcal{E}(k') = 2 - Va^2(k')^2, \tag{5.14}$$

which shows that at the top of the band the electron behaves like a particle with a *negative* effective mass $m^* = -\hbar^2/(2a^2 V)$.

We can intuitively understand the ka dependence of the effective mass as follows. For $ka \ll 1$, the wavelength of the electron in the quantum state $c_n = e^{ikan}$ is $\lambda = 2\pi/k$, and it

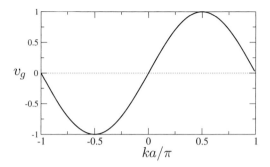

FIGURE 5.4. The group velocity of the electron, given by equation (5.16), in the tight binding model.

is much larger than the lattice spacing. In this state, the electron does not see details of the lattice and propagates as a free particle in a homogeneous medium. Its effective mass is positive. When ka increases, the lattice structure has an influence on the propagation of the electron. For $k = k_c = \pi/(2a)$ the wavelength of the electron is exactly four spatial periods, $\lambda = 4a$, so that the electron in this state is located on either even or odd lattice sites. Note that, the dispersion relation $\mathcal{E} = 2V \cos ka$ is *linear* in the wave vector around $k = k_c$, which is typical for waves. Therefore, the wave character of electron propagation becomes crucial.

5.1.3 Group Velocity

The k dependence of the effective mass can be understood also from Newton's equation for an electron,

$$F = \frac{\partial p}{\partial t} = \hbar \frac{\partial k}{\partial t},$$ (5.15)

where $p = \hbar k$ is called the *crystal momentum*.

The group velocity of a wave packet in a one-dimensional system is defined by the relation

$$v_g = \frac{1}{\hbar}\frac{\partial \mathcal{E}}{\partial k}.$$ (5.16)

In figure 5.4 we plot the group velocity $v_g = (2Va/\hbar)\sin ka$ for the one-dimensional tight binding model as a function of ka for the dispersion relation given by equation (5.8). Notice that the velocity decreases for $k > k_c = \pi/(2a)$. Thus, the acceleration is negative, i.e., opposite to the applied force, implying a negative mass. This means that, in this region of k-space, the lattice exerts such a large retarding force on the electron that it overcomes the applied force and produces a negative acceleration, and, therefore, a negative mass.

5.1.4 Density of States

The density of states was defined in section 4.7.1 by the equation

$$n(\mathcal{E}) = \frac{1}{2\pi} \frac{1}{\left|\frac{\partial \mathcal{E}}{\partial k}\right|}. \tag{5.17}$$

We remind the reader that $n(\mathcal{E})d\mathcal{E}$ is the number of allowed states per unit length of the system with energy inside the narrow interval $(\mathcal{E}, \ \mathcal{E} + d\mathcal{E})$. If the spin of the electron is considered, then $n(\mathcal{E})$ should be multiplied by 2, since each state can be occupied by two electrons, one with spin up and the second with spin down.

For the one-dimensional tight binding model, the relationship between the energy and the wave vector is given by the dispersion relation, equation (5.8). Therefore, the density of states can be easily calculated to be equal to

$$n'(\mathcal{E}) = Ln(\mathcal{E}) = \frac{L}{2Va\pi \sin ka} = \frac{N}{\pi\sqrt{4V^2 - \mathcal{E}^2}}. \tag{5.18}$$

We have used the fact that $L = Na$. The additional factor of 2 comes from the fact that for a given energy $\mathcal{E}$ we have two wave vectors, $+k$ and $-k$.

If we integrate $n'(\mathcal{E})$ over $\mathcal{E}$ from $-2V$ to $2V$ we find the result to be N, the number of particles in our one-dimensional system. This is expected, since the number of allowed values of the wave vector is exactly N.

We plot $n(\mathcal{E})$ as a function of the energy $\mathcal{E}$ in figure 5.5. We see that the density of states diverges in the vicinity of the band edges as

$$\frac{1}{\sqrt{|\mathcal{E} - \mathcal{E}_{\text{edge}}|}}, \tag{5.19}$$

exactly in the same way as in the case of the Kronig-Penney model (figures 4.7 and 4.8). These singularities are called *van Hove* singularities.

5.1.5 Fermi Energy and Fermi Vector

Let us assume that each site in the tight binding model has only one electron. So we have N electrons. At zero temperature, we would like to place the electrons in the lowest-energy eigenstates in order to minimize the total energy of our system. We have argued above that the energy band has N eigenstates, each of which can accept two electrons (one with spin up and one with spin down). So the electrons will fill exactly half the eigenstates of the energy band as shown in figure 5.5. Therefore, the Fermi energy $\mathcal{E}_F$ is equal to zero and the Fermi wave vector k_F is equal to $\pi/(2a)$.

The conclusion is that the simple one-dimensional tight binding model behaves as a metal, i.e., the density of states at the Fermi energy is nonzero. Close to the Fermi energy there are a lot of other eigenstates (spread over the system). This allows two typical important properties of metal: quick response to external perturbations and easy increase in temperature. If the density of states at the Fermi energy is equal to zero, then the system is either an insulator (large gap) or a semiconductor (small gap).

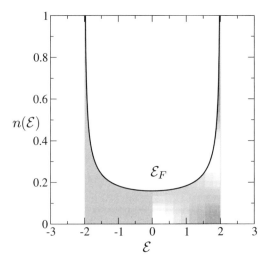

FIGURE 5.5. The density of states given by equation (5.18) for the tight binding model, as a function of energy $\mathcal{E}$. If the spin of the electron is considered, then $n(\mathcal{E})$ should be multiplied by 2, since each state can be occupied by two electrons, one with spin up and the second with spin down. Then, for the system with N electrons, one-half of the states are occupied and the Fermi energy is $\mathcal{E}_F = 0$.

5.2 The Transfer Matrix

Schrödinger's equation, given by equation (5.6), can be written in the matrix form

$$\begin{pmatrix} c_{n+1} \\ c_n \end{pmatrix} = \begin{pmatrix} (\mathcal{E} - \varepsilon_n)/V & -1 \\ 1 & 0 \end{pmatrix} \begin{pmatrix} c_n \\ c_{n-1} \end{pmatrix}. \tag{5.20}$$

To simplify our analysis, we rescale in this section the energy $\mathcal{E}$ and the site energies ε_n by a factor of V. We obtain

$$\begin{pmatrix} c_{n+1} \\ c_n \end{pmatrix} = \begin{pmatrix} \mathcal{E} - \varepsilon_n & -1 \\ 1 & 0 \end{pmatrix} \begin{pmatrix} c_n \\ c_{n-1} \end{pmatrix}. \tag{5.21}$$

The simple form of the relationship (5.21) inspires us to define the transfer matrix $\mathcal{M}_n$ for the tight binding model, given by equation (5.6), by the relation

$$\begin{pmatrix} c_{n+1} \\ c_n \end{pmatrix} = \mathcal{M}_n \begin{pmatrix} c_n \\ c_{n-1} \end{pmatrix} = \begin{pmatrix} \mathcal{E} - \varepsilon_n & -1 \\ 1 & 0 \end{pmatrix} \begin{pmatrix} c_n \\ c_{n-1} \end{pmatrix}. \tag{5.22}$$

The transfer matrix $\mathcal{M}_n$ has a very simple structure. Note, however, that this definition differs from the definition of the transfer matrix used in previous chapters. Indeed, $\mathcal{M}_n$

gives the relationship ship between two vectors,

$$
\begin{pmatrix} c_{n+1} \\ c_n \end{pmatrix} \quad \text{and} \quad \begin{pmatrix} c_n \\ c_{n-1} \end{pmatrix},
\tag{5.23}
$$

which contain the wave function of two nearest-neighbor sites. To derive the transfer matrix for the left-moving and right-moving waves on a given site n, we must express the two vectors in equation (5.22) in terms of left- and right-moving waves:

$$
c_n = A e^{inka} + B e^{-inka}.
\tag{5.24}
$$

We easily verify that the two vectors in equation (5.23) can be expressed as [115]

$$
\begin{pmatrix} c_n \\ c_{n-1} \end{pmatrix} = \mathbf{Q} \begin{pmatrix} A e^{inka} \\ B e^{-inka} \end{pmatrix}
\tag{5.25}
$$

and

$$
\begin{pmatrix} c_{n+1} \\ c_n \end{pmatrix} = \mathbf{Q} \begin{pmatrix} A e^{i(n+1)ka} \\ B e^{-i(n+1)ka} \end{pmatrix}.
\tag{5.26}
$$

The matrix $\mathbf{Q}$ is given by

$$
\mathbf{Q} = \begin{pmatrix} 1 & 1 \\ e^{-ika} & e^{+ika} \end{pmatrix}.
\tag{5.27}
$$

Inserting expressions (5.25) and (5.26) in equation (5.22), we obtain

$$
\begin{pmatrix} A e^{i(n+1)ka} \\ B e^{-i(n+1)ka} \end{pmatrix} = \mathbf{Q}^{-1} \mathcal{M}_n \mathbf{Q} \begin{pmatrix} A e^{inka} \\ B e^{-inka} \end{pmatrix} = \mathbf{M}_n \begin{pmatrix} A e^{inka} \\ B e^{-inka} \end{pmatrix},
\tag{5.28}
$$

where we have used the inverse matrix $\mathbf{Q}^{-1}$, given by

$$
\mathbf{Q}^{-1} = \frac{1}{2i \sin ka} \begin{pmatrix} e^{ika} & -1 \\ -e^{-ika} & 1 \end{pmatrix},
\tag{5.29}
$$

and have introduced the transfer matrix $\mathbf{M}_n$,

$$
\mathbf{M}_n = \mathbf{Q}^{-1} \mathcal{M}_n \mathbf{Q} = \mathbf{Q}^{-1} \begin{pmatrix} \mathcal{E} - \varepsilon_n & -1 \\ 1 & 0 \end{pmatrix} \mathbf{Q}.
\tag{5.30}
$$

The explicit form of the transfer matrix $\mathbf{M}_n$ reads

$$
\mathbf{M}_n = \begin{pmatrix} e^{+ika} \left[1 - \dfrac{\varepsilon_n}{2i \sin ka} \right] & -\dfrac{e^{+ika} \varepsilon_n}{2i \sin ka} \\[2ex] +\dfrac{e^{-ika} \varepsilon_n}{2i \sin ka} & e^{-ika} \left[1 + \dfrac{\varepsilon_n}{2i \sin ka} \right] \end{pmatrix}.
\tag{5.31}
$$

The transfer matrix $\mathbf{M}_n$ determines the relationship between the wave function at site na and $(n+1)a$, so that it is defined in accordance with the definition (1.12). Consequently, all properties of the transfer matrix discussed in chapter 1 are also valid for the transfer matrix $\mathbf{M}_n$. In particular, the transmission coefficient through the site n is given by $t = M_{22}^{-1}$, and the ratio

$$\left|\frac{r}{t}\right|^2 = |M_{12}|^2 = \frac{\varepsilon_n^2}{4\sin^2 ka}. \tag{5.32}$$

Note that the matrix $\mathbf{Q}$ does not depend on the site energy ε_n. Therefore, the relation (5.30) holds not only for the transfer matrix defined on a given site, but also for any product of transfer matrices. In particular, for the case of N sites, it is much easier to calculate the transfer matrix

$$\mathcal{M}^{(N)} = \mathcal{M}_N \mathcal{M}_{N-1} \cdots \mathcal{M}_2 \mathcal{M}_1 \tag{5.33}$$

than the corresponding product of matrices, $\mathbf{M}$. The resulting transfer matrix $\mathbf{M}^{(N)}$ can be obtained from the matrix $\mathcal{M}^{(N)}$ as

$$\mathbf{M}^{(N)} = \mathbf{Q}^{-1} \mathcal{M}^{(N)} \mathbf{Q}. \tag{5.34}$$

The very simple structure of the transfer matrix $\mathcal{M}$ is the reason why $\mathcal{M}$ is used in various applications, and in numerical calculations, especially in studies of disordered materials, where the site energy ε_n varies randomly as a function of the site index n.

5.3 Transmission Coefficient

Consider now the system in which $\varepsilon_n \equiv 0$ for all $n \leq 0$ and $n > N$, but ε_n possess arbitrary values at sites $n = 0, 1, \ldots, N-1$. We want to determine the general formulas for the transmission coefficient T of the electron propagating through the sample represented by N sites, $n = 0, 1, \ldots, N-1$.

Consider an electron coming from the right side of the system. Then, left of the system, there is only a transmitted wave going to the left. Its wave function at the sites $n = 0$ and $n = -1$ can be written as

$$c_{-1} = e^{+ika}, \quad c_0 = 1. \tag{5.35}$$

The wave function on the right-hand side of the system is given by the superposition of incoming and reflected waves. We can use the transfer matrix $\mathcal{M}$ to express the wave function at sites N and $N-1$:

$$\begin{pmatrix} c_N \\ c_{N-1} \end{pmatrix} = \mathcal{M}_{N-1} \mathcal{M}_{N-2} \ldots \mathcal{M}_1 \mathcal{M}_0 \begin{pmatrix} c_0 \\ c_{-1} \end{pmatrix} = \mathcal{M}^{(N)} \begin{pmatrix} c_0 \\ c_{-1} \end{pmatrix}, \tag{5.36}$$

where $\mathcal{M}_n$ is the transfer matrix for the nth site, given by equation (5.22).

Now we multiply both sides of equation (5.36) by the matrix $\mathbf{Q}^{-1}$:

$$\mathbf{Q}^{-1}\begin{pmatrix} c_N \\ c_{N-1} \end{pmatrix} = \mathbf{Q}^{-1}\mathcal{M}^N\mathbf{Q}\mathbf{Q}^{-1}\begin{pmatrix} c_0 \\ c_{-1} \end{pmatrix}. \tag{5.37}$$

With the help of the relationship

$$\begin{pmatrix} Ae^{inka} \\ Be^{-inka} \end{pmatrix} = \mathbf{Q}^{-1}\begin{pmatrix} c_n \\ c_{n-1} \end{pmatrix} = \frac{1}{2i\sin ka}\begin{pmatrix} e^{ika}c_{n+1} - c_n \\ -e^{-ika}c_{n+1} + c_n \end{pmatrix}, \tag{5.38}$$

which follows from equation (5.25), we obtain that

$$\frac{1}{2i\sin ka}\begin{pmatrix} e^{ika}c_N - c_{N-1} \\ -e^{-ika}c_N + c_{N-1} \end{pmatrix} = \mathbf{M}^{(N)}\frac{1}{2i\sin ka}\begin{pmatrix} e^{ika}c_0 - c_{-1} \\ -e^{-ika}c_0 + c_{-1} \end{pmatrix}. \tag{5.39}$$

The right-hand side of equation (5.39) can be considerably simplified with the use of the initial values of c_0 and c_1, given by equation (5.35). We obtain that $e^{ika}c_0 - c_{-1} = 0$ and that $-e^{-ika}c_0 + c_{-1} = -e^{-ika} + e^{+ika} = 2i\sin ka$. Inserting these expressions into the r.h.s. of equation (5.39), we obtain

$$\frac{1}{2i\sin ka}\begin{pmatrix} e^{ika}c_N - c_{N-1} \\ -e^{-ika}c_N + c_{N-1} \end{pmatrix} = \begin{pmatrix} M_{12}^{(N)} \\ M_{22}^{(N)} \end{pmatrix} = \begin{pmatrix} rt^{-1} \\ t^{-1} \end{pmatrix}. \tag{5.40}$$

As $T = 1/|M_{22}^{(N)}|^2$, we finally obtain the expression for the transmission coefficient [149]

$$T = \frac{4\sin^2 ka}{|e^{-ika}c_N - c_{N-1}|^2}. \tag{5.41}$$

We also have the reflection coefficient

$$R = \left| \frac{e^{ika}c_N - c_{N-1}}{e^{-ika}c_N - c_{N-1}} \right|^2. \tag{5.42}$$

Expression (5.41) is often used in numerical simulations. For the case when the energies ε_n are chosen randomly, we cannot calculate the transmission and reflection coefficients analytically. Simple iteration procedure, equations (5.22) and (5.36), enable us to calculate the amplitudes c_N and c_{N-1} and obtain the transmission coefficient.

5.4 Single Impurity

Consider now the case of one individual scatterer located at the site $n = 0$,

$$\varepsilon_n = \delta_{n,0}\varepsilon_0. \tag{5.43}$$

The transfer matrix for site $n = 0$ is given by equation (5.31) with $\varepsilon_n = \varepsilon_0$. By comparing this with the general expression for the transfer matrix given by equation (1.95), we immediately obtain the ratio $|r/t|^2$ of reflection and transmission probabilities,

$$\left|\frac{r}{t}\right|^2 = \left|\frac{r'}{t'}\right|^2 = \frac{\varepsilon_0^2}{4\sin^2 ka}, \tag{5.44}$$

and the transmission coefficient through a single impurity is

$$T = \frac{1}{1 + |r/t|^2}. \tag{5.45}$$

To find the energy of the bound state, we follow the general arguments presented in section 1.4.3. We insert $k = i\kappa$ into the expression for the transfer matrix given by equation (5.31). The bound state is given by the solution of the equation $M_{22}(k = i\kappa) = 0$, and given by the expression

$$\sinh \kappa = \frac{\varepsilon_0}{2}. \tag{5.46}$$

For $\varepsilon_0 > 0$, the solution of equation (5.46), $\kappa_b^+ = \sinh^{-1}(\varepsilon_0/2)$, is real. The energy of a bound state is $\mathcal{E}_b^+ = 2\cosh \kappa_b^+ a = 2\sqrt{1 + \sinh^2 \kappa_b^+ a} = \sqrt{4 + \varepsilon_0^2}$. For $\varepsilon_0 < 0$, we use $\sinh(x + i\pi) = -\sinh x$, so that the solution of equation (5.46) is $\kappa_b^- = \sinh^{-1}|\varepsilon_0/2| + i\pi = \kappa_b^+ + i\pi$. We remind the reader that the real part of κ_b^- must be positive in order to have exponentially decaying solutions on both sides of the impurity. Since $\cosh(x + i\pi) = -\cosh x$, the energy of the bound state is now $\mathcal{E}_b^- = -2\cosh \kappa_b^- a = -\sqrt{4 + \varepsilon_0^2}$. Thus we obtain the bound energy [11]

$$\mathcal{E}_b^\pm = \pm 2\sqrt{1 + \varepsilon_0^2/4}, \tag{5.47}$$

where the sign $+$ $(-)$ corresponds to positive (negative) ε_0, respectively. We see that a bound state appears for any nonzero ε_0. Since $|\mathcal{E}_b| > 2$, the energy of the bound states always lies *outside* the transmission band $|E| < 2$. Note that, in contrast to the rectangular potential well (chapter 2), bound states exist for both signs of ε_0.

5.5 Transmission through Impurities

In this section, we will apply Chebyshev's formula, equation (1.106), for the calculation of the transmission coefficient in two special problems.

5.5.1 Transmission through N Identical Impurities

As the first case, we consider N identical impurities: $\varepsilon_n = \varepsilon_A$, $n = 0, 1, \ldots, N-1$. The model for $N = 6$ is shown in figure 5.6. With the use of Chebyshev's identity, equations (1.106) and (5.44), for the transmission and reflection amplitudes of one individual impurity, we obtain the transmission coefficient in the form

$$T_N = \frac{1}{1 + \left|\dfrac{r^2}{t^2}\right| \dfrac{\sin^2 Nqa}{\sin^2 qa}} = \frac{1}{1 + \dfrac{\varepsilon_A^2}{4\sin^2 ka} \dfrac{\sin^2 Nqa}{\sin^2 qa}}, \tag{5.48}$$

$e^{ikn} + r\,e^{-ikn}$ $t\,e^{ikn}$

○ ○ ○ ○ ○ ○ ○ ● ● ● ● ● ● ○ ○ ○ ○ ○ ○ ○ ○

$\varepsilon_n = 0$ $\varepsilon_n = \varepsilon_A$ $\varepsilon_n = 0$

FIGURE 5.6. Model for $N = 6$ identical impurities $\varepsilon_n = \varepsilon_A$.

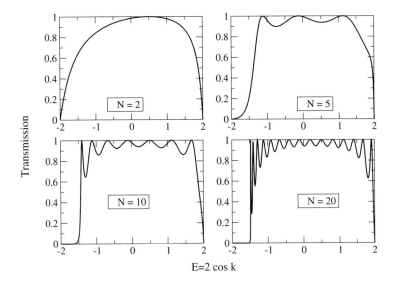

FIGURE 5.7. Transmission coefficient as a function of energy $E = 2\cos ka$ for a barrier of length $N = 2, 5, 10,$ and 20 with $\varepsilon_n = \varepsilon_A = 0.5$, shown in figure 5.6 for $N = 6$. Note the special values of energy for which $T = 1$.

where

$$2\cos qa = \mathcal{E} - \varepsilon_A. \tag{5.49}$$

We plot the transmission coefficient as a function of energy $\mathcal{E}$ in figure 5.7. Similarly to the transmission through N identical δ-function potential barriers, we find that the transmission coefficient oscillates as a function of the energy, and equals 1 when $\sin Nqa = 0$. Note that, the energy interval in which transmission is possible is restricted by two conditions: (1) the wave vector q given by equation (5.49) must be real, and (2) $|\mathcal{E}| \leq 2$ in order to have waves propagating left and right from the sample. Therefore, the transmission band is restricted to values of $-2 + \varepsilon_A \leq \mathcal{E} < 2$.

5.5.2 Transmission through $2N$ Alternating Impurities

In the second problem, we assume that our sample consists of $2N$ alternating impurities:

$$\varepsilon_n = \begin{cases} +\varepsilon_A, & n = 0, 2, \ldots, 2N-2, \\ -\varepsilon_A, & n = 1, 3, \ldots, 2N-1, \end{cases} \tag{5.50}$$

$$e^{ikn} + r\,e^{-ikn} \qquad\qquad\qquad t\,e^{ikn}$$

$$\varepsilon_n = 0 \qquad\qquad \varepsilon_n = (-1)^n \varepsilon_A \qquad\qquad \varepsilon_n = 0$$

FIGURE 5.8. Model for $2N = 8$ alternating impurities ε_n.

as shown schematically in figure 5.8. In this case, the spatial period of the system is $2a$ and the transfer matrix of two neighboring sites can be expressed as

$$\mathbf{M}^{+-} = \mathbf{M}^{+}\mathbf{M}^{-}, \tag{5.51}$$

where $\mathbf{M}^{+}$ ($\mathbf{M}^{-}$) is the transfer matrix of even (odd) sites and given by equation (5.31), with $\varepsilon_0 = \varepsilon_A$ and $\varepsilon_0 = -\varepsilon_A$, respectively. The transfer matrix for N pairs of impurities is then

$$\left[\mathbf{M}^{+-}\right]^{N}. \tag{5.52}$$

From the explicit form of the transfer matrix $\mathbf{M}$, given by equation (5.31), we easily find that

$$M_{12}^{+-} = e^{ika}\left[\varepsilon_A + i\,\frac{\varepsilon_A^2}{2\sin ka}\right]. \tag{5.53}$$

Then we can express the analytical expression for the transmission coefficient, $T^{+-} = 1/|M_{22}^{+-}|^2 = 1/(1 + |M_{12}^{+-}|^2)$, as

$$T^{+-} = \frac{1}{1 + \varepsilon_A^2 + \dfrac{\varepsilon_A^4}{4\sin^2 ka}}. \tag{5.54}$$

We have also

$$\operatorname{Tr}\mathbf{M}^{+-} = \operatorname{Tr}\mathcal{M}^{+-} = 2\cos 2qa = \mathcal{E}^2 - 2 - \varepsilon_A^2, \tag{5.55}$$

as can be easily verified from the form of the matrix $\mathcal{M}^{+-}$,

$$\mathcal{M}^{+-} = \mathcal{M}^{+}\mathcal{M}^{-} = \begin{pmatrix} \mathcal{E} - \varepsilon_A & -1 \\ 1 & 0 \end{pmatrix}\begin{pmatrix} \mathcal{E} + \varepsilon_A & -1 \\ 1 & 0 \end{pmatrix}$$

$$= \begin{pmatrix} \mathcal{E}^2 - \varepsilon_A^2 - 1 & \mathcal{E} - \varepsilon_A \\ \mathcal{E} + \varepsilon_A & -1 \end{pmatrix}. \tag{5.56}$$

The transmission coefficient of the system consisting of $2N$ alternating impurities is

$$T_N^{+-} = \frac{1}{1 + \left(\varepsilon_A^2 + \dfrac{\varepsilon_A^4}{4\sin^2 ka}\right)\dfrac{\sin^2 Nqa}{\sin^2 qa}}, \tag{5.57}$$

where qa can be obtained from equation (5.55).

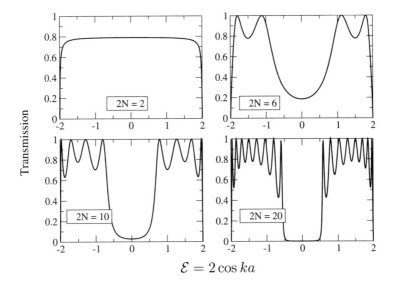

$$\mathcal{E} = 2\cos ka$$

FIGURE 5.9. The transmission coefficient as a function of energy $E = 2\cos ka$, for a barrier consisting of N impurities $\varepsilon_n = \varepsilon_A$ (n odd) and $\varepsilon_n = -\varepsilon_A$ (n even), $n = 1, 2, \ldots N$ shown in figure 5.8. $\varepsilon_A = 0.5$.

Figure 5.9 presents the energy dependence of the transmission coefficient (5.57) for different values of N. For $N = 1$ we have only one unit cell with two impurities $+\varepsilon$ and $-\varepsilon$. Notice that $T \simeq 0.8$, as can easily be obtained from equation (5.57) with $N = 1$. We see in figure 5.9 that, for large N, the system of alternating impurities splits the band structure of the original tight binding model. The original transmission band, $|\mathcal{E}| \leq 2$, splits into two bands, separated by a gap where the transmission is very small and decreases as the number of impurities, $2N$, increases. For $N \to \infty$, the width of the gap is $\approx 2\varepsilon_A = 1$. In transmission bands, the transmission coefficient oscillates as a function of energy and reaches maximum values $T = 1$ for some special energies.

5.6 Coupled Pendulum Analogy of the Tight Binding Model

Now we introduce a simple model from classical mechanics, namely, the system of N coupled linear pendulums as shown in figure 5.10. The model, introduced by Economou in his book about Green's functions [11], mimics all the typical properties of the tight binding model, as discussed in previous sections.

The displacement of the nth pendulum (we assume this displacement is along the x axis and is small) obeys Newton's equation

$$m\ddot{u}_n = F_B + F_{n+1} + F_{n-1}, \tag{5.58}$$

where

$$F_B = -\frac{mg}{\ell} u_n \tag{5.59}$$

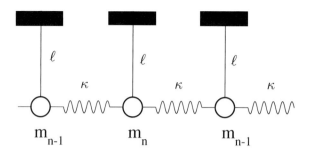

FIGURE 5.10. Model of N coupled pendulums.

is the gravitational force, together with the tension of the wire of the length ℓ, and

$$F_{n+1} = \kappa \left(u_{n+1} - u_n \right) \tag{5.60}$$

is the force acting on the nth pendulum from the $(n+1)$th pendulum, and

$$F_{n-1} = \kappa \left(u_{n-1} - u_n \right) \tag{5.61}$$

is the force acting on the nth pendulum from the $(n-1)$th pendulum. We assume that all pendulums have the same time dependence $e^{-i\omega t}$ so that $\ddot{u} = -\omega^2 u$. Substituting equations (5.59) and (5.60) into equation (5.58) and setting $\ddot{u} = -\omega^2 u$ and $\omega_0^2 \equiv g/\ell$, we obtain the following equation:

$$\left(m_n \omega_0^2 + 2\kappa - m\omega^2 \right) u_n - \kappa u_{n-1} - \kappa u_{n+1} = 0. \tag{5.62}$$

For simplicity, we assume that $m_n \equiv m$ for all n. Thus we have a periodic system. By using Bloch's theorem, we have

$$u_{n+1} = e^{ika} u_n \quad \text{and} \quad u_{n-1} = e^{-ika} u_n. \tag{5.63}$$

Substituting equations(5.63) into equation (5.62), we find that equation (5.62) is satisfied for any value of u_n, provided that the eigenfrequency satisfies the condition

$$\omega^2 = \omega_0^2 + 2\frac{\kappa}{m} \left[1 - \cos ka \right]. \tag{5.64}$$

This is called the dispersion relation.

The allowed values of k are obtained from the periodic boundary conditions

$$u_{N+1} = u_1, \tag{5.65}$$

which, together with Bloch's theorem $u_{N+1} = u_1 e^{ikNa}$, give the following relation:

$$Nka = 2\pi p, \quad p \text{ integer}, \tag{5.66}$$

or

$$k = \frac{2\pi}{a} \frac{p}{N} = \frac{2\pi}{L} p, \tag{5.67}$$

where $L = Na$ is the length of the system. Notice that the two integers p and p', which differ by an integer number of N, give the same solution. Therefore, the wave vector k

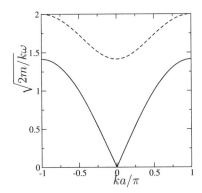

FIGURE 5.11. Frequency band of the system of coupled linear pendulums for two values of the frequency ω_0: $\omega_0 = 0$ (solid line) and $2m\omega_0^2/k = 2$ (dashed line).

can be restricted to the region

$$-\frac{\pi}{a} \leq k \leq \frac{\pi}{a}. \tag{5.68}$$

This region is called the first Brillouin zone. Since the number of nonequivalent values of k is N, the number of degrees of freedom of the system, we have obtained all the N eigenfrequencies of the system.

In figure 5.11 we plot ω versus ka. Notice that plotting ω versus ka outside the region $(-\pi, +\pi)$ does not give any additional information, since all the eigenfrequencies are already given in the first Brillouin zone. The graph in figure 5.11 is called the *reduced zone scheme*. If we allow k to take all values, then the graph is called a *repeated zone scheme*.

Also notice from figure 5.11 or from equation (5.64) that there is a region of allowed frequencies, $\omega_{min} \leq \omega \leq \omega_{max}$, where $\omega_{min} = \omega_0$ and $\omega_{max} = \sqrt{\omega_0^2 + 4\kappa/m}$, called a *band*. There are also regions that do not allow propagating solutions. These regions are called *gaps*. In this simple system of N coupled pendulums, the energy spectrum is given by a band, bounded by two gaps. This property of having bands and gaps as the frequency of the system changes is a general property of wave propagation in periodic systems.

Notice that the dispersion relation is not linear (as is the case for classical wave propagation in homogeneous systems) neither is it quadratic (as is the case of free particles). From equation (5.64), one can distinguish the following dispersion relations.

(1) For $k \simeq 0$ (that means that the frequency ω is very close to ω_{min}), we have

$$\omega - \omega_{min} \simeq \frac{1}{2}\frac{\hbar^2}{m^*}k^2, \tag{5.69}$$

where

$$m^* = m\frac{\hbar\omega_0}{\kappa a^2} \tag{5.70}$$

is called the effective mass. Equation (5.69) is generally obeyed very close to the minimum of the dispersion relation.

(2) For $k \simeq \pi/a$ we have

$$\omega - \omega_{max} \simeq \frac{1}{2} \frac{\hbar^2}{m^*} \left(k \pm \frac{\pi}{a} \right)^2, \tag{5.71}$$

where

$$m^* = -m \frac{\hbar \omega_0}{\kappa a^2}. \tag{5.72}$$

Notice that the effective mass very close to the maximum of the dispersion relation is negative.

(3) When $\omega_0 \sim 0$ and $\kappa \approx 0$, we have

$$\omega = c|k|, \quad \text{where} \quad c = \sqrt{\frac{\kappa a^2}{m}} \tag{5.73}$$

is the velocity of sound inside the medium.

For a homogeneous medium the velocity of sound is given by $c = \sqrt{B/\rho}$ where B is the bulk modulus and ρ is the density. In the present case we have $B = \kappa a$ and $\rho = m/a$. The case with $\omega_0 = 0$ is really important, since it describes the motion of ions inside a one-dimensional solid.

The group velocity is defined as $v_g = \partial \omega/\partial k$. We see that a wave packet with $k = \pm \pi/a$ (or $k = 0$ in the case of $\omega_0 \neq 0$) does not propagate in the system, i.e., the wave is a standing wave.

One can easily calculate the density of states by using the relation

$$\begin{aligned} n(\omega) &= 2n(k) \frac{1}{\dfrac{\partial \omega}{\partial k}} \\ &= 2 \frac{L}{2\pi} \frac{1}{\dfrac{\partial \omega}{\partial k}} = \frac{2N}{\pi} \frac{\omega}{\sqrt{(\omega_{max}^2 - \omega^2)(\omega^2 - \omega_{min}^2)}}. \end{aligned} \tag{5.74}$$

Notice that $n(\omega) \neq 0$ only inside the band and is zero in the gaps. Notice also that $n(\omega)$ very close to the band edges diverges as

$$n(\omega) \sim (\omega - \omega_{min})^{-1/2} \quad \text{or} \quad n(\omega) \sim (\omega - \omega_{min})^{-1/2}. \tag{5.75}$$

Equation (5.75) is a general characteristic of all one-dimensional systems.

5.7 Problems

Problems with Solutions

Problem 5.1

Prove that the Fermi energy $\mathcal{E}_F = 0$ and the Fermi wave vector $k_F = \pi/(2a)$ for a system of N noninteracting electrons in the one-dimensional tight binding model.

Solution. The Fermi energy is defined by the relation

$$\int_{-\infty}^{\mathcal{E}_F} n'(\mathcal{E}) \, d\mathcal{E} = N \tag{5.76}$$

and the Fermi vector is defined by

$$2 \sum_{k \leq k_F} 1 = 2 \left(\frac{L}{2\pi} \right) = \int_{-k_F}^{+k_F} dk = N. \tag{5.77}$$

The integral in equation (5.76) can be calculated analytically and gives

$$\frac{2N}{\pi} \left(\frac{\pi}{2} + \sin^{-1} \frac{\mathcal{E}_F}{2V} \right) = N. \tag{5.78}$$

The last equation is satisfied if $\mathcal{E}_F = 0$. An easier way to prove that $\mathcal{E}_F = 0$ is to use $k_F = \pi/2$ from equation (5.77) and then use the relation

$$\mathcal{E}_F = \mathcal{E}(k_F) = 2V \cos k_F a, \tag{5.79}$$

which is the relationship between the wave vector and the energy of the electron in the one-dimensional tight binding model.

Problem 5.2
Verify that

$$\text{Tr} \, \mathcal{M} = \text{Tr} \, \mathbf{M}, \tag{5.80}$$

The transfer matrices $\mathcal{M}$ and $\mathbf{M}$ are given by equations (5.22) and (5.31), respectively.

Solution. We start from the relation $\mathbf{M} = \mathbf{Q}^{-1} \mathcal{M} \mathbf{Q}$, given in equation (5.30). Using the explicit form of the matrices $\mathbf{Q}$ and $\mathbf{Q}^{-1}$, we get

$$M_{11} = \frac{1}{2i \sin ka} \left[\mathcal{M}_{11} e^{ika} + \mathcal{M}_{12} - \mathcal{M}_{21} + \mathcal{M}_{22} e^{-ika} \right] \tag{5.81}$$

and

$$M_{22} = \frac{1}{2i \sin ka} \left[-\mathcal{M}_{11} e^{-ika} - \mathcal{M}_{12} + \mathcal{M}_{21} + \mathcal{M}_{22} e^{ika} \right]. \tag{5.82}$$

Now, we sum the above equations and indeed obtain

$$\text{Tr} \, \mathbf{M} = M_{11} + M_{22} = \mathcal{M}_{11} + \mathcal{M}_{22} = \text{Tr} \, \mathcal{M}. \tag{5.83}$$

The relationship (5.80) follows directly from a theorem of linear algebra, which states that the trace and the determinant of any matrix $\mathbf{A}$ are invariant with respect to similarity transformations $\mathbf{A} \to \mathbf{Q}^{-1} \mathbf{A} \mathbf{Q}$.

Problem 5.3
Derive the following formula for the transmission amplitude t_N through N impurities in the one-dimensional tight binding model:

$$t_N = \frac{-2i \sin ka}{\mathcal{M}_{11} e^{-ika} + \mathcal{M}_{12} - \mathcal{M}_{22} e^{ika} - \mathcal{M}_{21}}. \tag{5.84}$$

Solution. The wave function on one side of this segment of N impurities can be expressed in terms of the wave function on the opposite side:

$$\begin{pmatrix} c_N \\ c_{N-1} \end{pmatrix} = \prod_{n=0}^{N-1} \mathcal{M}_n \begin{pmatrix} c_0 \\ c_{-1} \end{pmatrix} = \begin{pmatrix} \mathcal{M}_{11} & \mathcal{M}_{12} \\ \mathcal{M}_{21} & \mathcal{M}_{22} \end{pmatrix} \begin{pmatrix} c_0 \\ c_{-1} \end{pmatrix}. \tag{5.85}$$

This relationship can be written explicitly as

$$c_N = \mathcal{M}_{11}c_0 + \mathcal{M}_{12}c_{-1},$$
$$c_{N-1} = \mathcal{M}_{21}c_0 + \mathcal{M}_{22}c_{-1}. \tag{5.86}$$

Now suppose that the electron is approaching the system from the left. The wave function left of the system is of the form

$$\Psi_n = \begin{cases} e^{ikan} + r_N e^{-ikan} & \text{for} \quad n \leq 0, \\ t_N e^{ikan} & \text{for} \quad n \geq N-1, \end{cases} \tag{5.87}$$

where r_N is the reflection amplitude. In particular,

$$c_0 = 1 + r_N, \quad c_{-1} = e^{-ika} + r_N e^{ika}. \tag{5.88}$$

On the opposite side of the sample, there is only the transmitted wave, so that we have

$$c_N = t_N, \quad c_{N-1} = t_N e^{-ika}. \tag{5.89}$$

We need to solve the system of linear equations (5.86), (5.88), and (5.89). First, by inserting equation (5.89) in equation (5.86), we obtain

$$t_N = \mathcal{M}_{11}c_0 + \mathcal{M}_{12}c_{-1}, \quad t_N e^{-ika} = \mathcal{M}_{21}c_0 + \mathcal{M}_{22}c_{-1}. \tag{5.90}$$

Then, from equations (5.88) we obtain that $c_{-1} = e^{-ika} - e^{ika} + e^{ika}c_0 = -2i \sin ka + e^{ika}c_0$. Inserting this expression into equation (5.90), we get

$$t_N = \left[\mathcal{M}_{11} + \mathcal{M}_{12}e^{ika} \right] c_0 - 2i \sin ka \mathcal{M}_{12},$$
$$t_N e^{-ika} = \left[\mathcal{M}_{21} + \mathcal{M}_{22}e^{ika} \right] c_0 - 2i \sin ka \mathcal{M}_{22}. \tag{5.91}$$

Now, express c_0 using both equations (5.92),

$$\frac{t_N + 2i \sin ka \mathcal{M}_{12}}{\mathcal{M}_{11} + \mathcal{M}_{12}e^{ika}} = \frac{t_N e^{-ika} + 2i \sin ka \mathcal{M}_{22}}{\mathcal{M}_{21} + \mathcal{M}_{22}e^{ika}}. \tag{5.92}$$

After simple algebraic operations and with the help of det $\mathcal{M} = 1$, we obtain the required expression for the transmission amplitude t_N, given by equation (5.84).

Problem 5.4
Calculate the current density for the tight binding model.

Solution. The current density

$$j = \frac{\hbar i}{2m} \left[\psi(x) \frac{\partial \psi^*(x)}{\partial x} - \psi^*(x) \frac{\partial \psi(x)}{\partial x} \right] \tag{5.93}$$

is defined in section 1.5.1 by equation (1.114). In our special case, the wave function is defined only on the discrete set of sites $x = na$,

$$\psi(x) = \psi(na). \tag{5.94}$$

Therefore, we substitute the derivative by the expression

$$\frac{\partial \psi(x)}{\partial x} = \frac{1}{a}(\psi_{n+1} - \psi_n). \tag{5.95}$$

Inserting equation (5.95) in equation (5.93), we obtain

$$j = \frac{i\hbar}{2ma} \left[\psi_n \psi_{n+1}^* - \psi_n^* \psi_{n+1} \right]. \tag{5.96}$$

For the plane wave $\psi_n = A e^{ikan}$, we obtain the following expression for the current density:

$$j = \frac{\hbar}{m} \frac{\sin ka}{a} |A|^2. \tag{5.97}$$

The current density can be expressed in terms of the group velocity

$$v_g = \frac{2Va}{\hbar} \sin ka \tag{5.98}$$

and the effective mass $m^* = \hbar^2/(2a^2 V)$,

$$j = \frac{m^*}{m} v_g |A|^2. \tag{5.99}$$

Note that, the current density given by equation (5.97) converges to $j = \hbar k/m |A|^2$ in the limit of $ka \ll 1$. This is consistent with the expression for the current, given by equation (1.123).

Problem 5.5
Show that the Kronig-Penney model, discussed in chapter 4,

$$-\frac{\hbar^2}{2m} \frac{\partial^2 \Psi}{\partial x^2} + \sum_{n=-\infty}^{+\infty} \frac{\hbar^2}{2m} \frac{\beta_n}{a} \delta(x - na) \Psi(x) = E\Psi(x), \tag{5.100}$$

can be considered as a tight binding model with k-dependent potential

$$\varepsilon_n = \beta_n \frac{\sin ka}{ka}. \tag{5.101}$$

Solution. The easiest way is to compare the transfer matrix for the Kronig-Penney model, given by equation (4.7), with the transfer matrix for the tight-binding model, given by equation (5.31). One immediately sees that both matrices are formally equivalent, if the random potential ε_n is written in the form of equation (5.101).

Then, equation (5.100) is equivalent to the tight binding model

$$\Psi_{n+1} + \Psi_{n-1} = 2\cos ka \Psi_n - \beta_n \frac{\sin ka}{ka} \Psi_n. \tag{5.102}$$

$$e^{ikn} + r\,e^{-ikn} \qquad\qquad\qquad t\,e^{ikn}$$

○ ○ ○ ○ ○ ○ ○ ● ● ○ ○ ● ● ○ ○ ○ ○ ○ ○ ○ ○ ○

$$\varepsilon_n = 0 \qquad\qquad\qquad\qquad \varepsilon_n = 0$$

FIGURE 5.12. Model for $4N = 8$ alternating impurities, $\varepsilon_n = \varepsilon_A$ for $n = 1, 2$, $\varepsilon_n = -\varepsilon_A$ for $n = 3, 4$, and $\varepsilon_{n+4} \equiv \varepsilon_n$.

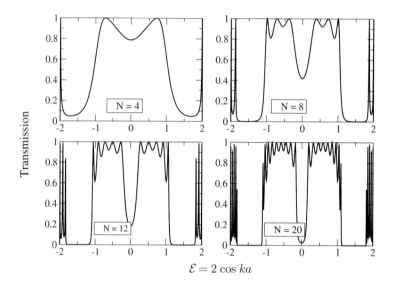

FIGURE 5.13. Transmission coefficient as a function of energy $E = 2\cos k$ for a barrier of N impurities $\varepsilon_n = \varepsilon_A$ $(n = 1, 2)$, $\varepsilon_n = -\varepsilon_A$ $(n = 3, 4)$, and $\varepsilon_{n+4} \equiv \varepsilon_n$ (figure 5.12) and $\varepsilon_A = 0.5$.

For the periodic case we have that all $\beta_n = \beta$. We immediately see that all the propagating solutions of equation (5.102) are given by

$$\cos qa = \cos ka - \frac{\beta}{2}\frac{\sin ka}{ka}. \tag{5.103}$$

This is exactly the same equation as equation (4.10).

Problems without Solutions

Problem 5.6
Show that for the case of $\varepsilon_n = 0$ the transfer matrix

$$\mathbf{M}_0 = \mathbf{Q}^{-1}\mathcal{M}_0\,\mathbf{Q} = \begin{pmatrix} e^{+ika} & 0 \\ 0 & e^{-ika} \end{pmatrix} \tag{5.104}$$

is diagonal.

Problem 5.7
Consider now the general problems of the transmission of electrons through N impurities located on sites $0, 1, \ldots, N-1$. Use formula (5.41) for the transmission coefficient and write a numerical program to calculate T for an arbitrary n dependence of ε_n. Apply the program to three special cases discussed in figures 5.6, 5.8 and 5.12.

Problem 5.8
Calculate numerically the transmission of an electron through the system of $4N$ impurities alternating with period 4:

$$\varepsilon_n = \begin{cases} +\varepsilon_A, & n = 0, 1, 4, 5, 8, 9, \ldots, \\ -\varepsilon_A, & n = 2, 3, 6, 7, \ldots, 4N-2, 4N-1, \end{cases} \tag{5.105}$$

shown schematically in figure 5.12. $\varepsilon_A = 0.5$. Compare your results with the data presented in figure 5.13.

Problem 5.9
Show that energies for which the transmission coefficient T shown in figure 5.9 equals 1 are given by

$$\mathcal{E}_n = \sqrt{2 + \varepsilon_A^2 - 2\cos\frac{n\pi}{N}}. \tag{5.106}$$

Explain why the number of maxima observed in figure 5.9 is smaller than $N-1$.

6 Tight Binding Models of Crystals

In this chapter we study how the spatial periodicity of the system influences the structure of the energy spectrum. We introduce two tight binding models, the first one with a period $\ell = 2a$, and the second one with a period $\ell = 4a$. We show that the spectrum of the allowed energies changes considerably when the spatial period of the lattice increases. The energy band splits into subbands separated from each other by gaps [23, 39]. Also, the wave function does not have the simple form of a plane wave, but possesses more complicated spatial structure, known as Bloch waves.

6.1 Periodic One-Dimensional System with Two Different Atoms

We first consider a one-dimensional solid with two different atoms, with site energies $+\varepsilon_A$ and $\varepsilon_B = -\varepsilon_A$, as shown in figure 6.1. For simplicity we assume that the nonzero matrix element $V_2 = \langle \Phi_A | H | \Phi_B \rangle$ is real. In the following discussion we normalize all the energies with V_2, i.e., we have $V_2 = 1$.

6.1.1 The Wave Function

It is instructive to calculate the site dependence of the wave function. To do so, let us write Schrödinger's equation

$$\mathcal{E} c_n = \varepsilon_n c_n + c_{n-1} + c_{n+1} \tag{6.1}$$

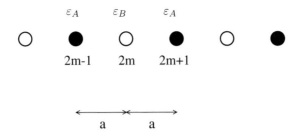

FIGURE 6.1. Periodic one-dimensional solid with two different atoms. The period of the system is $\ell = 2a$.

explicitly for two neighboring sites, $n = 2m$ and $n = 2m + 1$,

$$c_{2m-1} \quad +(-\varepsilon_A - \mathcal{E})c_{2m} \qquad\qquad +c_{2m+1} \qquad\qquad = 0,$$
$$c_{2m} \quad +(+\varepsilon_A - \mathcal{E})c_{2m+1} \quad +c_{2m+2} = 0. \tag{6.2}$$

We are looking for a solution of equation (6.2) of the form

$$c_{2m-1} = \frac{a}{\sqrt{1+a^2}} e^{ika(2m-1)},$$
$$c_{2m} = \frac{1}{\sqrt{1+a^2}} e^{ika(2m)}, \tag{6.3}$$

where a is a constant, which needs to be determined. It will become clear later why we choose this special form for the prefactor in front of the free exponentials in equations (6.3).

Inserting equations (6.3) into equations (6.2) we obtain from the first equation (6.2) that

$$\mathcal{E} + \varepsilon_A = 2a \cos ka, \tag{6.4}$$

and from the second equation (6.2) that

$$\mathcal{E} - \varepsilon_A = \frac{2}{a} \cos ka. \tag{6.5}$$

By multiplication of the last two equations, we obtain the energy of the electron,

$$\mathcal{E}^\pm = \pm\sqrt{\varepsilon_A^2 + 4\cos^2 ka}. \tag{6.6}$$

Now, we solve for the energy $\mathcal{E}$ given by equation (6.4) and substitute it into equation (6.5). We obtain

$$\left(a - \frac{1}{a}\right) \cos ka = \varepsilon_A, \tag{6.7}$$

which can be rewritten as

$$a^2 - \frac{\varepsilon_A}{\cos ka} a - 1 = 0. \tag{6.8}$$

Equation (6.8) has two solutions,

$$a_\pm = \frac{\varepsilon_A}{2\cos ka} \pm \sqrt{\left(\frac{\varepsilon_A}{2\cos ka}\right)^2 + 1}. \tag{6.9}$$

Note that, $a_+ = -a_-^{-1}$. Thus, we have two possible solutions, one with $a = a_+$ and the other with $a = a_- = -1/a_+$. Inserting $a = a_+$ and $a = a_- = -a_+^{-1}$ into equation (6.3), we find the following form of the wave functions:

$$c_n^+ : \begin{cases} c_{2m+1}^+ = \dfrac{e^{i2kma}}{\sqrt{1+a_+^2}} \times a_+ e^{ika}, \\[3mm] c_{2m}^+ = \dfrac{e^{i2kma}}{\sqrt{1+a_+^2}} \times 1, \end{cases} \tag{6.10}$$

and

$$c_n^- : \begin{cases} c_{2m+1}^- = \dfrac{e^{i2kma}}{\sqrt{1+a_+^2}} \times (-e^{ika}), \\[3mm] c_{2m}^- = \dfrac{e^{i2kma}}{\sqrt{1+a_+^2}} \times a_+. \end{cases} \tag{6.11}$$

Consider now the case when $\cos ka > 0$ (i.e., $-\pi/2 \le ka \le +\pi/2$). Then we have from equation (6.9) that $a_+ > 1$ (we remind the reader that $\varepsilon_A > 0$). That means that the electron with the wave function c_n^+, given by equation (6.10), prefers to stay on odd sites $n = 2m + 1$. Inserting $a = a^+$ into the expression (6.4), we find that the energy of the electron is $\mathcal{E}_+ = \sqrt{\varepsilon_A^2 + 4\cos^2 ka}$. In the same way, we see that the electron in the quantum state c_n^-, given by equation (6.11), prefers to stay at even sites $n = 2m$ of the lattice, and its energy is $\mathcal{E}_- = -\sqrt{\varepsilon_A^2 + 4\cos^2 ka}$.

In the case of $\cos ka < 0$, (i.e., either $-\pi \le ka \le -\pi/2$ or $+\pi/2 \le ka \le +\pi$), we obtain from equation (6.9) that $0 < a_+ < 1$. Therefore, the electron in the state with the wave function c_n^+ occupies mostly even sites. To find its energy, we multiply both sides of equation (6.9) by $\cos ka$. Since $\cos ka < 0$, we obtain

$$2a_+ \cos ka = +\varepsilon_A - \sqrt{\varepsilon_A^2 + 4\cos^2 ka}, \tag{6.12}$$

since $\sqrt{\cos^2 ka} = -\cos ka$. Inserting this expression in equation (6.4) we find that the energy of the electron is now $\mathcal{E} = \mathcal{E}_-$. Similarly, we find that the electron in the quantum state c_n^- possesses energy $\mathcal{E} = \mathcal{E}_+$. Thus, the two quantum states c_n^+ and c_n^- simply exchange their roles for negative $\cos ka$.

We see that an electron distinguishes between even and odd sites of the lattice. Its wave functions $c_n^+(k)$ and $c_n^-(k)$ possess spatial modulations within one spatial period $\ell = 2a$. The wave functions $c_n^+(k)$ and $c_n^-(k)$ represent the simplest form of the Bloch states.

The energy of an electron in quantum states c_n^+ and c_n^- is plotted in figure 6.2. The figure shows that the quantum states of the electron are invariant with respect to the transformation $ka \to ka + \pi$. Therefore, we can restrict our use of the wave vector to

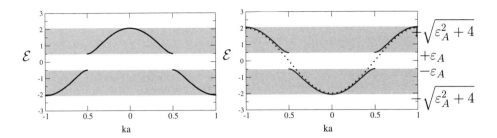

FIGURE 6.2. Energy of the electron in quantum states c_n^+ (right) and c_n^- (right). The energy spectrum of the infinite crystal consists of two transmission bands, given by equation (6.6). For comparison, we show in the right figure also the band of the unperturbed tight binding model with a spatial period of $\ell = a$ (dotted line).

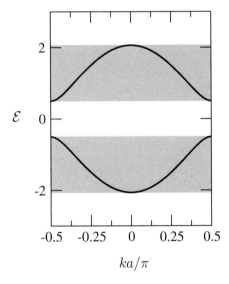

FIGURE 6.3. Reduced energy zone scheme for the periodic lattice model with spatial periodicity $\ell = 2a$. The lower band represents the energy of an electron located mostly on even sites with negative site energy $\varepsilon_{2m} = -\varepsilon_A$. The upper band contains electrons located mostly on odd sites. The two bands are separated by a gap of width $2\varepsilon_A$.

the first Brillouin zone,

$$-\frac{\pi}{2a} \leq k \leq +\frac{\pi}{2a}, \tag{6.13}$$

and construct the reduced energy zone scheme, shown in figure 6.3.

The expressions for the wave functions given by equations (6.10) and (6.11) enable us also to understand the origin of the gap in the energy spectrum. We see that, when $ka \to (\pi/2)^-$, $a_+ \to \infty$, so that the two wave functions c_n^+ and c_n^- occupy only even (odd) sites, respectively. The difference $\mathcal{E}_+ - \mathcal{E}_- = 2\varepsilon_A$ represents the width of the gap.

In the limit of $\varepsilon_A \to 0$, we get $a_+ = 1$ and $a_- = -1$, and we recover from equations (6.10) and (6.11) the solution for the periodic tight binding model with $c_n^+(k) = e^{ikna}$ and $c_n^-(k) = e^{i(k+\pi)na}$. The energy of both states c_n^+ and c_n^- becomes $\pm 2\cos ka$, plotted

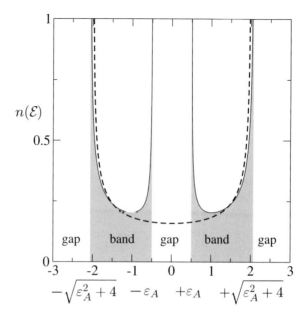

FIGURE 6.4. Density of states for the two-atom tight binding model. The dashed line is the density of states of the periodic tight binding model, discussed in chapter 5.

in the right panel of figure 6.2. We see that the gap at $ka = \pi/2$ closes, in agreement with equation (6.6), and we come back to the tight binding model discussed in chapter 5.

6.1.2 The Density of States

The density of states per unit length,

$$n(\mathcal{E}) = \frac{1}{2\pi} \frac{1}{\left|\dfrac{\partial \mathcal{E}}{\partial k}\right|}, \tag{6.14}$$

is plotted in figure 6.4. We see that the density of states consists of two bands of allowed energies. These two bands are separated by a gap. In the vicinity of the band edges, $n(\mathcal{E})$ diverges as $|\mathcal{E} - \mathcal{E}_{\text{edge}}|^{-1/2}$.

Note that, the band edges correspond exactly to the wave vectors where $|\cos 2ka| = 1$. Thus, we came again to the typical property of the one-dimensional density of states, namely, that it diverges at the band edge of each band.

Note also that, the two energy bands are not centered on the corresponding energies ε_A and $-\varepsilon_A$ of the atomic orbitals, but on the bonding and antibonding energies $\mathcal{E}_a = +\sqrt{1 + \varepsilon_A^2}$ and $\mathcal{E}_b = -\sqrt{1 - \varepsilon_A^2}$ of the molecule AB (problem 6.1).

If we assume that each atom, A and B, can be separated into an ion and an electron, then we have $N = L/a$ electrons. Each energy band has $N/2$ eigenstates. Each eigenstate can accept only two electrons with opposite orientation of spins. Thus, at zero

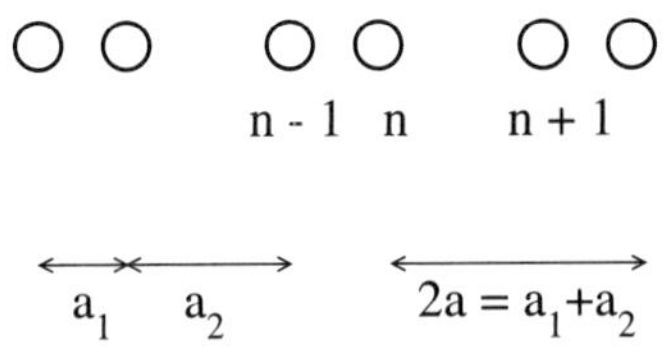

FIGURE 6.5. Tight binding model with two different distances between neighboring atoms.

temperature, we will have the lower band fully occupied, while the upper band will remain empty. The lower occupied band is called the valence band, while the upper empty band is called the conduction band. In this case we have an ionic insulator.

The energy gap $\mathcal{E}_g = \varepsilon_A - \varepsilon_B = 2\varepsilon_A$ for ionic solids is relatively large (of the order of 10 eV). We call a solid a semiconductor if the energy gap is smaller than 2.5–3 eV.

6.2 Periodic Model with Different Distances between Neighboring Atoms

A small gap in the band structure can also be introduced if we modify the periodic tight binding model with only one type of atom. In figure 6.5, we show a model in which all odd atoms have been moved by a small distance, while the positions of the even atoms remain unchanged. This model possesses the spatial period $2a = a_1 + a_2$, where a_1 and a_2 are the two distances between neighboring atoms. We have that the overlap integral between the small-distance (a_1) atoms is V_1 and for the large distance (a_2) is V_2, and we expect that $V_1 > V_2$. Then, Schrödinger's equations for the two neighboring sites can be written as follows:

$$
\begin{aligned}
V_1 c_{n-1} \quad -\mathcal{E} c_n \quad +V_2 c_{n+1} \quad &= 0, \\
V_2 c_n \quad -\mathcal{E} c_{n+1} \quad +V_1 c_{n+2} &= 0.
\end{aligned}
\tag{6.15}
$$

Now we insert $c_{n-1} = e^{-i2ka} c_{n+1}$ and $c_{n+2} = e^{i2ka} c_n$, as expected from Bloch's theorem, and obtain

$$
\begin{aligned}
-\mathcal{E} c_n \quad +(V_1 e^{-i2ka} + V_2) c_{n+1} \quad &= 0, \\
(V_2 + V_1 e^{+i2ka}) c_n \quad -\mathcal{E} c_{n+1} &= 0.
\end{aligned}
\tag{6.16}
$$

Equations (6.16) represent a system of two linear equations for unknown parameters c_n and c_{n+1}. Such a system has a nonzero solution only when its determinant is zero. The reader can easily find that the energy bands in this case are given by

$$
\mathcal{E}_\pm = \pm\sqrt{(V_1 - V_2)^2 + 4 V_1 V_2 \cos^2 ka},
\tag{6.17}
$$

so that the energy gap in this case is given by

$$
\mathcal{E}_{\text{gap}} = 2|V_1 - V_2|,
\tag{6.18}
$$

and the energy bands span the range

$$-(V_1 + V_2) \leq \mathcal{E} \leq -|V_1 - V_2| \quad \text{and} \quad |V_1 - V_2| < \mathcal{E} < V_1 + V_2. \tag{6.19}$$

The lower band will be full of electrons, while the higher band is empty. So we obtained either an insulator (if $\mathcal{E}_{\text{gap}}$ is large) or a semiconductor. Therefore, we see that a parallel equal displacement of half of the atoms, shown in figure 6.5, transforms a solid that has metallic behavior to a semiconductor. This instability in one-dimensional systems is called the *Peierls instability* [23] and is experimentally observed, for instance, in one-dimensional polyacetylene chains.

6.3 Periodic One-Dimensional System with Two Different Atoms and Spatial Period $\ell = 4a$

Now, consider a system with a spatial period $\ell = 4a$. The site energies ε_n possess the following values:

$$\varepsilon_1 = \varepsilon_2 = +\varepsilon_A,$$

$$\varepsilon_3 = \varepsilon_4 = -\varepsilon_A, \tag{6.20}$$

$$\varepsilon_n \equiv \varepsilon_{n+4}, \quad n = 1, 2, \ldots.$$

Following the arguments of section 6.1, we find the wave function of the electron propagating in the tight binding model with spatial periodicity $\ell = 4a$. We set $n = 4m + a$, $a = 1, 2, 3, 4$, and seek a wave function of the form

$$c_n(k) = e^{i4mka} u_n(k). \tag{6.21}$$

with the function $u_n(k)$ periodic with period 4 so that

$$u_{n+4}(k) = u_n(k). \tag{6.22}$$

Therefore, it is sufficient to solve Schrödinger's equation, in its tight binding form, only for four neighboring sites.

We insert the wave function (6.21) into the tight binding equation and obtain a system of linear equations for the unknown parameters $u_n(k)$:

$$\mathbf{A}u = 0, \tag{6.23}$$

where $\mathbf{A}$ is a 4×4 matrix of the form

$$\mathbf{A} = \begin{pmatrix} -\mathcal{E} + \varepsilon_A & 1 & 0 & e^{-i4ka} \\ 1 & -\mathcal{E} + \varepsilon_A & 1 & 0 \\ 0 & 1 & -\mathcal{E} - \varepsilon_A & 1 \\ e^{i4ka} & 0 & 1 & -\mathcal{E} - \varepsilon_A \end{pmatrix}. \tag{6.24}$$

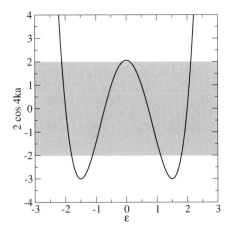

FIGURE 6.6. Trace of the matrix $\mathcal{M}^{(4)}$ as a function of energy $\mathcal{E}$. The shaded area highlights the interval in which $|\mathrm{Tr}\ \mathcal{M}^{(4)}| \leq 2$.

The system of linear equations given by equation (6.24) has a solution only if

$$\det \mathbf{A} = 0. \tag{6.25}$$

After some simple algebra, we obtain an explicit form of the determinant of the matrix $\mathbf{A}$, which is the following:

$$\det \mathbf{A} = (\mathcal{E}^2 - \varepsilon_A^2)^2 - 4\mathcal{E}^2 + 2 - 2\cos 4ka = 0. \tag{6.26}$$

Since $\det \mathbf{A}$ is a polynomial of the 4th order in $\mathcal{E}$, we have for each value of ka exactly four eigenenergies. Solving the system of linear equations given by (6.24), we can find the corresponding eigenvectors $u_n(k)$. We will not present the explicit form of the eigenvectors $u_n(k)$; we note only that, the matrix $\mathbf{A}$ given by equation (6.24) is invariant with respect to the following transformation of the wave vector ka:

$$ka \rightarrow ka - \frac{\pi}{2}. \tag{6.27}$$

Therefore, we can restrict ourselves to the reduced interval of wave vectors

$$-\frac{\pi}{4} \leq ka \leq \frac{\pi}{4}. \tag{6.28}$$

From equation (6.26) we obtain the result that the allowed energies are given by

$$2\cos 4ka = f(\mathcal{E}) = (\mathcal{E}^2 - \varepsilon_A^2)^2 - 4\mathcal{E}^2 + 2. \tag{6.29}$$

To obtain the band structure of the present model, we first plot in figure 6.6 the function $f(E)$ as a function of energy. We see that $f(E)$ has a local maximum at the energy

$$\mathcal{E}_1 = 0, \tag{6.30}$$

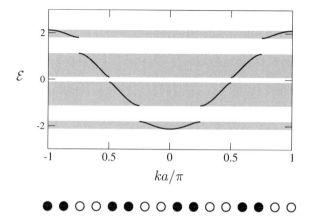

FIGURE 6.7. Band structure for the infinite periodic crystal with $\varepsilon_n = +\varepsilon_A$ for $n = 1, 2$, $\varepsilon_n = -\varepsilon_A$ for $n = 3, 4$, and $\varepsilon_{n+4} = \varepsilon_n$. The period of the lattice is 4. Four energy bands are shown. Their band edges are given by equation (6.32) with Tr $\mathcal{M}^{(4)}$ given by equation (6.68).

and two minima, located at

$$\mathcal{E}_2^{\pm} = \pm\sqrt{2 + \varepsilon_A^2}. \tag{6.31}$$

By inserting equations (6.30) and (6.31) in $f(\mathcal{E})$ given by equation (6.29), we immediately see that once $\varepsilon_A \neq 0$ then $|2\cos 4ka| > 2$ for all energies $\mathcal{E}_1$ and $\mathcal{E}_2^{\pm}$. Therefore, the energy spectrum always consists of four energy bands, separated by three gaps. So we have the result that the number of bands is equal to the lattice period.

The band edges can be obtained by solving the equation

$$|f(\mathcal{E})| = 2. \tag{6.32}$$

Equation (6.32) has the following solutions:

$$\varepsilon_1^- = -\sqrt{2 + 2\varepsilon_A + \varepsilon_A^2}, \quad \varepsilon_1^+ = -\sqrt{2 + \varepsilon_A^2 + 2\sqrt{1 + \varepsilon_A^2}},$$

$$\varepsilon_2^- = -\sqrt{2 + \varepsilon_A^2 - 2\sqrt{1 + \varepsilon_A^2}}, \quad \varepsilon_2^+ = -\sqrt{2 - 2\varepsilon_A + \varepsilon_A^2}, \tag{6.33}$$

$$\varepsilon_3^- = -\varepsilon_2^+, \quad \varepsilon_3^+ = -\varepsilon_2^-, \quad \varepsilon_4^- = -\varepsilon_1^+, \quad \varepsilon_4^+ = -\varepsilon_1^-,$$

where ε_i^- (ε_i^+) is the lowest (highest) edge of the ith band, respectively.

The band structure for $\varepsilon_A = 0.5$ is shown in figure 6.7 in the extended zone scheme and in figure 6.8 in the reduced zone scheme.

The density of states

$$n(\mathcal{E}) = \frac{1}{2\pi} \frac{1}{\left|\dfrac{\partial \mathcal{E}}{\partial k}\right|}, \tag{6.34}$$

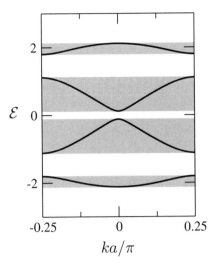

FIGURE 6.8. Reduced band zones for the tight binding model with spatial period $\ell = 4a$.

consists of four bands separated by gaps. Note that, $n(\mathcal{E})$ diverges again when the energy $\mathcal{E}$ approaches any band edge given by equation (6.33).

6.4 Reduced Zone Scheme

We can generalize our results presented in this chapter as follows.

For a system with lattice period $\ell = Na$, we find the band structure $\mathcal{E} = \mathcal{E}(k)$ from the solution of Schrödinger's equation for N neighboring sites. The wave function can be written in the form

$$c_n(k) = e^{ikna} u_n(k),$$ (6.35)

where the function $u_n(k)$ is periodic with period N: $u_{n+N}(k) = u_n(k)$.

From Schrödinger's equation, we see that both $\mathcal{E}(k)$ and $c_n(k)$ are periodic functions of k with period $2\pi/(Na)$. Therefore, it is sufficient to consider only a smaller interval of the wave vector, $-\pi/N \leq ka \leq +\pi/N$. Since we have N eigenenergies for each value of ka, we obtain N energy bands. Examples of the band structure for periods $\ell = 2a$ and $4a$ were shown in figures 6.3 and 6.8, respectively.

The wave function given by equation (6.35) has the form of a Bloch function for the discrete model. For a continuous model, Bloch's function has the form

$$\Psi_k(x) = e^{ikx} u_k(x)$$ (6.36)

where the function $u_k(x)$ is periodic with the period of the model, $u_k(x + \ell) = u_k(x)$.

For completeness, we show in figure 6.9 the reduced energy spectrum of the periodic Kronig-Penney model. In contrast to the tight binding model, the energy spectrum

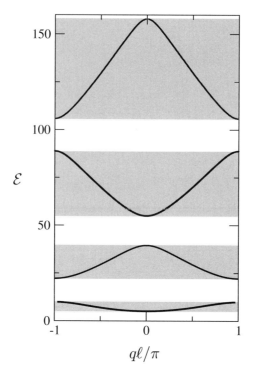

FIGURE 6.9. Energy spectrum of the Kronig-Penney model (figure 4.3) represented in the reduced zone scheme with $-\pi/\ell \leq k \leq +\pi/\ell$.

consists of an infinite number of energy bands. This is because of the spatial continuity of the model. Without the δ-function barriers, the model possesses the translational symmetry $x \rightarrow x + a$ for any a, even when a is infinitesimally small. The addition of the δ-function barriers breaks this symmetry. The new spatial symmetry is given by the distance ℓ between two δ-function barriers, which is "infinitely" larger than the original "period" of the model.

6.5 Problems

Problems with Solutions

Problem 6.1

Prove that the bonding and antibonding states of the AB molecule, with energies ε_A and $\varepsilon_B = -\varepsilon_A$, are given by $\mathcal{E}_b = -\sqrt{1 + \varepsilon_A^2}$ and $\mathcal{E}_a = +\sqrt{1 + \varepsilon_A^2}$. Prove that the ratio of the wave functions, c_B/c_A, for the bonding and antibonding states is given by

$$\frac{c_B}{c_A} = \mp \left[\varepsilon_A + \sqrt{1 + \varepsilon_A^2} \right], \tag{6.37}$$

where the sign $-$ $(+)$ is for the bonding (antibonding) state, respectively.

Solution. The Hamiltonian matrix of the system of two noninteracting atoms is diagonal,

$$\begin{pmatrix} \varepsilon_A & 0 \\ 0 & -\varepsilon_A \end{pmatrix}. \tag{6.38}$$

If two atoms are coupled, then the coupling V between the two atoms is given by the off-diagonal terms of the matrix (6.38), and the Hamiltonian matrix reads

$$\begin{pmatrix} \varepsilon_A & V \\ V & -\varepsilon_A \end{pmatrix}. \tag{6.39}$$

With $V = 1$, we find that the eigenvalues of the matrix (6.39) are $\mathcal{E}_b$ and $\mathcal{E}_a$ and that the corresponding eigenvectors have the form

$$u_{b,a} = C \begin{pmatrix} \pm(\varepsilon_A + \sqrt{1 + \varepsilon_A}) \\ 1 \end{pmatrix}, \tag{6.40}$$

where C is a normalization constant.

Problem 6.2
Using the transfer matrix for the tight binding model, find the band structure of the model with two different atoms.

Solution. The transfer matrix of one lattice period is given by the product of the transfer matrices of two neighboring sites:

$$\mathcal{M}^{(2)} = \mathcal{M}^+\mathcal{M}^- = \begin{pmatrix} \mathcal{E} - \varepsilon_A & -1 \\ 1 & 0 \end{pmatrix} \begin{pmatrix} \mathcal{E} + \varepsilon_A & -1 \\ 1 & 0 \end{pmatrix}. \tag{6.41}$$

One easily finds that

$$\mathcal{M}^{(2)} = \begin{pmatrix} \mathcal{E}^2 - \varepsilon_A^2 - 1 & \varepsilon_A - \mathcal{E} \\ \varepsilon_A + \mathcal{E} & -1 \end{pmatrix}. \tag{6.42}$$

Hence, the trace of the transfer matrix (6.42) equals

$$\mathrm{Tr}\,\mathcal{M}^{(2)} = \mathcal{E}^2 - \varepsilon_A^2 - 2. \tag{6.43}$$

If $\mathcal{E}$ is the energy of the propagating wave, then $|\mathrm{Tr}\,\mathcal{M}^{(2)}| < 2$, the eigenvalues of the matrix $\mathcal{M}^{(2)}$ are $e^{\pm i2ka}$ so that $\mathrm{Tr}\,\mathcal{M}^{(2)} = 2\cos 2ka$. By comparison of the last equation with equation (6.43), we obtain $\mathcal{E}^2 = \varepsilon_A^2 + 2 + 2\cos 2ka$, which can also be written as

$$\mathcal{E} = \pm\sqrt{\varepsilon_A^2 + 4\cos^2 ka}. \tag{6.44}$$

-1   0   +1

FIGURE 6.10. An impurity in the tight binding model with different
distances between neighboring atoms.

Problem 6.3

Find the bound states of the impurity in the tight binding model, defined by
equations (6.16). The impurity is shown in figure 6.10.

Solution. The wave function for the bound state must decrease exponentially on both
sides of the impurity. For sites $|n| > 1$, we write the equations for the wave func-
tion c_n. These equations have the same form as equations (6.16) with imaginary
wave vector $k = i\kappa$ $(\kappa > 0)$:

$$-\mathcal{E}_b c_n \quad +(V_2 + V_1 e^{2\kappa a})c_{n+1} \quad = 0,$$
$$(V_2 + V_1 e^{-2\kappa a})c_n \qquad\qquad -\mathcal{E}_b c_{n+1} \quad = 0.$$

(6.45)

From the requirement that the determinant of this system of equations must be zero, we
obtain the following relation between the bound energy $\mathcal{E}_b$ and wave vector ka:

$$\mathcal{E}_b = \pm\sqrt{(V_1 - V_2)^2 + 4V_1 V_2 \cosh^2 \kappa a}.$$

(6.46)

It is more convenient to write the above equation in the form

$$\mathcal{E}_b^2 = (V_1 e^{\kappa a} + V_2 e^{-\kappa a})(V_1 e^{-\kappa a} + V_2 e^{\kappa a}).$$

(6.47)

The site $n = 0$ must be treated separately. Consider first the symmetric bound state. Then
we have

$$c_1 = c_{-1},$$

(6.48)

and from Schrödinger's equation we get

$$-\mathcal{E}_b c_0 + 2c_1 V_1 = 0.$$

(6.49)

We need also the equation for the first site,

$$V_1 c_0 - \mathcal{E}_b c_1 + V_2 e^{-2\kappa a} c_0 = 0.$$

(6.50)

The last two equations have a solution only when

$$\mathcal{E}_b^2 = 2V_1 e^{-\kappa a}\left(V_1 e^{\kappa a} + V_2 e^{-\kappa a}\right).$$

(6.51)

We eliminate the energy $\mathcal{E}_b$ from the two equations (6.47) and (6.51) and obtain the
following equation for κa:

$$2V_1 e^{-\kappa a} = V_1 e^{-\kappa a} + V_2 e^{+\kappa a},$$

(6.52)

which has the solution

$$e^{2\kappa a} = \frac{V_1}{V_2}. \tag{6.53}$$

Inserting this result into equation (6.46), we obtain the energy of the bound state

$$\mathcal{E}_b^{\pm} = \pm\sqrt{2}\sqrt{V_1^2 + V_2^2}. \tag{6.54}$$

Since $\sqrt{2}\sqrt{V_1^2 + V_2^2} > |V_1 + V_2|$ for any values of V_1 and V_2, we obtain that these two band energies lie outside the bands given by equation (6.19). Note that, since $\kappa > 0$, we have the result that this symmetric bound state exists only when $V_1 > V_2$, otherwise equation (6.53) has no solution.

There is also another symmetric bound state with energy $\mathcal{E}_b = 0$. Inserting into equation (6.49) we have that $c_1 = 0$. With the help of Schrödinger's equation for even sites,

$$V_2 c_{2n-1} + V_1 c_{2n+1} = 0, \tag{6.55}$$

we immediately see that the wave function is zero at all odd sites,

$$c_{2n-1} \equiv 0. \tag{6.56}$$

From the equation for odd sites,

$$V_1 c_{2n} + V_2 c_{2n+2} = 0, \tag{6.57}$$

and from the requirement $c_{2n+2} = e^{-2\kappa a} c_{2n}$ we obtain the expression for κ,

$$V_1 + V_2 e^{-2\kappa a} = 0. \tag{6.58}$$

This equation has the solution

$$e^{\kappa a + i\pi/2} = \sqrt{\frac{V_2}{V_1}} \tag{6.59}$$

and

$$c_{2n+2} = -\frac{V_1}{V_2} c_{2n}. \tag{6.60}$$

Note that this solution exists only when $V_2 > V_1$.

There is also an antisymmetric bound state,

$$c_{-n} = -c_n \quad \text{and} \quad c_0 = 0. \tag{6.61}$$

From equation (6.50) we have that the energy of this bound state

$$\mathcal{E}_b = 0. \tag{6.62}$$

As in the previous case, we obtain that the wave function is nonzero only at odd sites, and κ satisfies the relation

$$e^{\kappa a + i\pi/2} = \sqrt{\frac{V_1}{V_2}} \tag{6.63}$$

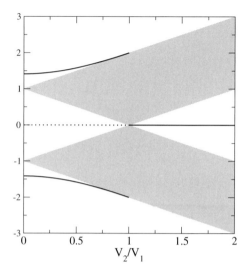

FIGURE 6.11. Spectrum of the periodic model with different distance between neighboring atoms (figure 6.5) as a function of V_2/V_1 (dashed area). Note that, there is no gap in the spectrum when $V_2 = V_1$. The solid lines are symmetric bound states, given by equation (6.54) for $V_2 < V_1$ and $\mathcal{E}_b = 0$ for $V_2 > V_1$. The dotted line is the antisymmetric bound state with $\mathcal{E}_b = 0$.

and

$$c_{2n+1} = -\frac{V_2}{V_1} c_{2n-1}. \tag{6.64}$$

Clearly, this solution exists only when $V_1 > V_2$.

In summary, when $V_1 > V_2$, there are three bound states. Two symmetric bound states have the energy $\mathcal{E}_b = \pm\sqrt{2}\sqrt{V_1^2 + V_2^2}$, and the antisymmetric bound state has the eigenenergy $\mathcal{E}_b = 0$. When $V_1 < V_2$, then we have only one symmetric bound state with the energy $\mathcal{E}_b = 0$. Figure 6.11 plots the energy of bound states as a function of ratio V_2/V_1.

Problem 6.4

Calculate the energy spectrum of the one-dimensional tight binding model shown in figure 6.12 with $\varepsilon_b = -\varepsilon_a$. Show that there are three energy bands in the spectrum of the model.

Solution. Using the spatial periodicity $\ell = 3a$, we write the Schrödinger equation for sites $n = 3m$, $3m + 1$, and $3m + 2$ as follows:

$$
\begin{aligned}
(-\mathcal{E} - \varepsilon_A)c_n &\quad + c_{n+1} \quad + e^{-i3ka}c_{n+2} = 0, \\
c_n &\quad + (-\mathcal{E} + \varepsilon_A)c_{n+1} \quad + c_{n+2} = 0, \\
e^{i3ka}c_n &\quad + c_{n+1} \quad + (-\mathcal{E} + \varepsilon_A)c_{n+2} = 0.
\end{aligned} \tag{6.65}
$$

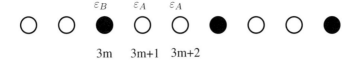

$$\varepsilon_B \qquad \varepsilon_A \qquad \varepsilon_A$$

3m   3m+1   3m+2

FIGURE 6.12. Periodic one-dimensional solid with two different atoms and period $\ell = 3a$.

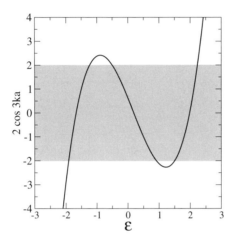

FIGURE 6.13. $2\cos 3ka$, given by equation (6.66).

The solution exists only when the determinant of this system of linear equations must be zero. This leads to a cubic equation for the energy,

$$-\mathcal{E}^3 + \mathcal{E}^2\varepsilon_A + \mathcal{E}(3 + \varepsilon_A^2) - \varepsilon_A^3 - \varepsilon_A + 2\cos 3ka = 0. \tag{6.66}$$

The plot of $2\cos 3ka$ versus energy $\mathcal{E}$ is shown in figure 6.13. We see that there are indeed three energy bands in which $|2\cos 3ka| \leq 2$. These bands correspond to propagating solutions of the model.

Problems without Solutions

Problem 6.5
Show that the two eigenfunctions $c_n^+(k)$ and $c_n^-(k)$ given by equations (6.10) and (6.11), respectively, are orthogonal to each other.

Problem 6.6
Use the technique explained in problem 6.3 to calculate all the bound states of the two-atom periodic model.

Problem 6.7

Using the transfer matrix for the tight binding model, find the band structure of the model with two different atoms, but with period 4. Calculate the trace of the transfer matrix of the segment of length $\ell = 4a$. The transfer matrix is a product of four transfer matrices for sites 1–4:

$$\mathcal{M}^{(4)} = \mathcal{M}^+\mathcal{M}^+\mathcal{M}^-\mathcal{M}^-, \tag{6.67}$$

where the transfer matrices $\mathcal{M}^+$ and $\mathcal{M}^-$ are given in equation (6.41). Show that

$$\text{Tr } \mathcal{M}^{(4)} = (\mathcal{E}^2 - \varepsilon_A^2)^2 - 4\mathcal{E}^2 + 2. \tag{6.68}$$

7 Disordered Models

In chapter 6 we studied the transmission of a quantum particle in an infinite periodic system. We found that the periodicity of the system creates bands and gaps in the energy spectrum. In the band, the particle moves freely throughout the sample for all allowed energies. This is due to the periodicity of the system, which enables successful interference of the back and forth scattered waves. In the band gap, there are no states at all and the transmission coefficient is zero.

We have also learned that a single impurity creates an isolated energy level which lies in the band gap. In this chapter we want to study what happens when the system contains many *different* impurities. By intuition we expect that the more impurities are included in the model, the less chance there is of the electron propagating long distances. The intuition is correct. There is an even stronger statement [11,111], namely, that in one-dimensional systems, the presence of impurities causes *localization* of electrons in a finite region of the lattice. In the limit of an infinite number of impurities, the transmission coefficient decreases exponentially with increasing system length [40,62,75,154]. This means that only bound (*localized*) states exists in disordered one-dimensional systems. The wave function of a localized state decreases exponentially with increasing distance on both sides of the localization center.

We will study how the transmission coefficient depends on the given spatial distribution of impurities, and find that the transmission coefficient becomes a statistical variable. Two samples, which differ in the random realization of the random energies, will have different transmission coefficients. A complete description of the transmission properties can be obtained only in terms of the *probability distribution* of the transmission and/or in terms of its mean value and higher statistical averages. It turns out that it is more suitable to deal with the *logarithm* of the transmission coefficient [45] because the mean value of the transmission coefficient is not a good representative quantity of the statistical ensemble.

We will investigate the transmission properties of two random models. The first one is the random tight binding model, which is also known as the *Anderson model* [44]. In this model, disorder is represented by randomly chosen site energies ε_n. Due to the disorder, we are not able to calculate the transmission coefficient analytically. We will only present our results of numerical simulations. The second model that we will discuss in this chapter is the random Kronig-Penney model [45]. In this model, the randomness is introduced by the irregular positions of the δ-function barriers. With the use of the transfer matrix technique, we can analytically calculate the transmission coefficient and prove that the electron is indeed localized.

7.1 Random Tight Binding Model

Consider the tight binding model defined in chapter 5, but now the energies ε_n on each site n are chosen randomly. Such a model is called the Anderson model. The time-independent Schrödinger equation of the Anderson model can be written as

$$\mathcal{E}c_n = \varepsilon_n c_n + c_{n+1} + c_{n-1}. \tag{7.1}$$

In what follows we will define the mean value and the variance of the random variable. The mean value $\langle \varepsilon \rangle$ is defined by averaging over the statistical ensemble. If we have N realizations of energies ε_i, then the mean value is simply given by

$$\langle \varepsilon \rangle = \frac{1}{N} \sum_{i=1}^{N} \varepsilon_i. \tag{7.2}$$

If we know the probability distribution $p(\varepsilon)$, then we can also calculate the mean values by the relation

$$\langle \varepsilon \rangle = \int d\varepsilon \, p(\varepsilon) \varepsilon. \tag{7.3}$$

Similarly, the variance var ε is defined by the relation

$$\text{var } \varepsilon = \langle \varepsilon^2 \rangle - \langle \varepsilon \rangle^2. \tag{7.4}$$

The square root of the variance, $\sqrt{\text{var } \varepsilon}$, gives us the width of the distribution $p(\varepsilon)$. For a given probability distribution, most of the realizations of random energies ε lie in the interval

$$\langle \varepsilon \rangle - \sqrt{\text{var } \varepsilon} \, < \, \varepsilon \, < \, \langle \varepsilon \rangle + \sqrt{\text{var } \varepsilon}. \tag{7.5}$$

For simplicity, we always assume that the mean value of random energies is zero:

$$\langle \varepsilon_n \rangle = 0. \tag{7.6}$$

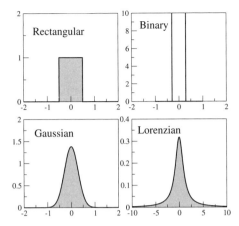

FIGURE 7.1. Four probability distributions of random energies. All have zero mean value, $\langle \varepsilon \rangle = 0$. For the rectangular, binary, and Gaussian distributions, given by equations (7.8), (7.9), and (7.10), respectively, we choose values of the parameter W such that all three distributions have the same variance $\langle \varepsilon^2 \rangle = 1/12$. Note that the variance is infinite in the case of a Lorenzian distribution, given by equation (7.11).

We also assume that any two random energies ε_n are statistically independent of each other. This means that for a given sample the value of the random energy ε_n on a given site n does not depend on the values of random energies on other sites. Mathematically, this means that

$$\langle \varepsilon_n \varepsilon_{n'} \rangle = 0. \tag{7.7}$$

Assumption (7.7) simplifies considerably the analysis of random systems. Nevertheless, recently it became clear that the one-dimensional models with a spatial correlation of random energies might have new interesting transport properties [76]. This problem is, however, beyond the scope of this book.

Figure 7.1 shows four probability distributions that are used in the analysis of disordered systems. The most used distribution of random energies ε_n is the so-called rectangular distribution, defined as

$$p(\varepsilon_n) = \begin{cases} W^{-1}, & -W/2 < \varepsilon_n < W/2, \\ 0 & \text{otherwise.} \end{cases} \tag{7.8}$$

The parameter W ($W = 1/2$ in the case shown in figure 7.1) determines the width of the distribution. Since higher values of W mean that the random energies are dissipated in a broader interval, we will call W the *strength of the disorder*. Other distributions shown in figure 7.1 are the binary distribution

$$p_{\text{bin}}(\varepsilon_n) = \frac{1}{2} \left[\delta(\varepsilon_n - W) + \delta(\varepsilon_n + W) \right], \tag{7.9}$$

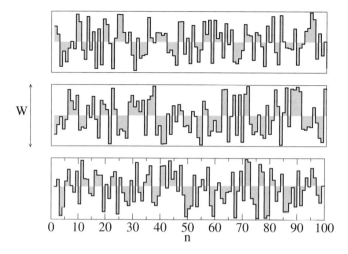

FIGURE 7.2. Three realizations of the random energies ε_n on a lattice of length $N = 100$ sites. Although the strength of the disorder, W, is the same for all three samples, the structure of each sample is considerably different from the structures of the others. Also, we expect that the three samples will have different transmission coefficients T. The random energies are distributed with the rectangular distribution given by equation (7.8).

the Gaussian distribution

$$p_g(\varepsilon_n) = \frac{1}{\sqrt{2\pi}\,W}\exp -\frac{\varepsilon_n^2}{2\,W^2},\tag{7.10}$$

and the Lorenzian distribution

$$p_l(\varepsilon_n) = \frac{W/\pi}{\varepsilon_n^2 + W^2}.\tag{7.11}$$

In all these distributions, the parameter W determines the width of the distribution. Note that the variance var ε of the Lorenzian distribution is infinite, since the integral

$$\int_{-\infty}^{+\infty} d\varepsilon \; p_L(\varepsilon)\varepsilon^2 \tag{7.12}$$

diverges.

Figure 7.2 shows three realizations of random energies ε_n in a system of $N = 100$ sites. We see that, although the random energies have the same probability distribution [rectangular distribution, given by equation (7.8)], the three samples shown in figure 7.2 differ considerably from each other. Hence, it is not sufficient to characterize the sample by macroscopic parameters, like length and the strength of the disorder. If we measure the transmission coefficient of the electron propagating through a disordered sample, then the results will depend not only on the macroscopic parameters, such as the disorder strength, W, the number of sites, N, and the energy of the electron, but also on the realization of the random energies in the sample. Two samples with the same macroscopic parameters, but different realization of random energies, will give us two

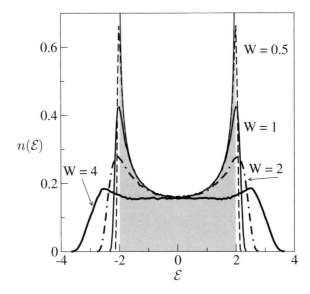

FIGURE 7.3. Density of states of the periodic tight binding model compared with the density of states of the disordered Anderson model with different strengths of the disorder, W. The band of energies for the periodic model, shown as a shaded area, becomes broader due to the disorder. The singularities of the density of states at the band edges, typical for periodic one-dimensional systems, disappear in the disordered case.

different values of the transmission coefficient. These two values can differ by many orders of magnitude.

7.1.1 The Density of States of a Disordered System

We learned in section 5.1 (see figure 5.5) that a periodic model possesses a transmission band with well-defined band edges at $\mathcal{E} = \pm 2$. We also know that random impurities create bound states in the gap region of the energy spectrum. When the number of impurities increases, more impurity eigenstates appears in the gap. In the limit of an infinite number of impurities, we expect the density of impurity states to become continuous and the energy band to become broader than the band of the periodic system. This is confirmed by the results of numerical calculations shown in figure 7.3. The disorder indeed broadens the density of states $n(\mathcal{E})$. Also, the singularities of $n(\mathcal{E})$ vanish at the band edges.

To calculate the density of states shown in figure 7.3, we calculated numerically all the eigenenergies of the Hamiltonian of a disordered lattice with a length of 1000 sites. The random energies were chosen from the rectangular probability distribution defined by equation (7.8). For each value of the disorder strength W, we repeated the calculation for 1000 different disordered samples and averaged the results obtained to get the smooth curves shown in the figure 7.3.

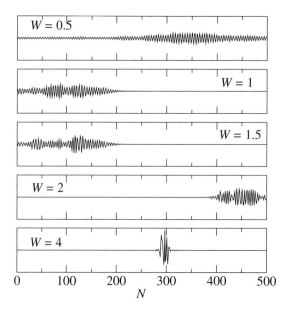

FIGURE 7.4. Wave function of the electron in a disordered tight binding model with a rectangular probability distribution of random energies given by equation (7.8). For each value of disorder strength W, we diagonalized the tight binding Hamiltonian, given by equation (7.1), and found the eigenenergy closest to $E = 0.5$. We see that with increasing disorder the electron becomes localized in certain parts of the lattice. The localized electron is not able to propagate through the lattice so that the transmission coefficient of an electron in a localized state is very small and eventually zero. The length of the system $N = 500$.

Note that, the density of state itself does not give us any information about the transmission coefficient. In contrast with periodic systems, in disordered systems we can have states localized in space. Such states do not contribute to the current through the sample.

7.1.2 Wave Functions of Disordered Systems

We have seen in chapter 5 that, if the energy of the electron lies in the interval $-2 < \mathcal{E} < 2$, the electron propagates freely in the periodic system. Its wave function oscillates as a function of the site index n. Disorder breaks the translational invariance of the model, as shown in figure 7.2. To see how the disorder influences the transmission through the system, it is instructive to find the space dependence of the wave eigenfunctions of disordered tight binding models. To do so, we diagonalize numerically the Hamiltonian matrix of a given size, say $N = 500$, and calculate all the eigenenergies and eigenfunctions.

We show in figure 7.4 the site dependence of the eigenfunction for five different values of disorder W, for a given energy $\mathcal{E} = 0.5$. One easily sees how the disorder causes the localization of the wave function. For small values of disorder ($W = 0.5$), the wave function is spread over the entire system, similarly to the periodic crystal. We expect the transmission coefficient T to be still sufficiently high, in spite of the randomness.

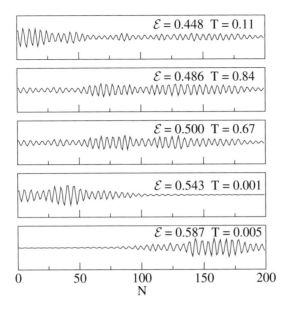

FIGURE 7.5. Five eigenfunctions of the electron in a disordered system of length $N = 200$ sites. The strength of the disorder is $W = 1$. We choose five neighboring eigenstates from the energy spectrum of the system. Although the eigenstates lie close to each other in the energy spectrum, the wave functions might have a completely different spatial extent. It is natural to expect that the transmission coefficient is larger for the eigenstate with energy $\mathcal{E} = 0.500$ than for the eigenstate with $\mathcal{E} = 0.543$. The transmission coefficients of these states indeed differ from each other by three orders of magnitude.

For larger values of W, however, the electron wave function becomes strongly localized in a certain region of the lattice, as is shown in figure 7.4.

In figure 7.5 we show the site dependence of the wave function of five neighboring eigenenergies of a random system. The form of the wave function varies from one eigenstate to the neighboring one in an unpredictable way. The spatial extent of the wave function is very sensitive to the actual configuration of random site energies and to the energy of the electron.

7.1.3 Transmission Coefficient

Contrary to periodic systems, we cannot calculate the transmission coefficient of the disordered samples analytically. Fortunately, numerical simulation of the transmission coefficient is easy with the help of the formula (5.41),

$$T = \frac{4 \sin^2 ka}{|e^{-ika}c_N - c_{N-1}|^2}. \tag{7.13}$$

derived in section 5.3. Here, ka is the wave vector of the electron with energy $\mathcal{E} = 2 \cos ka$ in a perfect system, and c_n is the wave function of the electron on the site n. We remind the reader that equation (7.13) gives the transmission coefficient of an electron propagating

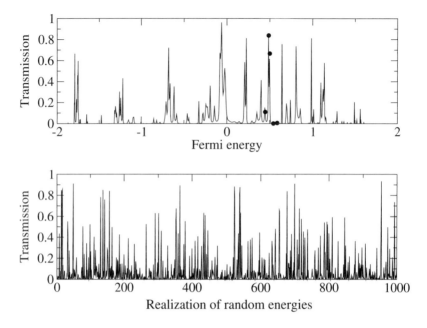

FIGURE 7.6. Top panel: Transmission coefficient of an electron propagating in a disordered lattice as a function of the energy $\mathcal{E}$ of the electron. All data were calculated for only one configuration of random site energies distributed according to the rectangular probability distribution with $W = 1$. The length of the system is $N = 200$ sites. The five dots indicate the transmission for five eigenenergies, discussed in figure 7.5. Bottom panel: Transmission coefficient of an electron with fixed energy $\mathcal{E} = 0.5$, calculated for 1000 different samples which have the same length and the same strength of the disorder W, but differ in the realization of the site energies ε_n.

through a disordered sample of length N. Outside the disordered region, the system is described by the tight binding model discussed in chapter 5.

Starting with $c_0 = 1$ and $c_{-1} = e^{-ika}$, we use the Schrödinger equation (7.1) to calculate the wave function at sites $N - 1$ and N, and find the transmission coefficient for a disordered one-dimensional system of length $L = Na$.

In this section, we present numerical data for the transmission coefficient, obtained with the use of equation (7.13). We will see that the transmission coefficient is a statistical variable, so that we need not only the mean value $\langle T \rangle$, but also higher averages, or the entire probability distribution of the transmission coefficient $p(T)$ [12, 41, 57, 81, 82, 103, 104, 109, 115–117] Also, we will see that the transmission coefficient decreases exponentially with increasing length of the disordered system.

Statistical Properties

In contrast to the periodic models, in which the particle moves freely when the energy of the electron lies inside the transmission band, the transmission coefficient of disordered systems strongly depends on the actual configuration of the disorder in the lattice. For a given sample, the transmission coefficient depends also on the energy $\mathcal{E}$ of the electron. This is shown in figure 7.6.

Figure 7.6 shows all the peculiar properties of the transmission in disordered systems: First, we see that the $\mathcal{E}$-dependence of the transmission is not smooth. There are some energies for which the transmission is very close to 1, and for other energies the transmission is smaller by a few orders of magnitude. The transmission strongly fluctuates when the energy changes by a very small value. This can be understood from the analysis of the wave functions shown in figure 7.5, since the transmission is determined by the spatial extent of the wave function.

The transmission coefficient is also very sensitive to the specific realization of the random energies. We show in the lower panel of figure 7.6 the transmission coefficient of an electron with a fixed energy $\mathcal{E} = 0.5$, propagating through 1000 samples, which differ only in the realization of randomness. Again, we see that the transmission coefficients of the two samples, which are macroscopically equivalent (they have the same strength of the disorder W), might differ by many orders of magnitude [43,68,104,109,110,129].

The above properties are consequences of the wave character of the electron. There are always some special configurations of random energies for which the transmission amplitudes of all possible trajectories inside the structure interfere positively, and the resulting transmission coefficient T is close to 1. Such configurations are, however, very sensitive to small changes of either the configurations of the random energies (in fact, changing only one random energy might cause the transmission coefficients to change considerably) or the energy $\mathcal{E}$ of the electron. Consequently, the transmission coefficient of the electron propagating in a disordered system is a random variable. To characterize the transmission properties of a random system completely, we need to consider the entire probability distribution $p(T)$ of the transmission coefficient.

System Length Dependence

As shown in figure 7.7, the transmission coefficient decreases when the system length increases. This is in agreement with our analysis of the wave function. The transmission is close to 1, only if the wave function spreads over the entire system. This is more probable when the system is sufficiently short, but becomes highly improbable when the length of the system, $L = Na$, is larger than a particular length ξ. The length ξ is called the *localization length* [75]. As shown in figure 7.8, the mean value of the *logarithm* of the transmission coefficient, $\langle \ln T \rangle$, decreases linearly with increasing system length and is given by the following relation [45]:

$$\langle \ln T \rangle = -\frac{2L}{\xi} \quad (L = Na). \tag{7.14}$$

By looking at figure 7.7, we see that the relation (7.14) holds only for the mean value of the transmission coefficient. For a given realization of random site energies, the electron still has a chance to have a large transmission even when the system is much longer than the localization length: $L \gg \xi$. The reason for such a big transmission is again the wave character of the electron propagation. Of course, the probability of such a rare event is small and decreases with increasing L.

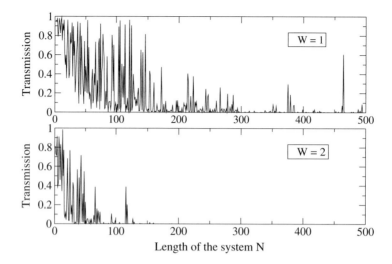

FIGURE 7.7. Transmission coefficient as a function of the system length $L = Na$ of a disordered system for two different strengths of the disorder, $W = 1$ and 2. The transmission decreases as the length of the system increases. Nevertheless, for a given realization of random energies, it might happen that T is very close to 1 even for a system that is much longer than the localization length, ξ.

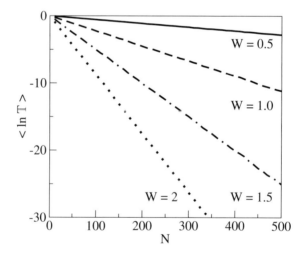

FIGURE 7.8. Mean value of the logarithm of the transmission coefficient $\langle \ln T \rangle$ as a function of the length of the disordered model (7.1) with energy $\mathcal{E} = 0.5$, and with the rectangular probability distribution of the disorder. The mean value $\langle \ln T \rangle$ decreases linearly with increasing length of the system, L, i.e., $\langle \ln T \rangle = -2L/\xi$, in agreement with equation (7.14). ξ is the localization length. We fit the numerical data and obtain that $\xi \approx 357$, 89, 40, and 23 for $W = 0.5$, 1, 1.5, and 2, respectively.

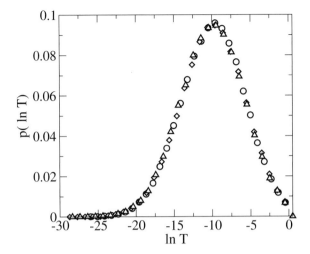

FIGURE 7.9. Probability distribution of the logarithm of the transmission coefficient $\langle \ln T \rangle$ of an electron propagating in a disordered lattice. The energy of the electron is $E = 0.5$. Three samples were considered, with different strengths of disorder $W = 1, 1.5,$ and 2. The length of the system is $N = 450, 200,$ and 115, respectively. With this choice of parameters $\langle \ln T \rangle$ has the same value for all three samples.

Probability Distribution

We show in figure 7.9 the probability distribution of the logarithm of the transmission coefficient for three different strengths of the disorder. We see that the distribution is similar to a Gaussian distribution centered around $\langle \ln T \rangle$ with a typical width measured by the variance, $\text{var} \ln T \le 2\langle \ln T \rangle$.[1]

Note that in figure 7.9 we plot the *logarithm* of the transmission coefficient, $\ln T$, not the transmission coefficient itself. The reason for this choice lies in the statistical properties of the transmission coefficient, which will be explained later. Here we only show the form of the probability distribution $P(\ln T)$ calculated for three samples (figure 7.9). We want to stress that, although we calculated $P(\ln T)$ for systems with different strengths of the disorder and different lengths, we obtained in all three cases the same distribution $P(\ln T)$. This indicates that the only parameter that determines the entire probability distribution is $\langle \ln T \rangle$, which is proportional to the ratio L/ξ. To verify this hypothesis, we use the values of the localization length, obtained in figure 7.8 and obtain $L/\xi = 450/89$, $200/40$, and $115/23$ for disordered systems with disorder strength $W = 1, 1.5,$ and 2, respectively. The hypothesis of the universality has been verified numerically in many different statistical ensembles of disordered systems.

7.1.4 Localization Length

In the previous section we observed that the transmission coefficient, T, decreases exponentially as a function of the length of the disordered system. The exponential

---

[1] var x is defined by equation (7.4).

decrease of the transmission is due to the fact that the electrons are localized in a certain region of the disordered lattice. Outside this region the wave function decreases as

$$|\Psi(Na)| \propto e^{-Na/\xi}. \tag{7.15}$$

This behavior is similar to the wave function of a bound state localized around a single impurity in a periodic model.

The localization length ξ can be defined through the so-called *Lyapunov exponent* $\gamma(E)$ [51,62,75,81,131,153], which characterizes the asymptotic behavior of a particular solution of the disordered model. For example, for the one-dimensional tight binding model, defined by equation (7.1),

$$\mathcal{E}c_n = \varepsilon_n c_n + c_{n+1} + c_{n-1}, \tag{7.16}$$

one can examine the asymptotic behavior of a particular solution $c_n(E)$ for a fixed energy $\mathcal{E}$. Starting with initial values c_0 and c_1 such that $c_0^2 + c_1^2 \neq 0$, we can iterate equation (7.16) or its matrix form, equation (5.22),

$$\begin{pmatrix} c_{n+1} \\ c_n \end{pmatrix} = \mathcal{M}_n \mathcal{M}_{n-1} \cdots \mathcal{M}_1 \begin{pmatrix} c_1 \\ c_0 \end{pmatrix}, \quad \mathcal{M}_n = \begin{pmatrix} \mathcal{E} - \varepsilon_n & -1 \\ 1 & 0 \end{pmatrix}. \tag{7.17}$$

The Lyapunov exponent $\gamma(E)$ is then defined by the relation

$$\gamma(\mathcal{E}) = \lim_{N \to \infty} \frac{1}{2N} \ln \left(c_N^2 + c_{N+1}^2 \right), \tag{7.18}$$

and the localization length is given by

$$\frac{a}{\xi} = \mathrm{Re}\, \gamma(\mathcal{E}). \tag{7.19}$$

Ishii [75] showed that $\mathrm{Re}\, \gamma(\mathcal{E}) = 0$ only for extended states. If $\mathrm{Re}\, \gamma(E) > 0$ then either the wave function is localized or the energy $\mathcal{E}$ lies inside the energy gap.

Thouless [153] proved the following relation between the *imaginary part* of the Lyapunov exponent and the density of states of a one-dimensional disordered system:

$$n(\mathcal{E}) = \frac{2}{\pi} \frac{\partial}{\partial \mathcal{E}} \, \mathrm{Im}\, \gamma(\mathcal{E}). \tag{7.20}$$

In the limit of weak disorder, $W \ll 1$, we can find also an analytical formula for for the Lyapunov exponent $\gamma(\mathcal{E})$. The derivation of the formula is presented in section 7.3.1. Then, from equation (7.19) we find the localization length [51,79,81,82,153]

$$\frac{\xi}{a} \approx \frac{2}{\langle \varepsilon^2 \rangle} \left(4 - \mathcal{E}^2 \right). \tag{7.21}$$

This estimation of the localization length agrees very well with our numerical data for the wave function (figure 7.4) and for the transmission coefficient (figure 7.8). It shows also that the localization length is nonzero for nonzero disorder W. This means that, there is no metallic regime in one-dimensional disordered systems.

7.1.5 Nonlocalized States

The absence of the metallic state is a common property of disordered systems with dimension $d \leq 2$. In dimension $d = 3$, the disorder must be larger than a certain critical value, called the critical disorder W_c, in order to obtain localization [44]. For smaller disorder strength, $W < W_c$, a three-dimensional system is metallic. The electron wave function is spread over the entire sample, and the system possesses finite electric conductivity σ. The metal-insulator transition was studied intensively both theoretically and numerically. In numerical works, the transfer matrix was used to calculate either the Lyapunov exponents [87, 101, 130] or the conductance [140]. These results confirm that the localization can be described within the *scaling theory of localization* [40, 45]. More detailed analysis of this theory is given in the reviews [87, 92, 103].

Two-dimensional systems deserve special attention. Although there is no metallic state in random systems of free electrons, the behavior of the model changes qualitatively when a strong magnetic field is applied to the system [74, 159], or if the electron can change the orientation of its spin when hopping from one site to the neighboring one [46, 63]. We will not discuss these situations here.

Although the one-dimensional disordered system does not exhibit metallic behavior in the limit of infinite system size, its transport properties are very similar to those of metals if the length of the system is much smaller than the localization length,

$$L \ll \xi. \tag{7.22}$$

In this limit, the electron propagates almost freely from one side of the sample to the other, and the transmission coefficient T varies between 0 and 1, depending on the actual configuration of the disorder in the sample. We show in figure 7.10 typical probability distributions of the transmission coefficient for a weakly disordered one-dimensional system. If the length of the system is much smaller than the localization length, then the system is almost transparent. In experiments, we can obtain, with rather large probability, that $T \sim 1$. When $L \sim \xi/2$, we see that the distribution $p(T)$ is almost flat. In experiments, we can obtain any value of T between 0 and 1 with equal probability. Only when the length of system is comparable to or larger than the localization length ξ does the transmission coefficient start to decrease, and for $L \gg \xi$ we observe the localized regime.

Thus, the transport properties of a disordered system depend not only on the disorder of the sample itself, but also on its length. By increasing the length of the system, one can change the character of the propagation from the propagating regime, where the transmission coefficient is close to 1, to the localized regime, characterized by an exponential decrease of the transmission coefficient.

The unusual properties of the electronic transmission caused by the quantum character of the electron propagation can be observed also in the metallic regime [12, 29, 49, 87, 92, 103, 109, 156]. Weak localization corrections to the transmission lead to the logarithmic size dependence of the transmission [87, 92, 103]. Even more surprising is the universality of statistical fluctuations of the transmission [93, 108, 129].

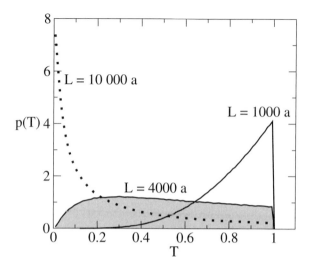

FIGURE 7.10. Probability distribution of the transmission coefficient for very weakly disordered systems. The strength of the disorder is $W = 0.1$ and the energy of electron is $E = 0.5$. For these parameters, we obtain from equation (7.21) the localization length $\xi = 9000a$. For samples much shorter than the localization length, the mean value of the transmission coefficient is close to 1 ($\langle T \rangle = 0.82$ for $N = 1000$), but $\langle T \rangle$ decreases when the system becomes longer. For $L = 10000a$, $\langle T \rangle$ is only 0.24. For $L = 4000a$, $\langle T \rangle = 0.504$ and the distribution is almost flat. This means that in experiment we obtain any value of the transmission coefficient, $0 < T < 1$, with the same probability. Note that all distributions terminate at $T = 1$, which is the maximum value of the transmission coefficient.

Since the localization is caused by the interference of scattered components of the waves, we expect that the same effects occur when classical—electromagnetic [69, 70, 151] or acoustic [91]—waves propagate in disordered media.

7.2 Random Kronig-Penney Model

In section 7.1 we studied the disordered tight binding model, or Anderson model. Because of the randomness, most of the information about the transmission properties of the disordered system was obtained numerically. In this section we study the model of randomly distributed barriers (for instance, δ-function barriers). This model allows, under certain limits, derivation of an analytical expression for the mean value of the transmission and reflection coefficients. We will see that the most important properties of the transmission of disordered systems, namely, the exponential decrease of the transmission coefficient as the length of the system increases, and the fluctuations of the transmission, can be described analytically. The results obtained are very important for the interpretation of the experimental data for the transmission of electrons through meso- and nanoscopic wires.

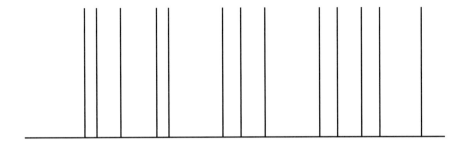

FIGURE 7.11. Random Kronig-Penney model. Horizontal lines indicate the positions of of the δ-function barriers. All barriers are identical, but the distance between two neighboring barriers is chosen randomly.

We will consider a one-dimensional model with randomly distributed δ-function impurities. The δ-function potential barrier was studied in detail in chapter 3. However, we do not need to describe the barrier in detail, we just need to know its transfer or scattering matrix. We consider all the barriers to be equal to each other (figure 7.11). For simplicity we also assume that the model possesses time-reversal symmetry. Therefore, each barrier is completely described by its transmission and reflection amplitudes t and r.

Besides the transmission and reflection coefficients, we are interested also in the Landauer resistance [89], which is defined as the ratio of the reflection and transmission coefficients,

$$\rho = \frac{\pi\hbar}{e^2} \frac{R}{T} = \frac{\pi\hbar}{e^2} \frac{1-T}{T}, \tag{7.23}$$

where e is the electric charge of the electron. We will discuss the Landauer resistance (7.23) in section 7.3.2. Now, we stress only that the Landauer resistance is given by the transmission coefficient T of as electron propagating in a one-dimensional system. So from the transmission coefficient we can find ρ. For our convenience, we omit in this section the multiplicative factor $\pi\hbar/e^2$. Then the Landauer resistance reads

$$\rho = \frac{R}{T} = \frac{1-T}{T} = \frac{R}{1-R}. \tag{7.24}$$

This means that the resistance is given in units of $e^2/(\pi\hbar)$.

7.2.1 Non-Ohmic Resistance

Composition Law of Two Scatterers
Consider two barriers defined through their transfer matrices $\mathbf{M}_1$ and $\mathbf{M}_2$, respectively. We calculate the Landauer resistance of the system consisting of these two barriers.

First, we calculate the transmission coefficient T_{12}. By using the composition law, given by equation (1.76), we obtain the transfer matrix of two scatterers at a distance ℓ,

$$\mathbf{M}_{12} = \mathbf{M}_2 \begin{pmatrix} e^{ik\ell} & 0 \\ 0 & e^{-ik\ell} \end{pmatrix} \mathbf{M}_1, \tag{7.25}$$

where $\mathbf{M}_1$ and $\mathbf{M}_2$ are the transfer matrices of the first and second impurities, respectively. We use the general form of the transfer matrix in terms of the transmission and reflection amplitudes given by equation (1.60). For instance, the transfer matrix $\mathbf{M}_1$ has the form

$$\mathbf{M}_1 = \begin{pmatrix} t_1' - \dfrac{r_1 r_1'}{t_1} & \dfrac{r_1}{t_1} \\ -\dfrac{r_1'}{t_1} & \dfrac{1}{t_1} \end{pmatrix}. \tag{7.26}$$

From equation (7.25), we obtain the transmission amplitude of the system of two barriers, t_{12}, as follows:

$$\frac{1}{t_{12}} = \frac{1}{t_1 t_2} \left(e^{ik\ell} - r_1' r_2 e^{-ik\ell} \right). \tag{7.27}$$

Now we express the reflection amplitudes r in terms of the reflection coefficient $R = |r|^2$ and the phase ϕ_r:

$$r_1' = \sqrt{R_1'} e^{i\phi_{r_1'}} \quad \text{and} \quad r_2 = \sqrt{R_2} e^{i\phi_{r_2}}. \tag{7.28}$$

Then, the transmission coefficient can be expressed in terms of the transmission and reflection coefficients of individual barriers as follows:

$$T_{12} = t_{12} t_{12}^*$$

$$= \frac{T_1 T_2}{\left(e^{ik\ell} - \sqrt{R_1' R_2} e^{-ik\ell + \phi_{r_1'} + \phi_{r_2}} \right) \left(e^{-ik\ell} - \sqrt{R_1' R_2} e^{+ik\ell - \phi_{r_1'} - \phi_{r_2}} \right)}$$

$$= \frac{T_1 T_2}{1 + R_1' R_2 - 2\sqrt{R_1' R_2} \cos\phi}, \tag{7.29}$$

where the phase ϕ is given by

$$\phi = \phi_{r_1'} + \phi_{r_2} - 2k\ell. \tag{7.30}$$

We remind the reader that we are considering only systems with time-reversal symmetry. Therefore, in what follows we use $R_1' = R_1$. If we assume that there is no absorption in our system, then $R_{12} = 1 - T_{12}$ and we obtain

$$R_{12} = 1 - T_{12} = \frac{R_1 + R_2 - 2\sqrt{R_1 R_2} \cos\phi}{1 + R_1 R_2 - 2\sqrt{R_1 R_2} \cos\phi}. \tag{7.31}$$

Finally, the Landauer resistance, defined by equation (7.24), is given by

$$\rho_{12} = \frac{R_{12}}{T_{12}} = \frac{R_1 + R_2 - 2\sqrt{R_1 R_2} \cos\phi}{T_1 T_2}. \tag{7.32}$$

All three quantities T_{12}, R_{12}, and ρ_{12} are random variables, since they are functions of the random distance ℓ between two scatterers. Now, we want to average over all possible realizations of ℓ. To do so, we assume that the random distance ℓ is always much larger than the wavelength of the electron. This can be easily satisfied if the mean distance $\langle \ell \rangle$ is much larger than the wavelength λ of the electron. Since $\lambda = 2\pi/k$, the last requirement can be written as

$$k\langle \ell \rangle \gg 1. \tag{7.33}$$

When the condition (7.33) is satisfied, then the phase $\phi = \phi_{r_1'} + \phi_{r_2} - 2k\ell$, given by equation (7.30), is very sensitive to small changes of the distance ℓ. We can assume that the phase ϕ is randomly distributed in the interval $(0, 2\pi)$. Under these conditions, the mean value of $\cos \phi$ is zero:

$$\int_0^{2\pi} d\phi \cos \phi = 0, \tag{7.34}$$

and we can easily average the Landauer resistance, given by equation (7.32). We obtain

$$\langle \rho_{12} \rangle = \frac{R_1 + R_2}{(1 - R_1)(1 - R_2)} = \frac{R_1(1 - R_2) + R_2(1 - R_1) + 2R_1R_2}{(1 - R_1)(1 - R_2)}, \tag{7.35}$$

where we have used the fact that $T_1 = 1 - R_1$ and $T_2 = 1 - R_2$. With the help of the definition of the Landauer resistance of each barrier, $\rho_1 = R_1/(1 - R_1)$ and $\rho_2 = R_2/(1 - R_2)$, the last equation can be rewritten in the simpler form

$$\langle \rho_{12} \rangle = \rho_1 + \rho_2 + 2\rho_1\rho_2. \tag{7.36}$$

Equation (7.36) is the starting point of our analysis of disordered systems.

Length Dependence of the Landauer Resistance

Consider now N identical scatterers, each characterized by the same transmission and reflection coefficients T_I and R_I. The scatterers are distributed *randomly* on a linear chain of length L. The mean distance between two scatterers satisfies the condition $\langle \ell \rangle = L/N \gg k^{-1}$, where k is the wave vector of the electron. We will use the composition relation, given by equation (7.36), to show how the resistance of the entire system depends on the number N of scatterers.

First, we note that in our derivation of (7.36) we have not used any information about the form or the spatial extent of the scatterers. In fact, we can assume that both scatterers represent large systems with their own spatial structure. The composition rule, given by equation (7.36), can be applied also to this particular problem. We need only to replace the Landauer resistances by their mean values, $\rho_1 \to \langle \rho_1 \rangle$ and $\rho_2 \to \langle \rho_2 \rangle$, where $\langle \cdots \rangle$ means the average over the randomness inside the scatterers 1 and 2, respectively. Thus, if $\langle \rho_1 \rangle$ represents the Landauer resistance of the one-dimensional system of N_1 scatterers and $\langle \rho_2 \rangle$ is the resistance of another system of N_2 scatterers, then, after substitution in equation (7.36), we obtain the resistance $\langle \rho_{12} \rangle$ of the system of $N = N_1 + N_2$ scatterers given by

$$\langle \rho_{12} \rangle = \langle \rho_1 \rangle + \langle \rho_2 \rangle + 2\langle \rho_1 \rangle \langle \rho_2 \rangle. \tag{7.37}$$

Due to the composition law given by equation (7.37), the total resistance of a system that consists of two parts does not equal the sum of the resistances of its constituents. This

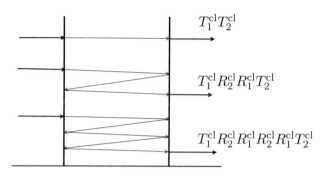

FIGURE 7.12. Schematic description of the first three scattering processes given in equation (7.38).

is a very important result, because this tells us that the Landauer resistance *does not* obey Ohm's law. In other words, the Landauer resistance $\langle \rho \rangle$ *is not proportional to the length of the system.*

Classical versus Quantum Scattering

A crucial assumption used in the derivation of the composition law, given by equation (7.37), is that of the wave character of electron propagation. Remember, we calculated the transmission and the reflection using formula (7.27), which combines the transmission and reflection *amplitudes* of two scatterers.

To understand the difference between quantum (wave) and classical (particle) propagation, let us calculate the Landauer resistance for classical particles. Then, instead of transmission and reflection *amplitudes,* we use the transmission and reflection *coefficients.* The classical transmission through two barriers is then given by the expression

$$T_{12}^{cl} = T_1^{cl} T_2^{cl} + T_1^{cl} R_2^{cl} R_1^{cl} T_2^{cl} + T_1^{cl} R_2^{cl} R_1^{cl} R_2^{cl} R_1^{cl} T_2^{cl} + \cdots \qquad (7.38)$$

The first term on the r.h.s. of equation (7.38) is a simple transmission through two scatterers. The second term is the contribution due to the process in which the electron is twice reflected between two scatterers and then transmitted (figure 7.12). There is an infinite number of similar contributions; the nth of them corresponds to the process in which the electron is n times reflected from the second barrier and n times reflected from the first barrier before it is transmitted through the second barrier. It is easy to sum all the contributions of such processes. We obtain

$$T_{12}^{cl} = \frac{T_1^{cl} T_2^{cl}}{1 - R_1^{cl} R_2^{cl}}. \qquad (7.39)$$

The reflection coefficient $R_{12}^{cl} = 1 - T_{12}^{cl}$ is given by the expression

$$R_{12}^{cl} = \frac{R_1^{cl} + R_2^{cl} - 2 R_1^{cl} R_2^{cl}}{1 - R_1^{cl} R_2^{cl}} \qquad (7.40)$$

(we used that $R_1^{cl} = 1 - T_1^{cl}$ and $R_2^{cl} = 1 - T_2^{cl}$). The Landauer resistance $\rho_{12}^{cl} = R_{12}^{cl}/T_{12}^{cl}$ is in the case of a classical particle given by

$$
\begin{aligned}
\rho_{12}^{cl} &= \frac{R_1^{cl} + R_2^{cl} - 2 R_1^{cl} R_2^{cl}}{T_1^{cl} T_2^{cl}} \\
&= \frac{R_1^{cl}(1 - R_2^{cl}) + R_2^{cl}(1 - R_1^{cl})}{(1 - R_1^{cl})(1 - R_2^{cl})} \\
&= \rho_1^{cl} + \rho_2^{cl}.
\end{aligned}
\tag{7.41}
$$

Thus, for a classical electron, we recover Ohm's law. This is not surprising, since by considering only the reflection coefficients and neglecting the phase of the wave function, we lost all the information about the disorder. Indeed, note that the length ℓ between two scatterers does not enter in the formula for the classical Landauer resistance.

7.2.2 Localization

Let us now derive the consequences of the non-Ohmic relation equation (7.37) for the length dependence of the Landauer resistance of a *quantum disordered system*. Consider a system of N scatterers. Each scatterer is given by its transmission and reflection coefficients T_I and R_I. The transmission and reflection coefficients of the whole system are T_N and R_N. We add to this system an additional scatterer with transmission and reflection coefficients T_I and R_I, respectively, and calculate the transmission and reflection coefficients for the resulting system of $N+1$ scatterers. This can be done using equation (7.29) with $t_1 \equiv t_N$, $r_1 \equiv r_N$, and $t_2 = t_I$. We obtain

$$
\rho_{N+1} = \frac{R_{N+1}}{T_{N+1}} = \frac{R_N + R_I - 2\sqrt{R_I R_N} \cos \phi_N}{(1 - R_N)(1 - R_I)}.
\tag{7.42}
$$

The phase ϕ_N is given by (7.30), where ℓ is now the distance between the system of N scatterers and the newly added scatterer. By averaging over ϕ_N, we again obtain that $\langle \cos \phi_N \rangle = 0$. Then, we obtain from equation (7.42) that

$$
\begin{aligned}
\langle \rho_{N+1} \rangle &= \frac{R_N + R_I}{(1 - R_N)(1 - R_I)} = \frac{R_N(1 + R_I) + R_I(1 - R_N)}{(1 - R_N)(1 - R_I)} \\
&= \frac{1 - R_I}{1 + R_I} \frac{R_N}{1 - R_N} + \frac{R_I}{1 - R_I}.
\end{aligned}
\tag{7.43}
$$

The last expression can be rewritten as a simple iterative equation

$$
\langle \rho_{N+1} \rangle = \Lambda \langle \rho_N \rangle + \frac{1}{2}(\Lambda - 1),
\tag{7.44}
$$

where

$$
\langle \rho_N \rangle = \frac{R_N}{1 - R_N}
\tag{7.45}
$$

is the Landauer resistance of the segment of N scatterers. The parameter Λ, defined as

$$
\Lambda = \frac{1 + R_I}{1 - R_I},
\tag{7.46}
$$

characterizes the properties of the individual scatterer. Note that, $\Lambda > 1$ for any value of R_I.

We will assume that the solution of equation (7.44) has the form $\langle \rho_n \rangle = a\Lambda^N + \beta$ [155]. Inserting this *Ansatz* into equation (7.44), we find $\beta = -1/2$. Then, for the initial condition $\rho_0 = 0 = a\Lambda^0 + \beta = a + \beta$, we obtain that $a = 1/2$. Thus, the solution of equation (7.44) can be written in the form

$$\langle \rho_N \rangle = \frac{1}{2} \left(\Lambda^N - 1 \right). \tag{7.47}$$

Since $\Lambda > 1$, equation (7.47) means that the mean resistance $\langle \rho_N \rangle$ of the system of N barriers *increases exponentially*, as a function of the number of scatterers, N: $\langle \rho_N \rangle \propto e^{N \ln \Lambda}$.

We can rewrite equation (7.47) in the more convenient form

$$\langle \rho(L) \rangle = \frac{1}{2} \left(e^{2L/\xi'} - 1 \right), \tag{7.48}$$

which shows that resistance increases exponentially with the length of the system. In equation (7.48) we introduce the characteristic length ξ' which depends upon the parameters of the individual scatterer as follows:

$$\xi' = \frac{2\langle \ell \rangle}{\ln \Lambda}. \tag{7.49}$$

Relation (7.48) states that the mean resistance of a random one-dimensional system increases exponentially, when the length of the system increases. This behavior is qualitatively and quantitatively different from the behavior of classical systems, where the resistance increases linearly, $\rho^{cl} \sim L$, with the system length. The exponential increase of the resistance is caused by the localization of the electrons inside the system. Only when the system is short in comparison to the characteristic length, $L \ll \xi'$, can we expand the right-hand side of equation (7.48) into a Taylor series, to obtain Ohm's law $\langle \rho_L \rangle = L/\xi'$.

The exponential increase of the Landauer resistance given by equation (7.48) agrees with our numerical data presented in section 7.1 (figure 7.8), where we found that the transmission coefficient of a one-dimensional disordered system decreases exponentially with increasing system length.

7.2.3 Statistical Properties of the Resistance

Since the resistance ρ is a statistical variable, we need also to know its variance. To do so, we repeat our calculations for the second moment $\langle \rho^2 \rangle$. Using the same method as for the mean value $\langle \rho \rangle$, we obtain that

$$\langle \rho_N^2 \rangle = \frac{1}{6}\Lambda_2^N - \frac{1}{2}\Lambda^N + \frac{1}{3}, \tag{7.50}$$

where Λ_2 is given by the relation

$$\Lambda_2 = \frac{1}{2} \left(3\Lambda^2 - 1 \right) \tag{7.51}$$

[155]. The derivation of equation (7.50) is given in problem 7.3. Using the relation $\Lambda = \exp 2L/\xi'$, with the length ξ' given by equation (7.49), we obtain

$$\langle \rho^2(L) \rangle = \frac{1}{6} e^{4L/\xi''} - \frac{1}{2} e^{2L/\xi'} - \frac{1}{6}, \tag{7.52}$$

where we have introduced another characteristic length by the relation

$$\Lambda_2 = e^{2L/\xi''}. \tag{7.53}$$

Since $\langle \rho_N^2 \rangle$ increases as $\left(\frac{3}{2}\Lambda^2\right)^N$ and the quantity $\langle \rho_N \rangle^2$ increases "only" as Λ^{2N}, in the limit $N \to \infty$ we obtain that

$$\langle \rho_N^2 \rangle \gg \langle \rho_N \rangle^2, \tag{7.54}$$

and the standard deviation, given by $\sqrt{\text{var } \rho_N}$, is much larger than the mean value:

$$\frac{\sqrt{\langle \rho_N^2 \rangle - \langle \rho_N \rangle^2}}{\langle \rho_N \rangle} \sim \left(\frac{3}{2}\right)^N \sim e^{N \ln(3/2)} \gg 1. \tag{7.55}$$

The ratio of the standard deviation to the mean value increases exponentially with the number N of barriers. This is a surprising result. It tells us that the mean value of the Landauer resistance is not a good quantity to describe the transmission properties of a disordered system. The reason is that the distribution of the resistance is very broad. It possesses a maximum value at values of $\rho_{\text{typ}} \ll \langle \rho \rangle$. The mean value is given mostly by very specific samples, which appear with a very small probability. We discuss the origin of this "absence of self-averaging" in section 7.3.3.

Since the mean value of the Landauer resistance happens to be not a good representative of the statistical ensemble, we must find another quantity which will better characterize the transport properties of the disordered system. A good candidate should have the mean value close to the most probable value. This means that the variance must be of the order of the mean value. We expect that such a quantity should be a linear function of the length of the system. Since $\langle \rho \rangle$ increases exponentially when L increases, it is tempting to consider the *logarithm* of the Landauer resistance. This idea is motivated also by our analysis presented in section 7.3.3, where we discuss two statistical variables x and $y = e^x$. We show that, even when the allowed values of the variable x are restricted to a finite interval, the probability distribution $p(y)$ of the variable y is very broad. The statistical properties of the random variable y remind us of the statistical properties of the Landauer resistance. This observation inspires us to investigate the statistical properties of the logarithm of the Landauer resistance.

Let us first consider the mean value of the logarithm of the transmission coefficient. From equation (7.29), it follows that

$$\ln T_{N+1} = \langle \ln T_N \rangle + \ln T_I - \ln \left[1 + R_N R_I - 2\sqrt{R_N R_I} \cos \phi \right], \tag{7.56}$$

where $\langle \cdot \rangle$ indicates that we have averaged over all phases between scatterers $1, 2, \ldots, N$. The phase ϕ is again given by equation (7.30). Assuming that all values of ϕ appear with the same probability, we average the expression on the r.h.s. of equation (7.56) over the

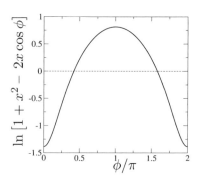

FIGURE 7.13. Plot of $\ln\left[1 + x^2 - 2x\cos\psi\right]$, $x = \sqrt{R_N R_I}$, for $x = 0.5$, as a function of ϕ. The integral of the function is zero as claimed in the text.

random phase ϕ. To do so, we need to calculate the integral

$$\int_0^{2\pi} d\phi \ln(a + b\cos\phi) = \pi\ln\left[\frac{1}{2}\left(a + \sqrt{a^2 - b^2}\right)\right].$$ (7.57)

We plot the function $\ln\left[1 + R_N R_I - 2\sqrt{R_N R_I}\cos\phi\right]$ in figure 7.13. In our particular case, we have $a = 1 + R_N R_I$ and $b = -2\sqrt{R_N R_I}$, so that

$$\frac{1}{2}\left[a + \sqrt{a^2 - b^2}\right] = \frac{1}{2}\left[1 + R_N R_I + |1 - R_N R_I|\right] = 1$$ (7.58)

(remember that $R_N < 1$ and $R_I < 0$ so that $1 - R_N R_I > 0$). Consequently, the r.h.s. of equation (7.57) is zero. We obtain from equation (7.56) the composition law

$$\langle \ln T_{N+1} \rangle = \langle \ln T_N \rangle + \ln T_I.$$ (7.59)

Note that $\ln T_I$ is the transmission of the individual scatterer.

The solution of equation (7.59) is given by

$$\langle \ln T_N \rangle = N \ln T_I,$$ (7.60)

as can be easily verified by inserting equation (7.60) into equation (7.59).

If we define the length of the system

$$L = N\langle \ell \rangle,$$ (7.61)

then the relation (7.60) can be written as

$$\langle \ln T(L) \rangle = -\frac{2L}{\xi}.$$ (7.62)

The parameter ξ is the *localization length* $\xi = 2\langle \ell \rangle / \ln T_I$ [45]. We can use the parameter $\Lambda = (1 + R_I)/(1 - R_I)$, defined by equation (7.46), to express $T_I = 1 - R_I = 2/(\Lambda + 1)$. The localization length ξ depends on Λ as follows:

$$\xi = \frac{2\langle \ell \rangle}{\ln\left[(\Lambda + 1)/2\right]}.$$ (7.63)

Using the relation $\rho = 1/T - 1$, we easily obtain the composition rule for the quantity f defined as

$$f = \ln(1+\rho) = -\ln T. \tag{7.64}$$

We immediately see that f satisfies the following equation:

$$\langle \ln(1+\rho) \rangle_{N+1} = \langle \ln(1+\rho) \rangle_N + \ln(1+\rho_I). \tag{7.65}$$

This means that

$$\langle \ln(1+\rho) \rangle_L = \frac{2L}{\xi}. \tag{7.66}$$

For a sufficiently long system, both sides of equation (7.66) are large so that we can neglect the 1 with respect to ρ on the l.h.s. We obtain that

$$\rho_{\text{typ}} = e^{\langle \ln \rho \rangle} = e^{2L/\xi}. \tag{7.67}$$

This expression must be compared with the mean resistance, given by equation (7.48). Since $\xi' < \xi$ for $\Lambda > 1$, we have

$$\langle \rho \rangle \gg \exp\langle \ln \rho \rangle. \tag{7.68}$$

The mean resistance $\langle \rho \rangle$ can be many orders of magnitude larger than the typical resistance.

It can be shown that the quantity f, defined by equation (7.64), possesses a rather narrow probability distribution with var $f \sim \langle f \rangle$. We do not present the calculations here. The reader can find all the details in the original paper [45]. Then, the probability distribution of $\ln \rho$ possesses a maximum for values close to the mean value $\langle \rho \rangle$ and the quantity ρ_{typ}, defined by equation (7.67). This is a good representative of the statistical ensemble.

Note that the results obtained are in very good agreement with our numerical results, presented in section 7.1. In particular, the distribution of $\ln T$, shown in figure 7.9, is close to a Gaussian distribution. The mean value $\langle \ln T \rangle$ lies close to the maximum of the distribution $p(\ln T)$, and the variance var $\ln T \sim \langle \ln T \rangle$. Because of the simple relationship between the quantity f and $\ln T$, we expect that the distribution of f has the same properties as the distribution $p(\ln T)$.

7.3 Supplementary Notes

7.3.1 Analytical Formulas for the Lyapunov Exponent

In this section we derive the formula for the Lyapunov exponent, given by equation (7.21), in the limit of weak disorder ($W \leq 1$) [51,75,79,153].

We first give an alternative definition of the Lyapunov exponent,

$$\gamma(\mathcal{E}) = \lim_{N \to \infty} \frac{1}{N} \sum_{n=1}^{N} \ln z_n. \tag{7.69}$$

In equation (7.69), we define the variable z_{n+1} by the relation

$$z_{n+1} = c_{n+1}/c_n. \tag{7.70}$$

To derive the expression (7.69) for $\gamma(E)$, we use the identity

$$c_{N+1}^2 + c_N^2 = \frac{c_{N+1}^2 + c_N^2}{c_N^2} \frac{c_N^2}{c_{N-1}^2} \cdots \frac{c_2^2}{c_1^2} \frac{c_1^2}{c_0^2} c_0^2. \tag{7.71}$$

Inserting this into the r.h.s. of equation (7.18), we obtain

$$\frac{1}{2N} \ln \left(c_N^2 + c_{N+1}^2 \right) = \frac{1}{2N} \sum_{n=1}^{N} \ln \frac{c_n^2}{c_{n-1}^2}$$
$$+ \frac{1}{2N} \ln \frac{c_{N+1}^2 + c_N^2}{c_N^2} + \frac{1}{2N} \ln c_0^2. \tag{7.72}$$

Since the last two terms give zero in the limit $N \to \infty$, we immediately get the relationship (7.69). Dividing both sides of equation (7.16) by c_n, we obtain the following iterative relation for the parameters z_n:

$$z_{n+1} = \mathcal{E} - \varepsilon_n - \frac{1}{z_n}. \tag{7.73}$$

Expression (7.69), together with (7.73), is very useful for the analysis of the Lyapunov exponent in the limit of small disorder.

Note that, in the absence of disorder, $\varepsilon_n = 0$, equation (7.73) has a simple solution

$$z_{n+1} = z_n = e^{ika}, \tag{7.74}$$

since $\mathcal{E} = 2 \cos ka$. We assume this solution to be only weakly affected by the disorder in the case of weak disorder. Therefore, we write

$$z_n = e^{ika} + u_n, \tag{7.75}$$

where the variable u_n is small. Since all the random energies are distributed with the same probability distribution, and since there is no correlation between any two energies on different sites, the mean value $\langle u \rangle$ and the mean of any the higher powers, $\langle u^m \rangle$, do not depend on the site index n. Therefore, we can introduce the parameters

$$\zeta_1 = \langle u_n \rangle \tag{7.76}$$

and

$$\zeta_2 = \langle u_n^2 \rangle. \tag{7.77}$$

Since we are interested only in the weak disorder limit, we neglect all the higher powers of the random energies. This means, we are interested only in the expression of the Lyapunov exponent to the order of W^2.

Inserting z_n, given by equation (7.75), into equation (7.73), we derive the following recursive relation for the quantities u_n:

$$u_{n+1} = \frac{e^{-ika}u_n + \varepsilon_n e^{ika} + \varepsilon_n u_n}{e^{ika} + u_n}. \tag{7.78}$$

To calculate the mean values ζ_1 and ζ_2 from equation (7.78), we use

$$\langle \varepsilon_n \rangle = 0. \tag{7.79}$$

We also use the fact that the random energies ε_n and $\varepsilon_{n'}$ are not correlated. That means that

$$\langle \varepsilon_n \varepsilon_{n'} \rangle = W^2 \delta_{nn'}. \tag{7.80}$$

Since u_n depends only on the random energies on sites $i < n$, the two quantities u_n and ε_n are also uncorrelated,

$$\langle \varepsilon_n u_n \rangle = 0. \tag{7.81}$$

We expand equation (7.78) into a power series of ε_n and u_n and neglect all the terms proportional to $\varepsilon_n u_n$ and all the terms which contain higher than the second power in ε and u_n. Hence we obtain

$$e^{ika}u_{n+1} = e^{-ika}u_n + \varepsilon_n e^{ika} - u_n^2 e^{-i2ka}. \tag{7.82}$$

By averaging over the random energies ε_n, we obtain

$$\zeta_1 e^{ika} = \zeta_1 e^{-ika} - \langle u^2 \rangle e^{-i2ka}, \tag{7.83}$$

which can be written as

$$\zeta_1 = -\frac{\langle u_n^2 \rangle e^{-i2ka}}{2i \sin ka} = -\frac{\zeta_2 e^{-i2ka}}{2i \sin ka}. \tag{7.84}$$

To find the mean value $\zeta_2 = \langle u^2 \rangle$, we square both sides of equation (7.82) and average. We obtain, after simple algebra,

$$\zeta_2 = \frac{\langle \varepsilon^2 \rangle e^{i2ka}}{2i \sin 2ka}. \tag{7.85}$$

Now, we calculate the weak disorder expansion of $\gamma(E)$. First, from equation (7.73), we obtain

$$\ln z_{n+1} = \ln \left[\mathcal{E} - \varepsilon_n - \frac{1}{z_n} \right]. \tag{7.86}$$

We need to expand the right-hand side of equation (7.86) in a power series. Inserting expression (7.75) into equation (7.86), we obtain

$$\begin{aligned} \ln z_{n+1} &= \ln \left[e^{ika} + e^{-ika} - \varepsilon_n - e^{-ika}(1 - u_n e^{-ika} + u_n^2 e^{-i2ka}) \right] \\ &= ika + \ln \left[1 - \varepsilon_n e^{-ika} + u_n e^{-i3ka} - u_n^2 e^{-i4ka} \right]. \end{aligned} \tag{7.87}$$

The logarithm on the right-hand side of equation (7.87) can be Taylor-expanded as $\ln(1 + x) = x - x^2/2 + \cdots$. Keeping only terms that are of the order of W^2, we obtain

$$\ln z_{n+1} = ika - \frac{\varepsilon_n^2}{2}e^{-i2ka} + u_n e^{-i3ka} - u_n^2 e^{-i4ka} - \frac{u_n^2}{2}e^{-i6ka}. \tag{7.88}$$

Since $\langle \ln z_{n+1} \rangle$ does not depend on the index n, from (7.69) we calculate that

$$\gamma(E) = \langle \ln z \rangle. \tag{7.89}$$

After some algebra with trigonometric functions, we calculate the final result,

$$\gamma(E) = ika + \frac{\langle \varepsilon^2 \rangle}{8 \sin^2 ka} = ika + \frac{\langle \varepsilon^2 \rangle}{2(4 - \mathcal{E}^2)}. \tag{7.90}$$

The real part of the r.h.s defines the localization length (7.21). The imaginary part contains no correction of the order of $\langle \varepsilon^2 \rangle$.

For a rectangular distribution of random energies, we have that $\langle \varepsilon^2 \rangle = W^2/12$. From equation (7.21), we obtain that the localization length of the electron, in a weakly disordered tight binding model, is given by

$$\xi(E) = \frac{1}{\operatorname{Re} \gamma(E)} = \frac{24(4 - \mathcal{E}^2)}{W^2}. \tag{7.91}$$

Comparing the numerical data, presented in figure 7.8, we see that, the analytical formula (7.91) agrees very well with the numerical data.

For readers who like mathematical problems, we note that the weak disorder expansion derived above does not work in the neighborhood of the band edge, when $|\mathcal{E}| \to 2$ because the leading term of the weak disorder expansion of the Lyapunov exponent, given by equation (7.90), is not small. A more detailed mathematical analysis [79,51] shows that the Lyapunov exponent behaves in the neighborhood of the band edge as $\gamma \propto \langle \varepsilon^2 \rangle^{2/3}$. Similarly, at the band center, $\mathcal{E} = 0$, one finds that the 4th-order term of the weak disorder expansion diverges. In general, the qth order of the weak disorder expansion must be treated very carefully in the neighborhood of all energies $\mathcal{E} = 2 \cos(\pi p/q)$ with p and q integers.

7.3.2 Electrical Conductance G

The law of electrical conductance in metals—Ohm's law—is $I = V/R$, where I is the current, V is the potential difference across the system, and R is the resistance of the wire. Suppose that L and A are, respectively, the length and the cross section of the wire. Then we have

$$J = I/A, \quad E = V/L, \quad \text{and} \quad R = \frac{L}{A}\rho, \tag{7.92}$$

where J is the current density (current per unit area), E the electric field, and ρ the electrical resistivity. The inverse of the resistivity,

$$\sigma = \frac{1}{\rho}, \tag{7.93}$$

is called the conductivity. If we substitute equations (7.92) and (7.93) into Ohm's law $I = V/R$, we obtain

$$J = \sigma E. \tag{7.94}$$

We also define the conductance G, equal to the inverse of the resistance, so that $G = 1/R = A/(L\rho) = \sigma A/L$. If L is the linear size of the sample in each direction, then $A = 1$, L, and L^2 in one-, two-, and three-dimensional systems, respectively. Consequently, we can write that

$$G = \sigma L^{d-2}, \tag{7.95}$$

where d is the dimension of the system. We want to express the conductivity σ in terms of the microscopic properties of the model. We start from the formula for the classical conductivity σ, given by the Drude model,

$$\sigma = \frac{ne^2\tau}{m}. \tag{7.96}$$

In equation (7.96), n is the number of electrons per unit volume, m is the electron mass, and τ is the collision time, defined as a time between two successive collisions of electrons. It is convenient to introduce the mean free path ℓ_p as the mean distance between two collisions. Then, the collision time may be expressed as

$$\tau = \ell_p/v_F, \tag{7.97}$$

where v_F is the Fermi velocity. By substituting τ, given by equation (7.97), into equation (7.96), we obtain that

$$\sigma = \frac{ne^2\ell_p}{mv_F} = \frac{ne^2\ell_p}{\hbar k_F}. \tag{7.98}$$

For a one-dimensional system, we have the number of electrons at the Fermi energy is

$$n = 2 \times \frac{L}{2\pi} \times 2k_F. \tag{7.99}$$

By substituting equation (7.99) into equation (7.98), we obtain that the conductivity is given by

$$\sigma = \frac{2e^2}{(2\pi)\hbar}\ell_p. \tag{7.100}$$

The expression (7.100) can be derived from the usual expression for the Drude conductivity.

The conductivity σ given by equation (7.100) is valid in the classical limit, where all quantum coherence effects have been neglected. In the limit of a perfect crystal and zero temperature, the conductivity is infinite and the resistivity $\rho = 1/\sigma$ is zero, as expected.

Now, we introduce two other formulas for the conductance, derived for the description of the *quantum* transport of electrons in a disordered one-dimensional system.

The first formula, called the Landauer formula [89, 90], gives the following expression for the conductance of disordered one-dimensional systems:

$$G = \frac{e^2}{\pi\hbar} \frac{T}{1-T},$$
(7.101)

where T is the transmission coefficient.

Another formula for the conductance, derived by Economou and Soukoulis [59], gives

$$G = \frac{e^2}{\pi\hbar} T.$$
(7.102)

There is a discrepancy between the two formulas. Let us now analyze how the two formulas describe the conductance in two limiting cases, namely, in the limit of strong disorder and in the limit of very weak disorder. For a strongly disordered one-dimensional system we have that $T = e^{-aL/\xi}$, where ξ is the localization length.[2] If the disorder is strong, ξ is small, and in the limit of $L \gg \xi$, both expressions for the conductance, (7.101) and (7.102), give the same value of the conductance $G = (e^2/\pi\hbar)e^{-aL/\xi}$. However, if the disorder is very small, the localization length ξ is large and in the limit of $\xi \gg L$, the transmission coefficient T approaches 1. Then equation (7.101) gives

$$G \simeq \frac{e^2}{\pi\hbar} \frac{\xi}{2L}.$$
(7.103)

Using the relation $\sigma = LG$, given by equation (7.95), we obtain the following expression for the conductance,

$$\sigma \simeq \frac{e^2}{2\pi\hbar}\xi = \frac{e^2}{2\pi\hbar} 2\ell_p.$$
(7.104)

In equation (7.104) we have used the fact that the localization length ξ is related to the mean free path ℓ_p by

$$\xi = 2\ell_p.$$
(7.105)

By this substitution, equation (7.104) gives the classical limit of the conductivity given by equation (7.100). We obtain that the Landauer formula (7.101) gives the classical limit of the conductivity and, in a perfect system, where the transmission coefficient $T \to 1$, gives that $G \to \infty$ and the resistance $R \to 0$.

On the other hand, equation (7.102) gives that in the limit of a perfect system the conductance approaches a universal constant $e^2/(\pi\hbar)$. This is a surprising result, since we have no disorder and $T = 1$. How can our one-dimensional system have a resistance equal to $\pi\hbar/e^2$?

---

[2] The coefficient $a = 1/4$ was obtained by detailed analysis of the probability distribution $p(T)$ [41].

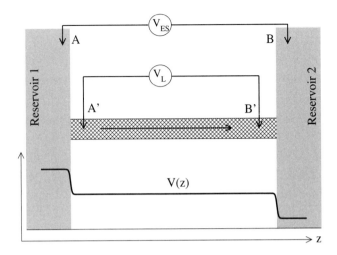

Figure 7.14. Measurement of the resistance. The electric current slows between the two reservoirs R_1 and R_2. If the sample is a perfect conductor, then the voltage between two reservoirs decreases at the interface between the sample and the reservoirs and is constant inside the sample. Therefore, $V_L = 0$ and $R_{\mathrm{wire}} = 0$ [8,9,19].

In order to understand the origin of the difference between expressions (7.101) and (7.102), let us describe an experiment in which the resistance is measured. A typical experimental setup is shown in figure 7.14. A sample is connected with two reservoirs R_1 and R_2. The reservoirs serve as a source of electrons. It is assumed that both reservoirs are big enough so that the electrons in the reservoirs are in thermodynamical equilibrium.

Now, suppose that there is a potential difference V between the two reservoirs. This causes current to flow from one reservoir to another. In order to determine the resistance, we must measure both the current and the voltage. While the measurement of the current is easy, we have two possibilities of the measurement for the voltage, as shown in figure 7.14. We measure the voltage either between points A and B or between points A' and B'. It turns out that these two possibilities correspond to the two formulas for the conductance, given by equations (7.101) and (7.102). The difference between the two formulas is due to the universal resistance $\pi\hbar/e^2$ of the contacts of a one-dimensional system. A drop in the voltage takes place exactly at the contacts of the wire with the two electrodes, as shown in figure 7.14.

The electric current between the points A and B is given by the ratio of the voltage drop V_0 across the points A and B to the resistance R_{AB}. The resistance R_{AB} is given by

$$R_{AB} = R_A + R_B + R_{\mathrm{wire}}, \tag{7.106}$$

where the sum of the two resistances $R_A + R_B = \pi\hbar/e^2$ and R_{wire} is the resistance of the one-dimensional wire. Therefore, when the wire is a perfect conductor, we have $R_{AB} = \pi\hbar/e^2$ and $R_{\mathrm{wire}} = 0$, as expected. When the wire is not a perfect conductor, then from equation (7.102) we have that $R_{AB} = \pi\hbar/(e^2 T)$ and from equation (7.106) we calculate

$$R_{\text{wire}} = R_{AB} - (R_A + R_B) = \frac{\pi\hbar}{e^2 T} - \frac{\pi\hbar}{e^2} = \frac{\pi\hbar}{e^2}\left[\frac{1}{T} - 1\right]$$

$$= \frac{\pi\hbar}{e^2}\frac{1-T}{T}, \tag{7.107}$$

which is equivalent to equation (7.101). Thus, there is no discrepancy between equations (7.101) and (7.102). Equation (7.101) gives the conductance of the one-dimensional wire measured by the voltmeter V_L in figure 7.14, while equation (7.102) gives the conductance of the one-dimensional wire measured by the voltmeter V_{ES}. The measurement V_{ES} is called a two-probe measurement since we measure both the voltage and the current at the same places, A and B. The experiment in which we measure V_L is called a four-probe experiment, because we need two probes to measure the current at A and B, and another two probes to measure the voltage at A' and B'. Thus, $G = e^2/\pi\hbar T$ is the outcome of a two-probe measurement and it includes the contact resistances. The result $G = (e^2/\pi\hbar)(T/1 - T)$, is the outcome of a four-probe measurement.

7.3.3 Probability Distributions

In section 7.2.3 we have seen that both the Landauer resistance and the transmission coefficient of the disordered one-dimensional system are statistical variables. Their probability distributions are extremely broad, so that it is more convenient to study the statistical properties of the logarithm of these variables than the variables themselves. Here, we present some useful formulas which allow us to derive the statistical properties of the Landauer resistance ρ and of the transmission coefficient T from the distributions of their logarithms.

Consider the statistical variable x, and assume it has a Gaussian distribution,

$$P(x) = \frac{1}{\sqrt{2\pi b}} \exp -\frac{(x-a)^2}{2b}. \tag{7.108}$$

It is evident that the mean value

$$\langle x \rangle = \int_{-\infty}^{+\infty} dx\, P(x)\, x = a, \tag{7.109}$$

and the variance

$$\langle x^2 \rangle - \langle x \rangle^2 = b. \tag{7.110}$$

Consider now another variable,

$$y = e^x. \tag{7.111}$$

It is easy to find that

$$\langle y^n \rangle = \int_{-\infty}^{+\infty} dx\, P(x)\, e^{nx} = \exp\left[an + \frac{bn^2}{2}\right]. \tag{7.112}$$

In the particular case of $n = 1$, we have

$$\langle y \rangle = \langle e^x \rangle = e^{a+b/2}. \tag{7.113}$$

On the other hand, we see from equation (7.109) that

$$e^{\langle x \rangle} = e^a. \tag{7.114}$$

Comparing equations (7.113) and (7.114), we see that

$$e^{\langle x \rangle} \neq \langle e^x \rangle. \tag{7.115}$$

The two expressions are equal only when $b = 0$, which means that the Gaussian distribution given by equation (7.108) is infinitesimally narrow.

We can calculate also the variance

$$\langle y^2 \rangle - \langle y \rangle^2 = e^{2a+2b} - e^{2a+b}. \tag{7.116}$$

Now we use the above mathematical formulas for analysis of the statistical properties of the transmission. In sections 7.1.3 and 7.2.3, we have learned that the parameter

$$f = -\ln T, \tag{7.117}$$

possesses a probability distribution similar to the Gaussian distribution, as shown in figure 7.9. We also know that the width of the distribution is of the same size as the mean value. Thus we can approximate the distribution $P(f)$ with a Gaussian distribution equation (7.108) with $b \approx 2a$. On the other hand, we know that the mean value, given by $\langle f \rangle = -\langle \ln T \rangle = 2L/\xi$, increases linearly with the length L.

Now, we consider a new statistical variable

$$y = e^f = e^{-\ln T}. \tag{7.118}$$

With the use of equations (7.115) and (7.116) we easily find that the mean value

$$\langle e^f \rangle = e^{a+b/2} = e^{2L/\xi + 2L/\xi} \approx e^{4L/\xi} \tag{7.119}$$

increases with L much faster than the typical value, $\exp\langle -\ln T \rangle$:

$$e^{\langle f \rangle} = e^{2L/\xi}. \tag{7.120}$$

Also, the standard deviation $\sqrt{\mathrm{var}\, e^f}$ is given by

$$\sqrt{\langle e^{2f} \rangle - \langle e^f \rangle^2} = e^{a+b} \tag{7.121}$$

and increases with L much faster than the mean value $\langle e^f \rangle = e^{a+b/2}$. This means that the distribution $P(e^f)$ is very broad, and the mean value $\langle e^f \rangle$ is not a representative value of the transmission. Since $f = \ln(1 + \rho)$, we conclude that the mean value of the Landauer resistance is not a good variable for characterization of the transmission in disordered systems.

Another example of a non-Gaussian statistical ensemble is the following. Consider the statistical ensemble of $N_{\text{stat}} = 10^3$ numbers ζ_i. Suppose that most (say, 995) of ζ_i are of the order of 10^{-6}, and the remaining five possess values 0.7, 0.2, 0.09, 0.005, and 0.0004. If we choose one number randomly from the ensemble, we get "almost always" (more precisely, with a probability of 99.5%) a small number of the order of 10^{-6}. We conclude that the typical value of ζ is 10^{-6}. However, the mean value, calculated as

$$\langle \zeta \rangle = \frac{1}{N_{\text{stat}}} \sum_i \zeta_i = \frac{1}{1000} [0.7 + 0.2 + 0.09 + 0.005 + \cdots] \approx 0.001, \tag{7.122}$$

is three orders of magnitude larger than the typical value! It is then evident that the mean value does not give us any relevant information about the statistical ensemble. Instead of studying the numbers ζ_i, it is more convenient to concentrate on $x = \ln \zeta$, which might possess a good probability distribution.

We have seen in section 7.1 that the typical value of the transmission coefficient T of a long disordered system has a very small value. However, there are a few samples with a much larger transmission coefficient T than the typical values. The statistical properties of the transmission coefficient T seem to be similar to those of the parameter ζ discussed above. This is the reason, that it is more convenient to study the statistical properties of the logarithm, $\ln T$.

7.4 Problems

Problems with Solutions

Problem 7.1
Consider $\varepsilon_n = 0$ for all n. Show that, for $|\mathcal{E}| > 2$,

$$\gamma(\mathcal{E}) = \begin{cases} +\lambda & \text{for } c_1/c_0 \neq \exp(-\lambda), \\ -\lambda & \text{for } c_1/c_0 = \exp(-\lambda), \end{cases} \tag{7.123}$$

where $\lambda = \cosh^{-1} |\mathcal{E}|/2$.

Solution. The transfer matrix $\mathbf{M}$ for the tight binding model is given by equation (5.22). For $\mathcal{E} = 2 \cos k$, it can be written as

$$\mathbf{M} = \begin{pmatrix} 2 \cos k & -1 \\ 1 & 0 \end{pmatrix} \quad \text{for} \quad |\mathcal{E}| < 2 \tag{7.124}$$

and

$$\mathbf{M} = \begin{pmatrix} 2 \cosh \lambda & -1 \\ 1 & 0 \end{pmatrix} \quad \text{for} \quad |\mathcal{E}| > 2. \tag{7.125}$$

We find c_{n+1} and c_n from the relation

$$\begin{pmatrix} c_{N+1} \\ c_N \end{pmatrix} = \mathbf{M}^N \begin{pmatrix} c_1 \\ c_0 \end{pmatrix}. \tag{7.126}$$

It is easy to show that the eigenvalues of the matrix (7.125) are e^λ and $e^{-\lambda}$, and the corresponding eigenvectors are

$$e_1 = \begin{pmatrix} e^\lambda \\ 1 \end{pmatrix} \quad \text{and} \quad e_2 = \begin{pmatrix} e^{-\lambda} \\ 1 \end{pmatrix} \tag{7.127}$$

and

$$\mathbf{M}e_1 = e^\lambda e_1, \qquad \mathbf{M}e_2 = e^{-\lambda} e_2. \tag{7.128}$$

Any starting vector can be expressed as a linear combination of the eigenvectors e_1 and e_2:

$$\begin{pmatrix} c_1 \\ c_0 \end{pmatrix} = a e_1 + \beta e_2. \tag{7.129}$$

Inserting (7.129) into (7.126), we have

$$\begin{pmatrix} c_{N+1} \\ c_N \end{pmatrix} = \mathbf{M}^N [a e_1 + \beta e_2] = a e^{\lambda N} e_1 + \beta e^{-\lambda N} e_2. \tag{7.130}$$

Consequently, both c_{N+1} and c_N contain exponentially increasing terms. Since $e^{\lambda N} \gg 1$, $e^{-\lambda N}$ all the other terms in the expressions for c_{N+1} and c_N can be neglected for $a \neq 0$ in the limit $N \to \infty$. This leads to equation (7.123).

The case $a = 0$ is special, since we do not have any exponentially increasing term in (7.130). We have from (7.129) that $c_1 = e^{-\lambda} c_0$. Only terms that exponentially decrease remain in (7.130), so that $\gamma(\mathcal{E}) = -\lambda$.

Problem 7.2
Use equation (7.21) for the estimation of the localization length for the case of a rectangular probability distribution of the random energies given by equation (7.8) and $\mathcal{E} = 0.5$. Compare equation (7.21) with the numerical data presented in the caption to figure 7.9 for the mean value of $\ln T$.

Solution. For $\mathcal{E} = 0.5$, we have $\sqrt{4 - \mathcal{E}^2} = \sqrt{15}/2$. For the rectangular probability distribution, we have $\langle \varepsilon^2 \rangle = W^2/12$. Inserting these data in equation (7.21) we obtain

$$\xi(W, E = 0.5) = \frac{90}{W^2}. \tag{7.131}$$

Thus, we have $\xi = 360, 90, 45$, and 22.5 for $W = 0.5, 1, 2$, and 4, respectively. These results are in very good agreement with those presented in section 7.1.3 and in the caption of figure 7.9.

Problem 7.3

Derive the relation (7.50) for the second moment of the resistance [155].

Solution. We start with the equation (7.42)

$$\rho_{N+1} = \frac{R_{N+1}}{T_{N+1}} = \frac{R_N + R_I - 2\sqrt{R_1 R_N}\cos\phi_N}{(1-R_N)(1-R_I)},$$
(7.132)

which can be rewritten in the form

$$\rho_{N+1} = \Lambda\rho_N + \frac{\Lambda-1}{2} - \sqrt{(\Lambda^2-1)(\rho_N + \rho_N^2)}\cos\phi_N.$$
(7.133)

To calculate the second moment of the resistance, we square both sides of equation (7.133) and obtain

$$\rho_{N+1}^2 = \rho_N^2\left[\Lambda^2 + (\Lambda^2-1)\cos^2\phi\right]$$

$$+\rho_N\left[\Lambda(\Lambda-1) + (\Lambda^2-1)\cos^2\phi_N\right] + \frac{1}{4}(\Lambda-1)^2$$

$$+\text{terms proportional to }\cos\phi_N.$$
(7.134)

Now, we average over the random phase ϕ_N, and use that $\langle\cos\phi_N\rangle = 0$ and $\langle\cos^2\phi_N\rangle = (1/2\pi)\int_0^{2\pi} d\phi\,\cos^2\phi = 1/2$. Inserting these values into equation (7.134), we obtain

$$\langle\rho_{N+1}^2\rangle = \Lambda_2\langle\rho_N^2\rangle + (\Lambda_2-\Lambda)\langle\rho_N\rangle + \frac{1}{4}(\Lambda-1)^2$$
(7.135)

where $\Lambda_2 = (3\Lambda^2-1)/2$.

We also know that $\langle\rho_N\rangle = (\Lambda^N-1)/2$. Substituting this result into equation (7.135), we obtain that

$$\langle\rho_{N+1}^2\rangle = \Lambda_2\langle\rho_N^2\rangle + \frac{1}{2}(\Lambda_2-\Lambda)\Lambda^N - \frac{1}{2}(\Lambda_2-\Lambda) + \frac{1}{4}(\Lambda-1)^2.$$
(7.136)

We are looking for a solution of equation (7.135) in the form of

$$\langle\rho_N^2\rangle = a\Lambda_2^N + \beta\Lambda_1^N + \gamma.$$
(7.137)

Inserting this form into equation (7.136) we obtain

$$a\Lambda_2\Lambda_2^N + \beta\Lambda\Lambda^N + \gamma = a\Lambda_2\Lambda_2^N + \beta\Lambda_2\Lambda^N + \gamma\Lambda_2$$

$$+\frac{1}{2}(\Lambda_2-\Lambda)\Lambda^N$$

$$+\frac{1}{2}(\Lambda_2-\Lambda) + \frac{1}{4}(\Lambda-1)^2.$$
(7.138)

This equation holds for any N. This is possible only if the coefficients of Λ_2^N, Λ^N on both sides of equation (7.138) are equal to each other. So we compare the terms proportional to Λ_2^N, Λ^N on both sides of the equation. We easily find that the coefficients of Λ_2^N are equal to each other for any a so that we need another relation to estimate a. This will be completed later. Comparison of terms of Λ^N gives us the following relationship:

$$\beta\Lambda = \beta\Lambda_2 + \frac{1}{2}(\Lambda_2-\Lambda)$$
(7.139)

so that $\beta = -1/2$. All remaining terms in equation (7.137) give us the equation for γ,

$$\gamma = \gamma\Lambda_2 = \frac{1}{4}(\Lambda-1)^2 - \frac{1}{2}(\Lambda_2 - \Lambda). \tag{7.140}$$

With the use of the relation $\Lambda_2 = (3\Lambda^2 - 1)/2$ we find, after simple algebra, that $\gamma = 1/3$. To obtain the parameter a, we use the initial condition $\rho_0 \equiv 0$, which gives also that $\langle \rho_0^2 \rangle = 0$, so that $a + \beta + \gamma = 0$. This gives us $a = 1/6$.

The relationship (7.137) with the obtained parameters a, β, and γ determines the N dependence of $\langle \rho_N^2 \rangle$ and proves the validity of equation (7.50).

Problems without Solutions

Problem 7.4
Find the second moment $\langle \varepsilon_n^2 \rangle$ for the probability distributions given by equations (7.8) and (7.10).

Problem 7.5
Write a computer program to diagonalize the tight binding Hamiltonian with random diagonal elements ε_n. Use a random number generator to generate the random energies ε. Use a diagonalizing subroutine from LAPACK or the Numerical Recipes package. Use the results to calculate the density of states, shown in figure 7.3 and the spatial form of eigenfunctions, shown in figures 7.4 and 7.5.

Problem 7.6
Write a program to calculate numerically the transmission coefficient, T, given by equation (7.13) for a one-dimensional disordered tight binding model. Assume that the random energies ε_n are distributed with a rectangular probability distribution given by eq equation (7.8) and they are statistically independent. Use formula (5.41) with initial values (5.35) to calculate numerically the energy dependence of the transmission coefficient for a system of length $N = 100$. The parameter W, which measures the strength of the disorder, is $W = 1$.

Problem 7.7
Use the same program to calculate numerically the length dependence of the transmission coefficient for $W = 0.2$, 1, and 2. The energy is $\mathcal{E} = 0.5$ and the system size $N = 20$, 100, and 500.

Problem 7.8
Repeat the calculations of the previous two exercises for various realizations of random energies. Using 10000 random realizations, plot the probability distribution of quantities $p(T)$ and $p(\ln T)$. Show how the mean transmission $\langle T \rangle$ and mean value of the logarithm of the transmission $\langle \ln T \rangle$ depend on N.

Problem 7.9
Following the solution of the problem 7.1, show that for $|\mathcal{E}| < 2$

$$\gamma(\mathcal{E}) = ik \quad \text{for} \quad \mathcal{E} = 2\cos k. \tag{7.141}$$

Problem 7.10
Show that the lengths ξ and ξ', given by equations (7.63) and (7.49), respectively, satisfy the relation $\xi > \xi'$.

Problem 7.11
Suppose that the statistical variable x follows a Gaussian probability distribution

$$P(x) = \frac{1}{\sqrt{2\pi b}} \exp -\frac{(x-a)^2}{2b}. \tag{7.142}$$

Using the identity

$$P(x)dx = P(y)dy \tag{7.143}$$

derive the probability distribution for the variable $y = e^x$ and calculate their moments $\langle y^n \rangle$. Compare your results with those shown in section 7.3.3.

8 Numerical Solution of the Schrödinger Equation

The one-dimensional Schrödinger equation

$$-\frac{\hbar^2}{2m}\frac{\partial^2}{\partial x^2}\Psi + V(x) = E\Psi(x) \tag{8.1}$$

can be solved analytically only for a few elementary problems [15]. In most of the applications, we have to find the transmission and the energy spectrum numerically. In this chapter, we describe the simplest numerical algorithm for solution of the Schrödinger equation, and discuss the accuracy of the results obtained. We describe a simple numerical algorithm that enables us to treat various scattering problems numerically. Applying this algorithm to the simplest problem—that of the free particle— enables us to discuss the accuracy of the numerical algorithm and to estimate the numerical error of our simulations. Finally, we apply the numerical procedure to the calculations of the transmission coefficient of the rectangular potential.

8.1 Numerical Procedure

In numerical applications [32], we divide the interval L on which we want to solve the equation (8.1) into N steps. We introduce a new length unit Δ,

$$\Delta = L/N, \tag{8.2}$$

and calculate the wave function only at a discrete set of points $x = n\Delta$. We define

$$c_n = \Psi(x = n\Delta) \tag{8.3}$$

and

$$V_n = V(x = n\Delta). \tag{8.4}$$

First, we need to rewrite the *differential* equation (8.1) into a *difference* equation for the wave amplitudes c_n. To do so, we approximate the first derivative of the wave function as

$$\frac{\partial \Psi(x)}{\partial x} \approx \frac{\Psi(x + \Delta/2) - \Psi(x - \Delta/2)}{\Delta}. \tag{8.5}$$

The second derivative is then

$$\frac{\partial^2 \Psi(x)}{\partial x^2} \approx \frac{\Psi(x + \Delta) + \Psi(x - \Delta) - 2\Psi(x)}{\Delta^2} = \frac{c_{n+1} + c_{n-1} - 2c_n}{\Delta^2}. \tag{8.6}$$

Inserting expression (8.6) into the Schrödinger equation (8.1) we obtain the equation for the wave amplitudes c_n,

$$c_{n+1} + c_{n-1} = \left[2 + \frac{2m\Delta^2}{\hbar^2}(V_n - E) \right] c_n. \tag{8.7}$$

It is useful to rewrite equation (8.7) in matrix form,

$$\begin{pmatrix} c_{n+1} \\ c_n \end{pmatrix} = \mathcal{M}_n \begin{pmatrix} c_n \\ c_{n-1} \end{pmatrix} = \begin{pmatrix} \tilde{E} - \tilde{V}_n & -1 \\ 1 & 0 \end{pmatrix} \begin{pmatrix} c_n \\ c_{n-1} \end{pmatrix}, \tag{8.8}$$

where

$$\tilde{E} = 2 - \frac{2m\Delta^2}{\hbar^2} E \tag{8.9}$$

and

$$\tilde{V}_n = \frac{2m\Delta^2}{\hbar^2} V_n. \tag{8.10}$$

Equation (8.7) can be easily solved numerically. Before we apply the above algorithm to the calculation of the transmission coefficient, we need to estimate the accuracy of the numerical procedure.

8.2 Accuracy of Numerical Data

Consider first the problem of propagation of the free particle: $V_n \equiv 0$ for all n. As the solution of the Schrödinger equation is $\Psi(x) = e^{ikx}$, with the wave vector k given by energy $E = \hbar^2 k^2 / (2m)$, we expect that equation (8.7) has a solution $c_n = e^{ik\Delta n}$. Inserting this into equation (8.7) we have on the left-hand side of equation (8.7)

$$c_{n+1} + c_{n-1} = e^{ik\Delta(n+1)} + e^{ik\Delta(n-1)} = 2\cos k\Delta e^{ik\Delta n}, \tag{8.11}$$

and the right-hand side of equation (8.7) is given by

$$[2 - \Delta^2 k^2] e^{ik\Delta n}. \tag{8.12}$$

Comparing the two expressions, we obtain

$$\cos k\Delta = 1 - \frac{1}{2}\Delta^2 k^2 \qquad (8.13)$$

This is never true, since the right-hand side of equation (8.13) consists only of the first two terms of the Taylor expansion of the cosine. Nevertheless, relation (8.13) represents a good approximation of the cosine if

$$k\Delta \ll 1. \qquad (8.14)$$

In terms of the wavelength of the electron, $\lambda = 2\pi/k$, condition (8.14) reads

$$\Delta \ll \frac{\lambda}{2\pi}. \qquad (8.15)$$

The inequality (8.15) has a clear physical interpretation: we can use the discrete approximation (8.13) only if the number of discrete points *within* one electron wavelength is sufficiently large so that we have enough discrete points within one wavelength to recover the spatial form of the wave.

To estimate the inaccuracy of the proposed procedure, we use the matrix equation (8.8) for zero potential $V_n = 0$. Then the 2×2 matrix $\mathcal{M}$ has eigenvalues $e^{\pm iq\Delta}$ with q given by

$$2\cos q\Delta = 2 - k^2\Delta^2. \qquad (8.16)$$

The corresponding eigenvectors of the matrix **M** are

$$e_1 = \begin{pmatrix} 1 \\ e^{-iq\Delta} \end{pmatrix} \quad \text{and} \quad e_2 = \begin{pmatrix} 1 \\ e^{+iq\Delta} \end{pmatrix}. \qquad (8.17)$$

For small Δ, q is close to k:

$$q = k + \frac{1}{\Delta}\cos^{-1}\left(1 - \frac{k^2\Delta^2}{2}\right). \qquad (8.18)$$

When solving equation (8.7), we choose the starting vector,

$$\begin{pmatrix} c_0 \\ c_{-1} \end{pmatrix} = ae_1 + be_2, \qquad (8.19)$$

and calculate

$$\begin{pmatrix} c_M \\ c_{M-1} \end{pmatrix} = \mathcal{M}^M \begin{pmatrix} c_0 \\ c_{-1} \end{pmatrix}. \qquad (8.20)$$

After M iterations, we obtain

$$\begin{pmatrix} c_M \\ c_{m-1} \end{pmatrix} = ae^{+iq\Delta M}e_1 + be^{-iq\Delta M}e_2, \qquad (8.21)$$

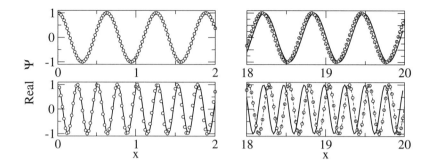

FIGURE 8.1. Wave function c_n calculated numerically for the case of a free particle (open circles) compared with the *exact analytical solution* of the Schrödinger equation $\Psi(x) - e^{ikx}$ (thin line). The discretization difference $\Delta = 0.02$. In the top panels: $k = 10$ and the wavelength $\lambda = 2\pi/k \approx 0.628$. The dashed line is $\cos qx$ with $q = 1/\Delta \cos^{-1}(1 - k^2\Delta^2/2) = 10.0167$. Numerical data are sufficiently accurate. Right top panel: Even after 1000 iterations ($x = 20$) the numerical data do not differ considerably from the exact values. Bottom panels: Same for $k = 10\sqrt{6} = 24.494897$. The wavelength $\lambda = 2\pi/k \approx 0.256$. In this case, the difference $q - k = 0.25$ ($q = 24.7467$). We see that we get numerical data with satisfactory accuracy only for short system. For $x \sim 20$ we see that numerical data are completely different from the exact solution. To get more accurate results, we need to use smaller values of Δ.

which should be compared with the *exact* solution

$$c_M^{\text{exact}} = \Psi(x = M\Delta) = e^{+ik\Delta M}. \tag{8.22}$$

The difference between the analytical and numerical solutions has two origins: First, due to the difference between k and q, the wavelength of the numerical solution differs from $2\pi/k$. It is clearly visible in figure 8.1 which compares the exact analytical solution of the free-electron problem with results of numerical calculations for two energies of the electron. We also see that the difference is larger for higher energy E.

The second source of the inaccuracy of numerical results is not so obvious. As k differs from q, the starting vector

$$\begin{pmatrix} c_0 \\ c_{-1} \end{pmatrix} = \begin{pmatrix} 1 \\ e^{-ik\Delta} \end{pmatrix} \tag{8.23}$$

is *not* an eigenvector of the matrix $\mathcal{M}$. That means that $b \neq 0$ in equation (8.19). In numerical calculations, we therefore obtain not only the wave propagating to the right, but also the wave propagating to the left. The amplitude of the left-going wave is small for $k\Delta \ll 1$. It can be obtained from equation (8.19). Inserting the vector (8.23) into (8.19), we obtain two equations for the unknown parameters a and b,

$$1 = a + b, \quad e^{-ik\Delta} = e^{-iq\Delta}a + e^{+iq\Delta}b, \tag{8.24}$$

which give

$$b = \frac{e^{-ik\Delta} - e^{-iq\Delta}}{2i\sin q\Delta}. \tag{8.25}$$

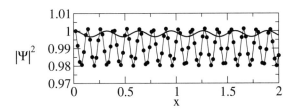

FIGURE 8.2. Numerically calculated $|\Psi(x)|^2$ as a function of x for a free particle with $k = 10$ (solid line) and $10\sqrt{6}$ (circles). For the analytical solution of the free-particle problem we have $|\Psi(x)| \equiv 1$. Small oscillations shown in the figure are due to the nonaccurate estimation of the starting vector equation (8.19) in the numerical procedure. Note that this inaccuracy is observable already at the beginning of the iteration and does not depend on the number of iteration steps.

For instance, for the case discussed in figure 8.1 we have $b = 0.0025 - 0.000167i$ for $k = 10$ and $b = 0.0046 + 0.0025i$ for $k = 10\sqrt{6}$. Nevertheless, the fact that $b \neq 0$ is responsible for the small spatial oscillations of the absolute value of the wave function $|c_n|$ shown in figure 8.2.

8.3 Numerical Data for Transmission

Consider an interval $-2a \leq x \leq 2a$ of length $L = 4a$ and a rectangular potential of height V_0 and width $\ell = 2a$ located in the middle of this interval. Using N discretization points inside the barrier, we have $\Delta = 4a/N$. Equation (8.7) can be written as

$$c_{n+1} = c_{n-1} + \left[2 + 16 \frac{\beta^2}{N^2} \frac{V_n - E}{V_0} \right] c_n, \qquad (8.26)$$

where $\beta = \sqrt{2ma^2}/\hbar$ is given by equation (2.6).

In the numerical procedure, we assume that the electron is coming from the left side only. Then, on the right side of the potential, we have only particles propagating to the right,

$$\Psi_R(x) = Ce^{+ikx}, \qquad (8.27)$$

with constant $C = 1$. Then we can start the iteration procedure with

$$c_N = 1 \quad \text{and} \quad c_{N-1} = e^{-ik\Delta}, \qquad (8.28)$$

where k is the wave vector of the electron $k = \sqrt{2mE}/\hbar$.

Equation (8.7) can be then iterated for $n = N - 2, N - 1, \ldots, 2, 1$. On the left side of the potential, the solution consists of the superposition of the incoming and reflected waves:

$$\Psi(x) = Ae^{ikx} + Be^{-ikx}. \qquad (8.29)$$

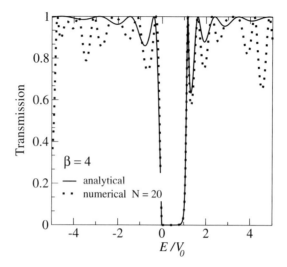

FIGURE 8.3. Transmission coefficient as a function of E/V_0 and for $\beta = 4$. The solid line is the analytical result shown already in figure 2.3. The dotted line is the numerical solution of equation (8.26) with only $N = 20$ discretization points inside the barrier. The results are wrong, especially for large energies. This indicates that the number of discretization points is not sufficient. Data for $N = 200$ are also shown but they are not distinguishable from the analytical results.

Because we know all the wave amplitudes c_n, we can calculate the parameters A and B. For instance, c_1 and c_2 can be expressed as

$$c_1 = Ae^{ik(N-1)\Delta} + Be^{-ik(N-1)\Delta}, \quad c_2 = Ae^{ik(N-2)\Delta} + Be^{-i2k(N-2)\Delta}. \tag{8.30}$$

Solving this system of equations gives us A and B as

$$A = \frac{c_2 - e^{-ik\Delta}c_1}{2i\sin k\Delta}e^{-lk(N-2)\Delta}, \quad B = \frac{c_2 - e^{+ik\Delta}c_1}{2i\sin k\Delta}e^{\,lik(N-2)\Delta}. \tag{8.31}$$

Using the definition of the transmission and reflection coefficients, we obtain

$$T = \frac{1}{|A|^2} \quad \text{and} \quad R = \frac{|B|^2}{|A|^2}. \tag{8.32}$$

In figure 8.3 we show the transmission as a function of E/V_0 for $\beta = 4$. To show how the discretization influences the accuracy of the results obtained, we use first a small number of discretization points, $N = 20$. For this case, we have

$$k^2\Delta^2 = 16\frac{\beta^2}{N^2}\frac{E}{V_0} = \frac{256}{400} = 0.64\frac{E}{V_0}, \tag{8.33}$$

which is small only for very small energy E. Comparison with the exact analytical solution (shown as a solid line in figure 8.3) indeed shows that the accuracy of our numerical data is good only for small E.

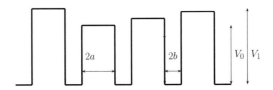

FIGURE 8.4. System of rectangular barriers of random potential height V_i.

In the case of nonzero potential V, we have to satisfy, besides the condition (8.14), another condition

$$k'\Delta \ll 1. \tag{8.34}$$

This is automatically satisfied for $V_0 > 0$ because $k' < k$. However, for $V_0 < 0$, we have $k' > k$ so that the condition (8.34) might not be satisfied. For a given Δ we expect that the accuracy of the numerical results for the transmission would be worse for $V_0 < 0$ than for $V_0 > 0$. This is indeed the case, as can be seen in figure 8.3. Transmission data are quite good for $V_0 > 0$ and $E / V_0 < 1.5$ but for $V_0 < 0$ they are already bad for $E/|V_0| = 0.5$.

Figure 8.3 shows also our numerical data for the discretization with $N = 200$. This discretization gives $k^2\Delta^2 = 0.0064 \ E/V_0$, which is already sufficient for our purposes. Indeed, no difference between numerical and analytical data is observable in figure 8.3.

8.4 Problems

Problem 8.1
Write a numerical program for the calculation of the transmission and reflection amplitudes for a periodic array of N rectangular potential barriers shown in figure 4.12. Compare your results with the analytical formula based on the formula for the transfer matrix, equation (4.67), and Chebyshev's formula for the transmission through N identical barriers, equation (1.106).

Problem 8.2
Consider a system of N rectangular barriers. All barriers have the same width $2a$, but the potential V_i of the ith barrier is chosen randomly from the interval (V_0, V_1) (figure 8.4). Using the analytical expression for the transfer matrix, equation (4.67), calculate numerically the transmission and the reflection of the present system for energy $E > V_1$, $V_0 < E < V_1$, and $0 < E < V_0$. Plot the transmission as a function of N for various choices of V_0 and V_1.

Problem 8.3
Do the same for an arrangement of rectangular barriers of the same height V_0 and width $2a$ but with a randomly chosen distance between barriers $2b_i$ (figure 8.5). Consider $b_0 < b_i < b_1$. Show how the results depend on the relation between the wavelength λ of the electron and the mean barrier distance $\langle b \rangle$.

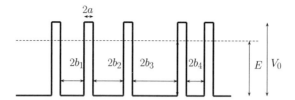

FIGURE 8.5. System of rectangular barriers separated by random distance b_i.

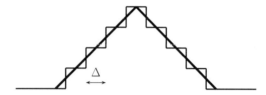

FIGURE 8.6. Approximation of the triangular barrier by system of narrow rectangular barriers of width Δ.

Problem 8.4

Another way to formulate the numerical calculation of the transmission is the following. Consider an arbitrary potential $V(x)$, $0 < x < \ell$. We can calculate transmission and reflection through this potential barrier as follows: (1) divide the interval ℓ into N equidistant subintervals of length $\Delta = \ell/N$. (2) In each interval $n\Delta < x < (n+1)\Delta$, approximate the potential $V(x)$ by constant $V_n = V(x = (n+1/2)\Delta)$ (figure 8.6). (3) In this way we transform the original potential to a system of $N+1$ potential steps. Each step is characterized by the transfer matrix M_s, equation (2.111). (4) Using the transfer matrices M_s and the transfer matrix for a constant potential, calculate the total transfer matrix **M**.

   Use the above algorithm to calculate numerically the transmission through a triangular potential shown in figure 8.6. Check how the accuracy of obtained result depend on the discretization step Δ.

9 Transmission and Reflection of Plane Electromagnetic Waves on an Interface

In this chapter, we will investigate the very basic phenomena of transmission and reflection of electromagnetic wave propagating through the interface between two media. From the requirements of the continuity of the tangential components of the electric and the magnetic fields, we derive the transfer matrix for a single interface between two media. Its elements determine the transmission and reflection amplitudes for both electric and magnetic fields.

Next, we study the behavior of the electromagnetic waves incident on the surface of a dielectric and a metal. We learn how different electromagnetic properties of these materials influence the transmission and reflection coefficients [2, 39]. We also derive the necessary conditions for *total transmission* of the electromagnetic wave through the interface and for *total reflection* from the interface.

We assume that the reader is already familiar with the theory of electromagnetic fields [2, 5, 16, 20, 26, 37]. In Appendix B we give a short summary of the relations for the electromagnetic waves, and we summarize the most important electromagnetic properties of conventional materials—dielectrics and metals. Finally, in chapter 16 we discuss properties of a new class of materials, the so-called left-handed materials.

9.1 Plane Wave at the Interface

Consider a plane wave striking the interface between two media, shown in figure 9.1. The electric field $\vec{E}$ and the magnetic field $\vec{H}$ are given by the expressions

$$\vec{E}(\vec{r}, t) = \vec{E}_0 e^{i[\vec{k}\cdot\vec{r} - \omega t]},$$
$$\vec{H}(\vec{r}, t) = \vec{H}_0 e^{i[\vec{k}\cdot\vec{r} - \omega t]},$$

(9.1)

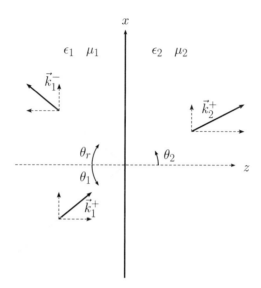

FIGURE 9.1. Wave vectors for an electromagnetic wave incident on a boundary separating two different media.

where $\vec{E}_0$ and $\vec{H}_0$ are the vectors that define the absolute value and the direction of $\vec{E}(\vec{r}, t)$ and $\vec{H}(\vec{r}, t)$, respectively. We know from the theory of electromagnetic fields that the fields $\vec{E}$ and $\vec{H}$ are mutually perpendicular.

The two media are characterized by different permittivities ε_1 and ε_2 and permeabilities μ_1 and μ_2. Since we discuss wave propagation, we assume that the permittivity and the permeability of the incoming medium, ε_1 and μ_1, are real. In most applications, we assume that ε_2 and μ_2 are real, too. Nevertheless, the formulation of the problem allows us to consider any complex value of ε_2 and μ_2.

We choose a coordinate system such that the interface between two media is perpendicular to the z-axis and the plane of incidence, which is the plane determined by the normal to the interface and the direction of the propagation of the electromagnetic wave, is perpendicular to the y-axis. The geometry of the problem is shown in figure 9.1. Thus, the wave vector has only two components,

$$\vec{k} = (k_x, 0, k_z).$$

(9.2)

We see that the formulation of the electromagnetic problem is more complex than that for propagation of the quantum particle, which was completely confined to be a one-dimensional problem. Since the electromagnetic waves propagate in a two-dimensional plane (only when the angle of incidence is zero, $\theta_1 = 0$, then we have one-dimensional propagation), we need to find additional information about the behavior of the electromagnetic wave at the boundary. We start with the condition of the continuity of tangential components of both the electric and the magnetic fields at the interface. The requirement of the continuity of the tangential components of both electric and magnetic fields at the interface, given by equation (B.70),

$$\vec{E}_{1t} = \vec{E}_{2t}, \quad \text{and} \quad \vec{H}_{1t} = \vec{H}_{2t}, \tag{9.3}$$

must be satisfied at any point x of the interface, the tangential component of the vector $\vec{k}$, k_x, must be the same for both media:

$$k_{1x} = k_{2x} = k_x. \tag{9.4}$$

The explicit form of the dispersion relation $n^2\omega^2/c^2 = k^2$ between the frequency and wave vector is given by

$$\frac{n_1^2\omega^2}{c^2} = k_1^2 = k_x^2 + k_{1z}^2, \quad n_1 = \sqrt{\mu_1\varepsilon_1}\,,$$

$$\frac{n_2^2\omega^2}{c^2} = k_2^2 = k_x^2 + k_{2z}^2, \quad n_2 = \sqrt{\mu_2\varepsilon_2}\,, \tag{9.5}$$

for media 1 and 2, respectively. Here, k_1 and k_2 are the absolute values of vectors $\vec{k}_1$ and $\vec{k}_2$, respectively.

In the case of planar waves, both components of the wave vector are real. We will, however, discuss also problems in which the z component of the wave vector k_z is complex. If both k_x and k_z are real, we can define an angle of incidence θ_1 and angle of refraction θ_2 by the relation

$$\tan\theta_1 = \frac{k_x}{k_{1z}} \tag{9.6}$$

and

$$\tan\theta_2 = \frac{k_x}{k_{2z}}, \tag{9.7}$$

respectively. From equation (9.4) we also have that

$$k_x = k_1 \sin\theta_1 = k_2 \sin\theta_2. \tag{9.8}$$

Using the dispersion relations (9.5) we finally obtain that

$$k_x = n_1 \frac{\omega}{c} \sin\theta_1 = n_2 \frac{\omega}{c} \sin\theta_2. \tag{9.9}$$

Taking the ratio of the the two equations in (9.9), we obtain a relation between angles of incidence and reflection, known as Snell's law [2],

$$\frac{\sin\theta_1}{\sin\theta_2} = \frac{n_2}{n_1}. \tag{9.10}$$

For the reflected wave, we have from equation (9.5) $k_1^{-2} = k_1^{+2}$. As the x-component of the wave vector is preserved, we obtain that

$$k_{1z}^- = -k_{1z}^+. \tag{9.11}$$

The negative sign is due to the opposite direction of propagation of the reflected wave. Consequently, the angle of the propagation of the reflected wave, $\theta_r = \tan^{-1}(k_x/k_{1z}^-)$, equals in absolute value the incident angle θ_1:

$$\theta_r = -\theta_1. \tag{9.12}$$

9.2 Transmission and Reflection Coefficients

Calculation of the transmission and reflection coefficients at the plane boundary will be discussed in two special wave polarizations. The first case is that in which the vector of the electric field, $\vec{E}$, of the incident wave is parallel to the boundary plane. This case is called transverse electric or TE polarization (we will use also the name s-*wave*). The second case is that in which the vector of the magnetic field, $\vec{H}$, of the incident wave is parallel to the boundary plane. This is called transverse magnetic or TM polarization. We will also call the TM polarized wave the *p-wave*.

We will discuss these two cases separately. Since any electromagnetic wave can be expressed as a linear combination of TE and TM waves, the results obtained provide us with the most general solution of the transmission problem.

The aim of this chapter is to determine the transmission and reflection coefficients for both TE and TM polarizations. We will use the well-known boundary conditions, given by equation (9.3), which require that the tangential components of the electric and magnetic fields are continuous as the boundary is crossed.

9.2.1 The TE Polarization

By definition, in the case of the TE polarization the electric intensity $\vec{E}$ is parallel to the interface. In the notation of figure 9.2 we see that vector $\vec{E}$ has only one component,

$$\vec{E} = (0, E, 0), \tag{9.13}$$

and vector $\vec{H}$ has two components,

$$H = (H_x, 0, H_z). \tag{9.14}$$

As shown in figure 9.2, each field contains components that propagate to the right (we denote them with the superscript $^+$) and components propagating to the left (denoted by the superscript $^-$).

The boundary condition (B.68) shown in Appendix B.5 states that the component of the electric intensity, parallel to the interface, is the same on both sides of the interface,

$$E_1^+ + E_1^- = E_2^+ + E_2^-. \tag{9.15}$$

The same holds for the intensity of the magnetic field,

$$H_{1x}^+ + H_{1x}^- = H_{2x}^+ + H_{2x}^-. \tag{9.16}$$

We use Maxwell's equation, given by equation (B.17) of Appendix B.1,

$$\vec{k} \times \vec{E} = \frac{\mu\omega}{c} \vec{H}, \tag{9.17}$$

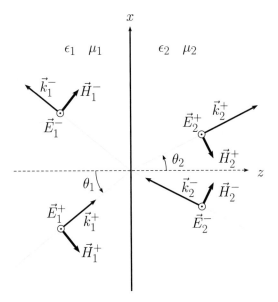

FIGURE 9.2. Scattering of electromagnetic waves at an interface. The TE (s) polarization is shown. This means that the vector of the electric field, $\vec{E}$, is perpendicular to the plane of incidence. Two incoming waves are shown: the wave ($\vec{E}_1^+$, $\vec{H}_1^+$) comes from the left, propagates in the medium (ε_1, μ_1), and scatters at the boundary, $z = 0$. The second incoming wave, ($\vec{E}_2^-$, $\vec{H}_2^-$) comes from the right. We use superscripts $^+$ and $^-$ for waves propagating to the right and to the left, respectively.

to express the magnetic field $\vec{H}$ in equation (9.16) in terms of $\vec{E}$,

$$\frac{\mu_1 \omega}{c} H_{1x}^+ = -k_{1z} E_1^+, \qquad \frac{\mu_1 \omega}{c} H_{1x}^- = k_{1z} E_1^-, \tag{9.18}$$

and similar relations for the field in the second medium. Here, k_{1z} is the z-component of the wave vector of the wave which propagates to the right. Note that, the r.h.s of the second equation (9.18) has the opposite sign because the wave E_1^- propagates in the opposite direction.

Inserting (9.18) in the boundary condition (9.16), we obtain

$$-\frac{k_{1z} c}{\mu_1 \omega} E_1^+ + \frac{k_{1z} c}{\mu_1 \omega} E_1^- = -\frac{k_{2z} c}{\mu_2 \omega} E_2^+ + \frac{k_{2z} c}{\mu_2 \omega} E_1^-. \tag{9.19}$$

Equation (9.19) together with equation (9.15) can be written in a matrix form,

$$\begin{pmatrix} 1 & 1 \\ -\dfrac{k_{1z}}{\mu_1} & +\dfrac{k_{1z}}{\mu_1} \end{pmatrix} \begin{pmatrix} E_1^+ \\ E_1^- \end{pmatrix} = \begin{pmatrix} 1 & 1 \\ -\dfrac{k_{2z}}{\mu_2} & +\dfrac{k_{2z}}{\mu_2} \end{pmatrix} \begin{pmatrix} E_2^+ \\ E_2^- \end{pmatrix}, \tag{9.20}$$

which express the electric field on one side of the interface in terms of the electric field on the other side of the interface. From equation (9.20) we obtain a relation between the electric fields on opposite sides of the interface,

$$\begin{pmatrix} E_2^+ \\ E_2^- \end{pmatrix} = \mathbf{M}^{(s)} \begin{pmatrix} E_1^+ \\ E_1^- \end{pmatrix} \tag{9.21}$$

with the transfer matrix

$$\mathbf{M}^{(s)} = \frac{1}{2} \begin{pmatrix} 1 + \dfrac{\mu_2}{\mu_1} \dfrac{k_{1z}}{k_{2z}} & 1 - \dfrac{\mu_2}{\mu_1} \dfrac{k_{1z}}{k_{2z}} \\ 1 - \dfrac{\mu_2}{\mu_1} \dfrac{k_{1z}}{k_{2z}} & 1 + \dfrac{\mu_2}{\mu_1} \dfrac{k_{1z}}{k_{2z}} \end{pmatrix}. \tag{9.22}$$

We derived the transfer matrix $\mathbf{M}^{(s)}$ for the general case when we have both the electromagnetic waves coming from the left and the right of the interface. Consider now the case when the electromagnetic wave E_1^+ is incident from the left and is scattered on the interface. Part of the wave, E_1^-, is reflected back, and another part, E_2^+, is transmitted through the barrier. Since no wave is incident from medium 2, we designate $E_2^- = 0$. Then, we obtain from equation (9.21) that

$$E_2^+ = M_{11}^{(s)} E_1^+ + M_{12}^{(s)} E_1^-,$$
$$E_2^- = 0 = M_{21}^{(s)} E_1^+ + M_{22}^{(s)} E_1^-. \tag{9.23}$$

We define the transmission and the reflection amplitudes for the electric field as

$$t_s = \frac{E_2^+}{E_1^+} \quad \text{and} \quad r_s = \frac{E_1^-}{E_1^+}. \tag{9.24}$$

Using the explicit form of the matrix $\mathbf{M}^{(s)}$, given by equation (9.22), we can solve the system of equations (9.23) to obtain

$$t_s = \frac{\det \mathbf{M}^{(s)}}{M_{22}^{(s)}} = \frac{2\mu_2 k_{1z}}{\mu_1 k_{2z} + \mu_2 k_{1z}} \tag{9.25}$$

(we used that $\det \mathbf{M}^{(s)} = \mu_2 k_{1z}/\mu_1 k_{2z}$), and

$$r_s = -\frac{M_{21}^{(s)}}{M_{22}^{(s)}} = \frac{\mu_2 k_{1z} - \mu_1 k_{2z}}{\mu_1 k_{2z} + \mu_2 k_{1z}}. \tag{9.26}$$

The transmission coefficient is given by the ratio of the energy flows in the two media. Using the relations between $\vec{E}$ and $\vec{H}$, we can express the energy flow perpendicular to the surface, given by equation (B.60), as

$$S_1 = \frac{c}{8\pi} \operatorname{Re}\left[\vec{E}_1^+ \times \vec{H}_1^{+*}\right] = \frac{c}{8\pi} \operatorname{Re}\left[E_1^+ H_{1x}^{+*}\right] = \frac{c^2}{8\pi} \frac{\operatorname{Re} k_{1z}}{\mu_1 \omega} |E_1^+|^2 \tag{9.27}$$

and

$$S_2 = \frac{c}{8\pi} \operatorname{Re}\left[\vec{E}_2^+ \times \vec{H}_2^{+*}\right] = \frac{c}{8\pi} \operatorname{Re}\left[E_2^+ H_{2x}^{+*}\right] = \frac{c^2}{8\pi} \frac{\operatorname{Re} k_{2z}}{\mu_2 \omega} |E_2^+|^2. \tag{9.28}$$

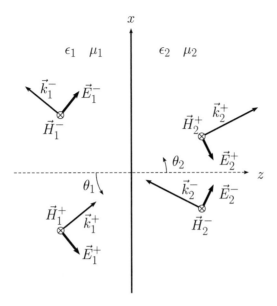

FIGURE 9.3. Scattering of an electromagnetic wave at an interface. The TM (p) polarization is shown. The wave comes from the left, propagates in the medium (ε_1, μ_1), and scatters at the boundary, $z = 0$. We use superscripts $^+$ and $^-$ for wave propagating to the right and to the left, respectively.

Since $E_2^+ = t_s E_1^+$, we finally obtain

$$T_s = \frac{S_2}{S_1} = \frac{\mu_1}{\mu_2} \frac{\operatorname{Re} k_{2z}}{\operatorname{Re} k_{1z}} |t_s|^2. \tag{9.29}$$

The reflection coefficient is

$$R_s = |r_s|^2. \tag{9.30}$$

9.2.2 The TM Polarization

For the TM polarization, shown in figure 9.3, we have the magnetic field vector $\vec{H}$ parallel to the interface and, therefore, it has only one component

$$\vec{H} = (0, H, 0). \tag{9.31}$$

The electric field vector $\vec{E}$ has two components

$$\vec{E} = (E_x, 0, E_z). \tag{9.32}$$

Following the same procedure used for the TE polarization we apply the same boundary conditions for the TM polarization case, too. We have then

$$H_1^+ + H_1^- = H_2^+ + H_2^- \tag{9.33}$$

and

$$E_{1x}^+ + E_{1x}^- = E_{2x}^+ + E_{2x}^-. \tag{9.34}$$

We use Maxwell's equation given by equation (B.18) of Appendix B.1,

$$\vec{k} \times \vec{H} = -\frac{\varepsilon\omega}{c}\,\vec{E}, \tag{9.35}$$

to relate the $\vec{E}$ and $\vec{H}$ components,

$$\frac{\varepsilon_1\omega}{c}E_{1x}^+ = k_{1z}H_1^+ \quad \text{and} \quad \frac{\varepsilon_1\omega}{c}E_{1x}^- = -k_{1z}H_1^-. \tag{9.36}$$

We obtain the relationship

$$\begin{pmatrix} 1 & 1 \\ \dfrac{k_{1z}}{\varepsilon_1} & -\dfrac{k_{1z}}{\varepsilon_1} \end{pmatrix} \begin{pmatrix} H_1^+ \\ H_1^- \end{pmatrix} = \begin{pmatrix} 1 & 1 \\ -\dfrac{k_{2z}}{\varepsilon_2} & +\dfrac{k_{2z}}{\varepsilon_2} \end{pmatrix} \begin{pmatrix} H_2^+ \\ H_2^- \end{pmatrix}, \tag{9.37}$$

which can be rewritten in the form

$$\begin{pmatrix} H_2^+ \\ H_2^- \end{pmatrix} = \tilde{\mathbf{M}}^{(p)} \begin{pmatrix} H_1^+ \\ H_1^- \end{pmatrix}. \tag{9.38}$$

The last equation expresses the magnetic field on one side of the interface in terms of the magnetic field on the opposite side of the interface. After simple algebra, we find that the form of the transfer matrix $\tilde{\mathbf{M}}^{(p)}$ reads as

$$\tilde{\mathbf{M}}^{(p)} = \frac{1}{2}\begin{pmatrix} 1 + \dfrac{\varepsilon_2}{\varepsilon_1}\dfrac{k_{1z}}{k_{2z}} & 1 - \dfrac{\varepsilon_2}{\varepsilon_1}\dfrac{k_{1z}}{k_{2z}} \\ 1 - \dfrac{\varepsilon_2}{\varepsilon_1}\dfrac{k_{1z}}{k_{2z}} & 1 + \dfrac{\varepsilon_2}{\varepsilon_1}\dfrac{k_{1z}}{k_{2z}} \end{pmatrix}. \tag{9.39}$$

Relations (9.38) and (9.39) solve the problem of transmission and reflection completely. We can define the transmission and the reflection amplitudes as

$$\tilde{t}_p = \frac{\det \tilde{\mathbf{M}}^{(p)}}{\tilde{M}_{22}^{(p)}} = \frac{2\varepsilon_2 k_{1z}}{\varepsilon_2 k_{1z} + \varepsilon_1 k_{2z}}, \tag{9.40}$$

$$\tilde{r}_p = -\frac{\tilde{M}_{21}^{(p)}}{\tilde{M}_{22}^{(p)}} = \frac{\varepsilon_2 k_{1z} - \varepsilon_1 k_{2z}}{\varepsilon_2 k_{1z} + \varepsilon_1 k_{2z}} \tag{9.41}$$

(we used that $\det \tilde{\mathbf{M}}^{(p)} = \varepsilon_2 k_{1z}/\varepsilon_1 k_{2z}$). We used the tilde to emphasize that $\tilde{t}$ is the transmission amplitude for the magnetic field.

The Poynting vectors in this case are given by

$$S_1 = \frac{c}{8\pi}\mathrm{Re}\left[\vec{E}_1^+ \times \vec{H}_1^{+*}\right] = \frac{c}{8\pi}\mathrm{Re}\left[\vec{E}_{1x}^+ H_1^{+*}\right] = \frac{c^2}{8\pi}\frac{\mathrm{Re}\,k_{1z}}{\varepsilon_1\omega}|H_1^+|^2 \tag{9.42}$$

with a similar expression for S_2. The transmission coefficient, given as the ratio of S_2/S_1, is given by

$$T_p = \frac{\varepsilon_1}{\varepsilon_2} \frac{\mathrm{Re}\, k_{2z}}{\mathrm{Re}\, k_{1z}} |\tilde{t}_p|^2 \tag{9.43}$$

and the reflection coefficient is given by

$$R_p = |\tilde{r}_p|^2. \tag{9.44}$$

It is easy to prove (problem 9.4) that, in the absence of absorption (Re $k_{2z} = k_{2z}$ and Re $k_{2z} = k_{2z}$), we have

$$T + R = 1 \tag{9.45}$$

for both the TE and TM polarization.

9.3 Interface between Two Dielectric Materials

In the previous section we calculated the transfer matrix $\mathbf{M}$ as well as the transmission and reflection amplitudes. Now, we will apply the general formulas in the case of a plane boundary between two dielectric materials. In the case of dielectric materials with only real dielectric constants ε_1 and ε_2, we have that both k_x and k_z are real. Thus, we can use the relations

$$k_{1z} = k_1 \cos\theta_1 \quad \text{and} \quad k_{2z} = k_2 \cos\theta_2 \tag{9.46}$$

to express the matrix $\mathbf{M}^{(s)}$, given by equation (9.22), in the form

$$\mathbf{M}^{(s)} = \frac{1}{2}\begin{pmatrix} 1 + z_{21}\dfrac{\cos\theta_1}{\cos\theta_2} & 1 - z_{21}\dfrac{\cos\theta_1}{\cos\theta_2} \\[2ex] 1 - z_{21}\dfrac{\cos\theta_1}{\cos\theta_2} & 1 + z_{21}\dfrac{\cos\theta_1}{\cos\theta_2} \end{pmatrix}, \tag{9.47}$$

where z_{21} is defined as

$$z_{21} = \frac{\mu_2}{\mu_1}\frac{k_1}{k_2}. \tag{9.48}$$

We can use also the definition of the refractive index $n = \sqrt{\varepsilon\mu}$ and impedance $z = \sqrt{\mu/\varepsilon}$ as well as the relation between k and frequency ω,

$$k_1 = \frac{n_1\omega}{c}, \quad k_2 = \frac{n_2\omega}{c}, \tag{9.49}$$

to express z_{21} in terms of the material parameters only,

$$z_{21} = \sqrt{\frac{\mu_2\varepsilon_1}{\mu_1\varepsilon_2}}. \tag{9.50}$$

The expression of the transmission amplitude for the s-polarization, given by equation (9.25) can be simplified to be

$$t_s = \frac{2z_{21} \cos \theta_1}{z_{21} \cos \theta_1 + \cos \theta_2}. \tag{9.51}$$

We used $\det \mathbf{M}^{(s)} = z_{21} \cos \theta_1 / \cos \theta_2$. The expression for the reflection amplitude, given by equation (9.26), is now given by

$$r_s = \frac{z_{21} \cos \theta_1 - \cos \theta_2}{z_{21} \cos \theta_1 + \cos \theta_2}. \tag{9.52}$$

In terms of incident and reflected angles, the transmission coefficient is given by the following expressions:

$$T_s = \frac{1}{z_{21}} \frac{\cos \theta_2}{\cos \theta_1} |t_s|^2 = \frac{4z_{21} \cos \theta_1 \cos \theta_2}{(z_{21} \cos \theta_1 + \cos \theta_2)^2}. \tag{9.53}$$

The reflection coefficient is simply given by

$$R_s = |r_s|^2. \tag{9.54}$$

Similarly, we can express the transmission and reflection amplitudes for the p-polarization, $\tilde{t}_p$ and $\tilde{r}_p$, in terms of the incident and reflected angles

$$\tilde{t}_p = \frac{2z_{12} \cos \theta_1}{z_{12} \cos \theta_1 + \cos \theta_2}, \tag{9.55}$$

$$\tilde{r}_p = \frac{z_{12} \cos \theta_1 - \cos \theta_2}{z_{12} \cos \theta_1 + \cos \theta_2}, \tag{9.56}$$

where

$$z_{12} = z_{21}^{-1} = \sqrt{\frac{\varepsilon_2 \mu_1}{\varepsilon_1 \mu_2}}. \tag{9.57}$$

The transmission coefficient is

$$T_p - z_{21} \frac{\cos \theta_2}{\cos \theta_1} |\tilde{t}_p|^2 - \frac{4z_{12} \cos \theta_1 \cos \theta_2}{(z_{12} \cos \theta_1 + \cos \theta_2)^2}. \tag{9.58}$$

All formulas derived in this section can be expressed in various forms used in the literature [2]. We present some other expressions in problems at the end of this chapter.

9.4 Interface between a Dielectric Material and a Metal

The formulas derived in section 9.2 can be used to derive the expressions for the specific interface between a dielectric (for instance air) and a metal. As discussed in appendix B.7, the metallic permittivity strongly depends on the frequency ω. Equation (B.87) gives the frequency dependence of the electrical permittivity of a metal [2, 20, 23, 26, 37],

$$\varepsilon_m = 1 - \frac{\omega_p^2}{\omega(\omega + i\gamma)}, \tag{9.59}$$

where ω_p is the plasma frequency and γ is the inverse relaxation time.

We calculate the transmission and the reflection coefficients of incident electromagnetic waves on a metallic interface in two limiting frequency regions. First we study the limit of low frequencies, $\omega \ll \gamma$, and then the case of high frequencies, $\gamma \ll \omega < \omega_p$. As shown in Appendix B.7, the electrical permittivity of the metal has completely different behavior for these two limits.

9.4.1 Low-Frequency Limit ($\omega \ll \gamma$)

We discuss here the low-frequency region $\omega \ll \gamma$. In this case the electrical permittivity of the metal can be approximately considered as purely imaginary

$$\varepsilon_2 = i\varepsilon_i, \tag{9.60}$$

where [equations (B.91) and (B.92)]

$$\varepsilon_i \approx \frac{\omega_p^2}{\omega\gamma} = \frac{4\pi\sigma}{\omega}, \tag{9.61}$$

and the conductivity $\sigma \approx \omega_p^2/(4\pi\gamma)$. The refractive index

$$n_2 = \sqrt{\varepsilon_2} = \frac{1}{\sqrt{2}}(1+i)\sqrt{\varepsilon_i} \tag{9.62}$$

is complex. From the dispersion relation of the electromagnetic wave inside the metal,

$$\frac{n_2^2\omega^2}{c^2} = k_x^2 + k_{2z}^2, \tag{9.63}$$

we obtain the wave vector k_{2z}, which possesses both real and imaginary parts,

$$k_{2z} = k_{2r} + i\kappa. \tag{9.64}$$

Consider for simplicity only the case of normal incidence, which means that $k_x = 0$. Then $k_2 = k_{2z} = n_2\omega/c$. Since the real and imaginary parts of the refractive index are equal to each other, we have that

$$\kappa = k_{2r} = \frac{\omega}{c}\sqrt{\frac{\varepsilon_i}{2}}. \tag{9.65}$$

We can easily calculate the reflection coefficient. From the general expression given by equation (9.26), we have

$$R = |r_s|^2 = \left|\frac{k_{1z} - k_{2r} - i\kappa}{k_{1z} + k_{2r} + i\kappa}\right|^2. \tag{9.66}$$

Now, we insert $k_1 = k_{1z} = \omega/c$ and $k_{2r} = \kappa$ from equation (9.65) in equation (9.66) and obtain

$$R = |r_s|^2 = \left|\frac{(1-\sqrt{\varepsilon_i/2}) - i\sqrt{\varepsilon_i/2}}{(1+\sqrt{\varepsilon_i/2}) + i\sqrt{\varepsilon_i/2}}\right|^2 = \frac{(1-\sqrt{\varepsilon_i/2})^2 + \varepsilon_i/2}{(1+\sqrt{\varepsilon_i/2})^2 + \varepsilon_i/2} = \frac{1 + \varepsilon_i - \sqrt{2\varepsilon_i}}{1 + \varepsilon_i + \sqrt{2\varepsilon_i}}. \tag{9.67}$$

As shown in Appendix B.7, typical values of ε_i are very large in the limit of $\omega \ll \gamma$. Therefore, we can expand equation (9.67) in power series of $\varepsilon_i^{-1/2}$. Neglecting all terms

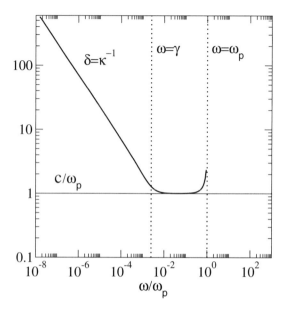

Figure 9.4. Frequency dependence of the skin depth, $\delta = \kappa^{-1}$, of a metal. In the region $\omega \ll \gamma$ the skin depth $\delta \propto \omega^{-1/2}$, in agreement with formula (9.71). Close to the plasma frequency, when $\gamma \ll \omega < \omega_p$, the skin depth is independent of the frequency and $\delta = c/\omega_p$, in agreement with equation (9.87).

higher than the first one, we obtain

$$R \approx 1 - 2\sqrt{\frac{2}{\varepsilon_i}}. \tag{9.68}$$

We find that the reflection of a metallic interface at frequencies $\omega \ll \gamma$ is close to unity.

Inside the metal, the electric field decreases exponentially as

$$E_2 \sim e^{-\kappa z}. \tag{9.69}$$

The electric field penetrates the metal up to a depth of length δ, called the skin depth,

$$\delta = \frac{1}{\kappa}. \tag{9.70}$$

With the use of expressions (9.65) and (9.61) we obtain that

$$\delta = \frac{c}{\sqrt{2\pi\sigma\omega}}. \tag{9.71}$$

By neglecting the frequency dependence of the conductivity, we observe the typical ω dependence of the skin depth as

$$\delta \sim \frac{1}{\sqrt{\omega}}, \tag{9.72}$$

shown in figure 9.4.

We can calculate also the Poynting vector inside the metal. We start with the general expression for the Poynting vector, equation (B.60),

$$S = \frac{c}{8\pi} \, \text{Re} \, [E_2 H_2^*]. \tag{9.73}$$

We need to express the electric and magnetic fields inside the metal. The electric field is related to the electric field E_1 of the incident wave by $E_2 = t_s E_1$. The transmission amplitude t_s can be easily calculated from equation (9.25). After simple algebra, t_s is given by

$$t_s = \frac{2}{1 + (1+i)\sqrt{\varepsilon_i/2}} \approx \frac{2}{(1+i)\sqrt{\varepsilon_i/2}}. \tag{9.74}$$

By the use of equation (9.18), we can express the magnetic field H_2 in terms of E_2,

$$H_2 = -\frac{ck_2}{\omega} E_2. \tag{9.75}$$

Inserting the above expressions in equation (9.73), and using $|t_s|^2 \approx 4/\varepsilon_i$, we obtain that the Poynting vector inside the metal is

$$S_2 = \frac{c^2}{8\pi\omega} \text{Re} \, k_2 |t_s|^2 |E_1|^2 = \frac{c}{8\pi} \sqrt{\frac{\varepsilon_i}{2}} \frac{4}{\varepsilon_i} |E_1|^2 \approx \frac{c}{8\pi} \frac{4}{\sqrt{2\varepsilon_i}} |E_1|^2. \tag{9.76}$$

Notice that inside the metal S_2 is not zero but decreases exponentially as a function of distance from the interface. This means that some portion of the energy passes through the interface inside the metal. Since no wave propagation is possible in a metal, the energy is absorbed in the surface layer of thickness equal to the skin depth δ.

Since the Poynting vector outside the metal is given by the expression

$$S_1 = \frac{c}{8\pi} |E_1|^2, \tag{9.77}$$

we can easily write down the expression for the transmission coefficient as

$$T = \frac{S_2}{S_1} = \frac{4}{\sqrt{2\varepsilon_i}}. \tag{9.78}$$

T represents the portion of the energy transfered through the interface. We have previously calculated the reflection coefficient R. Using the expression (9.68) we immediately see that

$$T + R = 1 \tag{9.79}$$

which guarantees the conservation of energy.

9.4.2 High-Frequency Limit $\omega \gg \gamma$

In the limit

$$\gamma \ll \omega < \omega_p, \tag{9.80}$$

we have that the real part of the permittivity of the metal, ε_r, is negative. Its absolute value $|\varepsilon_r|$ is much larger than the imaginary part $\varepsilon_i \ll |\varepsilon_r|$ [equation (B.99)]. The refractive index

n_m can be expressed as

$$n_m = \sqrt{\varepsilon_m} = [\varepsilon_r + i\varepsilon_i]^{1/2} \approx \sqrt{\varepsilon_r}\left[1 + i\frac{\varepsilon_i}{\varepsilon_r}\right]^{1/2}. \tag{9.81}$$

Since $\varepsilon_r < 0$, we have $\sqrt{\varepsilon_r} = i\sqrt{|\varepsilon_r|}$. Using the approximation $\sqrt{1+x} \approx 1 + x/2$ for $x \ll 1$, we obtain that

$$n_m \approx i\sqrt{|\varepsilon_r|}\left[1 - i\frac{\varepsilon_i}{2|\varepsilon_r|}\right] = i\sqrt{|\varepsilon_r|} + \frac{\varepsilon_i}{2\sqrt{|\varepsilon_r|}}. \tag{9.82}$$

With the help of the relations

$$\varepsilon_r \approx 1 - \frac{\omega_p^2}{\omega^2}, \tag{9.83}$$

given by equation (B.97) and $\varepsilon_i/|\varepsilon_r| \approx \gamma/\omega$ given by equation (B.99), we finally obtain

$$n \approx i\frac{\omega_p}{\omega} - \frac{\gamma\omega_p}{2\omega^2}. \tag{9.84}$$

Again, we consider only the case of normal incidence, so that the wave vector $k_2 = k_{2z} = k_{2r} + i\kappa$ and the real part of the wave vector, k_{2r}, is given by

$$k_{2r} = \frac{\omega}{c}\,\text{Re}\,n = \frac{\omega_p\gamma}{2\omega c}. \tag{9.85}$$

We see that the real part of the wave vector *decreases* when the frequency ω increases.

The imaginary part of k_2 is given by

$$\kappa = \frac{\omega}{c}\,\text{Im}\,n = \frac{\omega_p}{c}. \tag{9.86}$$

The skin depth in this case is given by

$$\delta = \frac{1}{\kappa} \approx \frac{c}{\omega_p}. \tag{9.87}$$

The skin depth does not depend on the frequency, as can be seen also in figure 9.4.

Note that

$$k_{2r} = \frac{\gamma}{2\omega}\kappa \ll \kappa. \tag{9.88}$$

The reflection coefficient can be calculated in the same way as in the previous section. Inserting k_{2r} and κ into equation (9.26), we obtain the reflection coefficient as

$$R = \left|\frac{k_1 - k_{2r} - i\kappa}{k_1 + k_{2r} + i\kappa}\right|^2 = \frac{\left(1 - \frac{\omega_p\gamma}{2\omega^2}\right)^2 + \frac{\omega_p^2}{\omega^2}}{\left(1 + \frac{\omega_p\gamma}{2\omega^2}\right)^2 + \frac{\omega_p^2}{\omega^2}}. \tag{9.89}$$

We Taylor-expand the last expression in powers of γ/ω ($\gamma/\omega \ll 1$). After some algebra, we get

$$R \approx 1 - \frac{2\omega_p\gamma}{\omega^2 + \omega_p^2}. \tag{9.90}$$

The transmission coefficient can be obtained directly from the general expression (9.29),

$$T = \frac{k_{2r}}{k_1} |t_s|^2,$$ (9.91)

where the transmission amplitude t_s is given by equation (9.25),

$$t_s = \frac{k_1}{k_1 + k_{2r} + i\kappa}.$$ (9.92)

We insert k_{2r} from equation (9.85) and κ from equation (9.86) in equation (9.91) and find, to the lowest order in γ, that

$$T = \frac{2\omega_p \gamma}{\omega^2 + \omega_p^2}.$$ (9.93)

Combination of equations (9.90) and (9.93) again confirms that $T + R = 1$.

9.5 Total Transmission

Now we will apply our general formulas for the transmission and reflection coefficients to find the conditions under which the interface is totally transparent. Since we have no absorption, we also have

$$T + R = 1.$$ (9.94)

The parameters for which total transmission appears can be obtained by solving the equations

$$r_s = 0$$ (9.95)

for the TE mode, and

$$r_p = 0$$ (9.96)

for the TM mode.

First, we calculate the conditions of zero reflectance of an interface between two regular dielectrics. The usual result for the Brewster angle will be recovered, where only the p (TM) polarization gives zero reflection. However, if we have an interface between a dielectric and a *left-handed medium*,[1] we can have zero reflection for both the TE and TM polarizations. This is a very important result, since we have perfect transmission for both polarizations of electromagnetic waves. This idea can be used to improve the performance of lasers. If the wells of a cavity are fabricated from a left-handed medium, no reflection will occur for either polarization and the efficiency of the laser will be improved dramatically.

---

[1] Left-handed materials will be discussed in chapter 16. For our purposes here we remind the reader that left-handed materials have both permittivity ε and permeability μ negative.

9.5.1 Brewster Angles

There are special angles of incidence for which the reflection amplitudes

$$r_s = \frac{z_{21} \cos \theta_1 - \cos \theta_2}{z_{21} \cos \theta_1 + \cos \theta_2} \quad \text{or} \quad \tilde{r}_p = \frac{z_{12} \cos \theta_1 - \cos \theta_2}{z_{12} \cos \theta_1 + \cos \theta_2} \tag{9.97}$$

given by equations (9.26) and (9.56), respectively, are zero. These angles are called Brewster angles. The TE (TM) wave incident on an interface with Brewster angle θ_{Bs} (θ_{Bp}) is totally transmitted through the interface.

It is easy to find the analytical expression for the Brewster angles for both TE and TM polarizations. For instance, for the TE polarization, the Brewster angle θ_{Bs} satisfies the equation

$$z_{21} \cos \theta_1 - \cos \theta_2 = 0. \tag{9.98}$$

We can write equation (9.98) as

$$z_{21}^2 \cos^2 \theta_1 = \cos^2 \theta_2, \tag{9.99}$$

and with the help of Snell's law, $n_1 \sin \theta_1 = n_2 \sin \theta_2$, which can also be written as

$$\frac{n_1^2}{n_2^2} \sin^2 \theta_1 = \sin^2 \theta_2, \tag{9.100}$$

we can derive an analytic expression for the Brewster angle. Remember that $z_{21} = \sqrt{\mu_2 \varepsilon_1 / \mu_1 \varepsilon_2}$, $n_1 = \sqrt{\varepsilon_1 \mu_1}$, and $n_2 = \sqrt{\varepsilon_2 \mu_2}$. Then, combining equations (9.99) and (9.100), we obtain an equation for the incident angle that gives zero reflection:

$$\frac{\mu_2}{\mu_1} \frac{\varepsilon_1}{\varepsilon_2} \cos^2 \theta_1 + \frac{\mu_1}{\mu_2} \frac{\varepsilon_1}{\varepsilon_2} \sin^2 \theta_1 = 1. \tag{9.101}$$

One immediately sees that, if $\mu_1 = \mu_2$, then equation (9.101) has only trivial solution $\varepsilon_1 \equiv \varepsilon_2$ (absence of the interface). Thus we see that for the s (TE) polarization no Brewster angle exists for an interface of two dielectrics. However, if $\mu_2 \neq \mu_1$, we solve equation (9.101) and get the Brewster angle θ_{Bs} as

$$\cos^2 \theta_{Bs} = \frac{\varepsilon_2 / \varepsilon_1 - \mu_1 / \mu_2}{\mu_2 / \mu_1 - \mu_1 / \mu_2}. \tag{9.102}$$

In the same way we obtain the expression for the Brewster angle θ_{Bp} for the p-(TM-)polarized wave,

$$\cos^2 \theta_{Bp} = \frac{\mu_2 / \mu_1 - \varepsilon_1 / \varepsilon_2}{\varepsilon_2 / \varepsilon_1 - \varepsilon_1 / \varepsilon_2}. \tag{9.103}$$

For nonmagnetic materials, $\mu_1 = \mu_2$, the formula (9.103) reduces to the well-known result

$$\tan \theta_{Bp} = \frac{n_2}{n_1}. \tag{9.104}$$

Total transmission of the TM wave is easily observable in any interface between two dielectrics.

If we want to have zero reflection for both TE- and TM-polarized waves at the same incidence angle θ_1, we must satisfy both equations

$$z_{21} \cos \theta_1 - \cos \theta_2 = 0 \tag{9.105}$$

and

$$z_{12} \cos \theta_1 - \cos \theta_2 = 0 \tag{9.106}$$

simultaneously. This is possible only when the two conditions

$$z_{21} = 1 \quad \text{and} \quad \theta_1 = \pm \theta_2 \tag{9.107}$$

are satisfied simultaneously. The second condition means that the absolute value of the index of refraction of both materials is the same,

$$|n_1| = |n_2|. \tag{9.108}$$

It is easy to check that both these requirements cannot be satisfied simultaneously for dielectric materials.

9.5.2 Total Transmission for Oblique Angles

There is also another possibility for achieving perfect transmission. We can choose the electromagnetic parameters of both media in such a way that the reflection amplitudes are zero *for any incident angle*.

From equation (9.101) we see that total transmission of the TE wave appears if

$$\frac{\mu_2 \, \varepsilon_1}{\mu_1 \, \varepsilon_2} = \frac{\mu_1 \, \varepsilon_1}{\mu_2 \, \varepsilon_2} = 1. \tag{9.109}$$

This result is quite general: total transmission for both polarizations is possible only through the interface between two media, which have both relative indices of refraction

$$n_{21} = \frac{n_2}{n_1} = \sqrt{\frac{\varepsilon_2 \mu_2}{\varepsilon_1 \mu_1}} \tag{9.110}$$

and the surface impedance

$$z_{21} = \frac{\mu_2 \varepsilon_1}{\mu_1 \varepsilon_2} \tag{9.111}$$

equal to unity. Both requirements are satisfied for the trivial case of two identical media, $\varepsilon_1 = \varepsilon_2$ and $\mu_1 = \mu_2$ (this means in fact the absence of any interface). However, another solution

$$\mu_1 = -\mu_2 \tag{9.112}$$

and

$$\varepsilon_1 = -\varepsilon_2 \tag{9.113}$$

is possible if the second medium is left handed. In particular, if we choose the first medium to be air, $\varepsilon_1 = \mu_1 = 1$, we obtain

$$\varepsilon_2 = -1 \quad \text{and} \quad \mu_2 = -1, \tag{9.114}$$

which is the case of an ideal left-handed medium. This important result states that all incident waves striking the surface between vacuum and a left-handed medium with $\varepsilon = -1$ and $\mu = -1$ will have zero reflection for any incident angle.

9.6 Total Reflection

Consider an interface between two dielectrics with $\mu_1 = \mu_2 = 1$. If the refractive index n_2 of the second medium is smaller than the refractive index of the first medium,

$$n_1 > n_2, \tag{9.115}$$

then we can find the critical angle of incidence, θ_{crit}, such that no transmission through the interface is possible for waves with incident angles $\theta > \theta_{\text{crit}}$. This observation follows directly from Snell's law,

$$\frac{\sin \theta_1}{\sin \theta_2} = \frac{n_2}{n_1}. \tag{9.116}$$

We see that, if $n_1 > n_2$, then the reflected angle θ_2 is always larger than θ_1. As the maximum value of the refracted angle is $\theta_2 = \pi/2$, it is evident that equation (9.116) has no real solution for θ_2, if

$$\sin \theta_1 > \sin \theta_{\text{crit}} = \frac{n_2}{n_1}. \tag{9.117}$$

From the previous equation we see that the critical angle θ_{crit} satisfies the equation

$$\frac{n_1}{n_2} \sin \theta_{\text{crit}} = 1. \tag{9.118}$$

Any incident wave with angle $\theta_1 > \theta_{\text{crit}}$ is totally reflected. One of the many application of the total internal reflection is the transmission of light through fibers.

Total reflection, however, does not mean that the field is exactly zero on the opposite side of the interface. To calculate the field in the second medium, we start with the dispersion relations. In the first medium we have the following dispersion relation

$$\frac{\omega^2}{c^2} \varepsilon_1 = k_x^2 + k_{1z}^2, \tag{9.119}$$

where

$$k_x = \frac{\omega}{c} \sqrt{\varepsilon_1} \sin \theta_1. \tag{9.120}$$

In the second medium, we have

$$k_{2z} = i\kappa_2, \quad \kappa_2 > 0, \tag{9.121}$$

and the dispersion relation is

$$\frac{\omega^2}{c^2} \varepsilon_2 = k_x^2 - \kappa_2^2. \tag{9.122}$$

By solving equations (9.119) and (9.122) for κ_2 and substituting equation (9.120), we obtain

$$\kappa_2^2 = \frac{\omega^2}{c^2} \left[\varepsilon_1 \sin^2 \theta_1 - \varepsilon_2 \right]. \tag{9.123}$$

It might be more suitable to express κ_2 in units of the wavelength of the incident wave,

$$\lambda_1 = \frac{2\pi}{k_1} = \frac{2\pi c}{\omega \sqrt{\varepsilon_1}}, \tag{9.124}$$

so that

$$\kappa_2 = \frac{2\pi}{\lambda_1} \sqrt{\sin^2 \theta_1 - \frac{\varepsilon_2}{\varepsilon_1}}. \tag{9.125}$$

Alternatively, we have

$$\kappa_2 = \frac{2\pi}{\lambda_2} \sqrt{\frac{\varepsilon_1}{\varepsilon_2} \sin^2 \theta_1 - 1}, \quad \lambda_2 = \frac{2\pi}{k_2} = \frac{2\pi c}{\omega \sqrt{\varepsilon_2}}. \tag{9.126}$$

Note that κ_2 is indeed real for $\theta_1 > \theta_{\text{crit}}$.

Now, we calculate the transmission and reflection amplitudes for the total reflection case. We used the general form of the transfer matrix given by equation (9.22) with $k_{2z} = i\kappa_2$. We have that $\det \mathbf{M}^{(s)} = k_{1z}/(i\kappa_2)$. From equations (9.25) and (9.26) we obtain

$$t_s = \frac{2i\kappa}{k_{1z} + i\kappa} \tag{9.127}$$

and

$$r_s = \frac{k_{1z} - i\kappa_2}{k_{1z} + i\kappa_2}. \tag{9.128}$$

Note that equations (9.127) and (9.128) are formally equivalent to the expressions (2.118) derived in chapter 2. We can also show that

$$R_s = |r_s|^2 = 1, \tag{9.129}$$

since r_s is the ratio of two complex numbers with the same absolute value.

Since $k_{2z} = k_2 = i\kappa_2$ is purely imaginary in this case, and the Poynting vector is proportional to the real part of k_{2z}, we immediately conclude that no energy is transferred through the interface. We conclude, in agreement with our result $R = 1$, that the energy of the incident wave is totally reflected back to the first medium.

9.6.1 Evanescent Waves

Since the transmission amplitude t_s is not zero, the electric field inside the second medium is nonzero,

$$E_2^+ = t_s E_1^+. \tag{9.130}$$

We remind the reader that continuity of the electric field at the interface requires that $\vec{E}_2 = (0, E_{2y}, 0)$. By using the Maxwell equation (B.17),

$$\vec{H}_2 = \frac{c}{\mu_2 \omega} \vec{k}_2 \times \vec{E}_2, \tag{9.131}$$

we can express the components of the magnetic field inside the second medium with respect to the electric field components,

$$H_{2z} = \frac{ck_{2x}}{\mu_2 \omega} E_{2y}, \qquad H_{2x} = \frac{ck_{2z}}{\mu_2 \omega} E_{2y}. \tag{9.132}$$

All the fields decrease exponentially inside the second medium, $\sim e^{-\kappa_2 z}$, along the propagation direction. Since $k_{2z} = i\kappa$ is purely imaginary, the Poynting vector has no component normal to the interface,

$$S_{2z} \equiv 0. \tag{9.133}$$

Therefore, the transmission coefficient $T = 0$. This is consistent with equation (9.129). However, the x-component of the Poynting vector is not zero:

$$S_{2x} = \frac{c}{8\pi} \text{Re} \left[E_{2y} H_{2z}^* \right] = \frac{c^2}{8\pi \mu_2 \omega} |t_s|^2 |E_1^+|^2 k_{1x} e^{-2\kappa z}. \tag{9.134}$$

S_{2x} represents the energy flow parallel to the interface.

We call such waves *evanescent waves*. They play an important role in the effect of tunneling (section 10.2.3). They are also responsible for the excitation of surface waves, as will be discussed in section 11.3.

9.7 Problems

Problems with Solutions

Problem 9.1
Show that

$$t_s - r_s = 1. \tag{9.135}$$

Solution. Relation (9.135) follows from the boundary conditions for the TE wave, given by equation (9.15), where we used $E_2^- = 0$ and $E_2^+ = t_s E_1^+$ and $E_1^- = r_s E_1^+$. The last two relationships follow from the definition of t_s and r_s, given by equations (9.24). Similarly, we obtain for the TM wave that

$$\tilde{t}_p - \tilde{r}_p = 1. \tag{9.136}$$

Problem 9.2
In section 9.2.2 we derived the transfer matrix $\tilde{\mathbf{M}}^{(p)}$ that expresses the magnetic fields H_2^+ and H_2^-, in terms of fields H_1^+ and H_1^-. Derive the matrix $\mathbf{M}^{(p)}$ that relates the electric fields of the TM-polarized wave scattered at the interface. Calculate the transmission coefficient in terms of the electric fields.

Solution. We use the relation between the amplitudes of the electric and magnetic fields, given by equation (B.25),

$$E = \sqrt{\frac{\mu}{\varepsilon}} H. \tag{9.137}$$

Inserting this in equation (9.38), we obtain

$$\mathbf{M}^{(p)} = \sqrt{\frac{\varepsilon_2 \, \mu_1}{\mu_2 \, \varepsilon_1}} \tilde{\mathbf{M}}^{(p)}. \tag{9.138}$$

We can also introduce the transmission and reflection amplitudes, $t_p = E_2^+ / E_1^+$, and $r_p = E_1^- / E_1^+$. Clearly,

$$t_p = \sqrt{\frac{\mu_2 \, \varepsilon_1}{\varepsilon_2 \, \mu_1}} \tilde{t}_p \quad \text{and} \quad r_p = \tilde{r}_p. \tag{9.139}$$

The transmission coefficient is given by

$$T_p = \frac{\mu_1}{\mu_2} \frac{\operatorname{Re} k_{2z}}{\operatorname{Re} k_{1z}} |t_p|^2. \tag{9.140}$$

[equation (9.29)]. Using the relation (9.139) between transmission amplitudes t_p and $\tilde{t}_p$, we see that expression (9.140) is identical with equation (9.43).

Problem 9.3
Use the data for the plasma frequency ω_p and for the inverse scattering time γ given in table B.1 to estimate the transmission coefficients for the electromagnetic wave at the interface between a vacuum and a metal.

Solution. In the low-frequency limit, we have that the transmission coefficient is given by equation (9.78),

$$T = 2\sqrt{\frac{2}{\varepsilon_i}} \approx 2\sqrt{2} \sqrt{\frac{\gamma \omega}{\omega_p^2}} \tag{9.141}$$

[we here inserted the frequency dependence of ε_i, given by equation (B.92) in equation (9.141)]. In the microwave region $\omega \sim 2\pi \times 10^{10}$ Hz, we use data from the table B.1 to estimate the ratios $\omega/\omega_p \sim 10^{-5}$ and $\gamma/\omega_p \sim 10^{-3}$. Inserting this in equation (9.141) we obtain

$$T \approx 2\sqrt{2} \times 10^{-4} \sim 10^{-3}. \tag{9.142}$$

so that approximately 0.1% of all energy of the incident electromagnetic wave is transmitted through the interface and absorbed in the metal. The rest of the energy is reflected back.

In the high-frequency limit, we have that ω is of the same order of magnitude as the plasma frequency ω_p. Thus, the transmission coefficient, given by equation (9.93),

$$T \approx \frac{2\omega_p \gamma}{\omega^2 + \omega_p^2} \approx \frac{\gamma}{\omega_p} \frac{2}{1 + (\omega/\omega_p)^2}. \tag{9.143}$$

Since $\gamma/\omega_p \sim 10^{-3}$ and $\omega < \omega_p$, we obtain again that $T \approx 0.001$.

Problems without Solutions

Problem 9.4
Show that in the absence of absorption

$$T_s + R_s = 1 \tag{9.144}$$

and

$$T_p + R_p = 1. \tag{9.145}$$

The above equations represent energy conservation. Since there is no absorption for real ε and μ, the wave is either transmitted through the interface, or reflected back.

Problem 9.5
Apply the general formulas for the transmission and reflection amplitudes for the special case with $\mu_1 = \mu_2 = 1$ and $\varepsilon_1 > 0$, $\varepsilon_2 > 0$, which corresponds to a dielectric material without a magnetic response. In this case, $z_{21} = n_1/n_2$, $z_{12} = n_1/n_2$ where $n_1 = \sqrt{\varepsilon_1}$ and $n_2 = \sqrt{\varepsilon_2}$ are the refractive indices of media 1 and 2, respectively. Show that in this case

$$
\begin{aligned}
t_s &= \frac{2n_1 \cos\theta_1}{n_1 \cos\theta_1 + n_2 \cos\theta_2}, \\[2mm]
r_s &= \frac{n_1 \cos\theta_1 - n_2 \cos\theta_2}{n_1 \cos\theta_1 + n_2 \cos\theta_2}, \\[2mm]
\tilde{t}_p &= \frac{2n_2 \cos\theta_1}{n_2 \cos\theta_1 + n_1 \cos\theta_2}, \\[2mm]
\tilde{r}_p &= \frac{n_2 \cos\theta_1 - n_1 \cos\theta_2}{n_2 \cos\theta_1 + n_1 \cos\theta_2}.
\end{aligned}
\tag{9.146}
$$

These are the well-known Fresnel equations [2, 39].

Problem 9.6
Plot the transmission and the reflection amplitudes, given by equation (9.146), as a function of the incident angle θ_1 for various values of the refractive indices n_1 and n_2 (figure 9.5).

Problem 9.7
For the case $\mu_1 = \mu_2 = 1$ and $\varepsilon_1 > 0$, $\varepsilon_2 > 0$ express the transmissions T_s and T_p as functions of incident angle θ_1:

$$T_s = \frac{2n_1 n_2 \cos\theta_1 \cos\theta_2}{(n_1 \cos\theta_1 + n_2 \cos\theta_2)^2} \tag{9.147}$$

and

$$T_p = \frac{2n_1 n_2 \cos\theta_1 \cos\theta_2}{(n_2 \cos\theta_1 + n_1 \cos\theta_2)^2} \tag{9.148}$$

(figure 9.6).

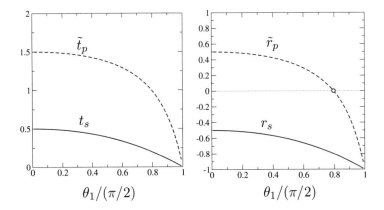

FIGURE 9.5. Transmission (left panel) and reflection (right panel) amplitudes, given by equation (9.146) as a function of incident angle θ_1 for the case with $n_1 = 1$ and $n_2 = 3$. Note that, $\tilde{r}_p = 0$ for $\theta_1 = \theta_{Bp}$.

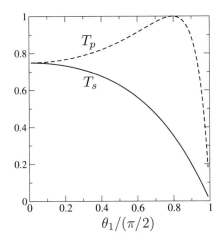

FIGURE 9.6. Transmission T_s and T_p, given by equations (9.147) and (9.148), respectively, as a function of incident angle θ_1 for the case with $n_1 = 1$ and $n_2 = 3$.

Problem 9.8
For the special case of normal incidence, $\theta_1 = 0$ and $\theta_2 = 0$. Note that, both E^+ and H^+ are parallel with the incident plane. Derive the formulas for the transmission and the reflection amplitudes:

$$
t_s = \frac{2n_1}{n_1 + n_2}, \quad \tilde{t}_p = \frac{2n_2}{n_1 + n_2},
$$

$$
r_s = -\tilde{r}_p = \frac{n_1 - n_2}{n_1 + n_2}.
$$

(9.149)

Note that the two reflection amplitudes r_s and $\tilde{r}_p$ have opposite signs.

Problem 9.9

Use Snell's law

$$\frac{n_1}{n_2} = \frac{\sin\theta_2}{\sin\theta_1} \tag{9.150}$$

and expressions (9.146) and (9.139) to derive another expression for the transmission and reflection amplitudes,

$$t_s = \frac{2\sin\theta_2\cos\theta_1}{\sin(\theta_2+\theta_1)}, \qquad r_s = \frac{\sin(\theta_2-\theta_1)}{\sin(\theta_2+\theta_1)},$$

$$t_p = \frac{2\sin\theta_2\cos\theta_1}{\sin(\theta_2+\theta_1)\cos(\theta_2-\theta_1)}, \qquad r_p = \frac{\tan(\theta_2-\theta_1)}{\tan(\theta_2+\theta_1)}. \tag{9.151}$$

Problem 9.10

For the special case $\mu_1 = \mu_2$, derive the relation for the Brewster angle θ_{Bp},

$$\tan\theta_{Bp} = \frac{n_2}{n_1} = \sqrt{\frac{\varepsilon_2}{\varepsilon_1}}. \tag{9.152}$$

Use Snell's law to show that, if $\theta_1 = \theta_{Bp}$, then

$$\theta_1 + \theta_2 = \pi/2. \tag{9.153}$$

Note that equation (9.153) follows directly from the expression (9.151) for r_p and from the requirement $r_p = 0$.

10 Transmission and Reflection Coefficients for a Slab

Consider now a slab of thickness ℓ with permittivity ε_2 and permeability μ_2, located between two semi-infinite media with electromagnetic parameters (ε_1, μ_1) and (ε_3, μ_3), respectively. We want to calculate transmission and reflection amplitudes for a plane wave arriving from the left for both TE and TM polarizations. As in section 9.2, we assume that the permittivity and permeability of incoming and outgoing media are real. This might not be true for the parameters of the slab.

Transmission through a planar slab is schematically shown in figure 10.1. We see that the problem is more complicated than that of a single interface. Not only do we have more parameters, but we also need to consider multiple scattering inside the slab, as shown in figure 10.2. Fortunately, as we have seen in chapter 1, the transfer matrix technique accounts for all the contributions of multiple scattering very efficiently. On the other hand, multiple scattering inside the slab gives us some interesting phenomena, such as the Fabry-Pérot resonances in dielectric slabs.

We first derive a general formula for the transfer matrix for a slab. Then we study in detail the transmission of an electromagnetic wave through a dielectric slab. We find that the problem of transmission through a dielectric slab is very similar to the problem of propagation of a quantum particle through a rectangular potential. We will use this analogy to explain various interesting phenomena, such as resonant transmission and tunneling of an electromagnetic wave through a slab of material which has permittivity smaller than the permittivity of the embedding medium. Then we discuss transmission through a metallic slab. Some applications of the formulas obtained will be given in chapter 16, where we analyze transmission of an electromagnetic wave through a slab of a left-handed medium.

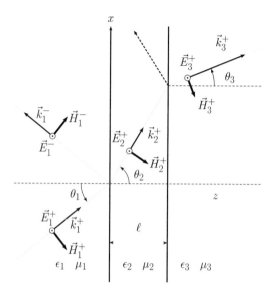

FIGURE 10.1. Schematic description of the transmission of an electromagnetic wave through a slab with width ℓ. The slab is characterized
by the parameters ε_2 and μ_2. The permittivity and the permeability of
the incoming and outgoing media are ε_1, μ_1 and ε_3, μ_3, respectively. The
orientations of the vectors of the electric and the magnetic fields are
shown for all waves propagating to the right and for the wave reflected
from the left boundary (E_1^-). The wave reflected from the right boundary
of the slab is not shown; only the direction of its propagation is depicted
by a dashed line. There is an infinite number of waves inside the slab.
Some are shown in figure 10.2.

10.1 Transmission and Reflection Amplitudes: TE and TM Modes

To find the transfer matrix for the propagation of wave through the slab, we use the
transfer matrix **M** for a single interface, derived in section 9.2, and the composition law
for transfer matrices, equation (1.76). In particular, for the TE polarization, we can write

$$\begin{pmatrix} E_3^+ \\ E_3^- \end{pmatrix} = \mathbf{M}^{12} \begin{pmatrix} e^{ik_{2z}\ell} & 0 \\ 0 & e^{-ik_{2z}\ell} \end{pmatrix} \mathbf{M}^{23} \begin{pmatrix} E_1^+ \\ E_1^- \end{pmatrix}, \tag{10.1}$$

where the matrices $\mathbf{M}^{12}$ and $\mathbf{M}^{23}$ are given by equation (9.47) and the diagonal matrix

$$\begin{pmatrix} e^{ik_{2z}\ell} & 0 \\ 0 & e^{-ik_{2z}\ell} \end{pmatrix} \tag{10.2}$$

is the transfer matrix for a homogeneous medium (ε_2, μ_2) between two interfaces.
Multiplication of the matrices leads to the relation

$$\begin{pmatrix} E_3^+ \\ E_3^- \end{pmatrix} = \mathbf{M}_{\text{slab}}^{(s)} \begin{pmatrix} E_1^+ \\ E_1^- \end{pmatrix}. \tag{10.3}$$

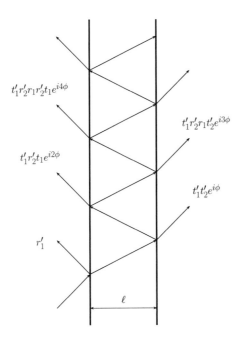

FIGURE 10.2. Multiple scattering of an electromagnetic wave inside a slab. The transmission and reflection amplitudes consist of an infinite number of contributions given by multiple scattering of the electromagnetic wave on the boundaries of the slab. $\phi = k\ell / \cos\theta_2$.

Explicit expressions for the elements of the matrix $\mathbf{M}^{(s)}_{\text{slab}}$ result from the matrix multiplications of equation (10.1):

$$M^{(s)}_{11} = \frac{1}{2}\left[1 + \frac{\mu_3}{\mu_1}\frac{k_{1z}}{k_{3z}}\right]\cos k_{2z}\ell + \frac{i}{2}\left[\frac{\mu_3}{\mu_2}\frac{k_{2z}}{k_{3z}} + \frac{\mu_2}{\mu_1}\frac{k_{1z}}{k_{2z}}\right]\sin k_{2z}\ell, \tag{10.4}$$

$$M^{(s)}_{12} = \frac{1}{2}\left[1 - \frac{\mu_3}{\mu_1}\frac{k_{1z}}{k_{3z}}\right]\cos k_{2z}\ell + \frac{i}{2}\left[\frac{\mu_3}{\mu_2}\frac{k_{2z}}{k_{3z}} - \frac{\mu_2}{\mu_1}\frac{k_{1z}}{k_{2z}}\right]\sin k_{2z}\ell, \tag{10.5}$$

and

$$M^{(s)}_{22} = \frac{1}{2}\left[1 + \frac{\mu_3}{\mu_1}\frac{k_{1z}}{k_{3z}}\right]\cos k_{2z}\ell - \frac{i}{2}\left[\frac{\mu_3}{\mu_2}\frac{k_{2z}}{k_{3z}} + \frac{\mu_2}{\mu_1}\frac{k_{1z}}{k_{2z}}\right]\sin k_{2z}\ell, \tag{10.6}$$

$$M^{(s)}_{21} = \frac{1}{2}\left[1 - \frac{\mu_3}{\mu_1}\frac{k_{1z}}{k_{3z}}\right]\cos k_{2z}\ell - \frac{i}{2}\left[\frac{\mu_3}{\mu_2}\frac{k_{2z}}{k_{3z}} - \frac{\mu_2}{\mu_1}\frac{k_{1z}}{k_{2z}}\right]\sin k_{2z}\ell. \tag{10.7}$$

Starting with the relation (10.3), we obtain the transmission and reflection amplitudes in the same way as for a single interface:

$$t_s = \frac{E_3^+}{E_1^+} = \frac{\det \mathbf{M}^{(s)}}{M^{(s)}_{22}}, \qquad r_s = \frac{E_1^-}{E_1^+} = -\frac{M^{(s)}_{21}}{M^{(s)}_{22}}, \tag{10.8}$$

and the corresponding transmission coefficient (9.29)

$$T_s = \frac{S_3}{S_1} = \frac{\mu_1}{\mu_3}\frac{\operatorname{Re} k_{3z}}{\operatorname{Re} k_{1z}}|t_s|^2 \tag{10.9}$$

and the reflection coefficient (9.30)

$$R_s = |r_s|^2. \tag{10.10}$$

Similarly, for the TM mode we obtain

$$\tilde{M}_{11}^{(p)} = \frac{1}{2}\left[1 + \frac{\varepsilon_3}{\varepsilon_1}\frac{k_{1z}}{k_{3z}}\right]\cos k_{2z}\ell + \frac{i}{2}\left[\frac{\varepsilon_3}{\varepsilon_2}\frac{k_{2z}}{k_{3z}} + \frac{\varepsilon_2}{\varepsilon_1}\frac{k_{1z}}{k_{2z}}\right]\sin k_{2z}\ell, \tag{10.11}$$

$$\tilde{M}_{12}^{(p)} = \frac{1}{2}\left[1 - \frac{\varepsilon_3}{\varepsilon_1}\frac{k_{1z}}{k_{3z}}\right]\cos k_{2z}\ell + \frac{i}{2}\left[\frac{\varepsilon_3}{\varepsilon_2}\frac{k_{2z}}{k_{3z}} - \frac{\varepsilon_2}{\varepsilon_1}\frac{k_{1z}}{k_{2z}}\right]\sin k_{2z}\ell, \tag{10.12}$$

and

$$\tilde{M}_{22}^{(p)} = \frac{1}{2}\left[1 + \frac{\varepsilon_3}{\varepsilon_1}\frac{k_{1z}}{k_{3z}}\right]\cos k_{2z}\ell - \frac{i}{2}\left[\frac{\varepsilon_3}{\varepsilon_2}\frac{k_{2z}}{k_{3z}} + \frac{\varepsilon_2}{\varepsilon_1}\frac{k_{1z}}{k_{2z}}\right]\sin k_{2z}\ell, \tag{10.13}$$

$$\tilde{M}_{21}^{(p)} = \frac{1}{2}\left[1 - \frac{\varepsilon_3}{\varepsilon_1}\frac{k_{1z}}{k_{3z}}\right]\cos k_{2z}\ell - \frac{i}{2}\left[\frac{\varepsilon_3}{\varepsilon_2}\frac{k_{2z}}{k_{3z}} - \frac{\varepsilon_2}{\varepsilon_1}\frac{k_{1z}}{k_{2z}}\right]\sin k_{2z}\ell. \tag{10.14}$$

Expressions (10.4)–(10.7) and (10.11)–(10.14) solve the problem of the transmission through a finite planar slab completely. Now we will discuss some special cases.

10.1.1 Slab Embedded in a Homogeneous Medium

The general formulas for the transfer matrix of a slab simplify considerably if the slab is embedded into one medium. In this special case, we have

$$\varepsilon_3 \equiv \varepsilon_1, \quad \mu_3 \equiv \mu_1 \tag{10.15}$$

and

$$k_{3z} = k_{1z}, \tag{10.16}$$

so that the incoming and outgoing plane waves propagate in the same direction. We present the matrix elements for the TE mode,

$$M_{11}^{(s)} = \cos k_{2z}\ell + \frac{i}{2}\left[\frac{\mu_1}{\mu_2}\frac{k_{2z}}{k_{1z}} + \frac{\mu_2}{\mu_1}\frac{k_{1z}}{k_{2z}}\right]\sin k_{2z}\ell, \tag{10.17}$$

$$M_{12}^{(s)} = \frac{i}{2}\left[\frac{\mu_1}{\mu_2}\frac{k_{2z}}{k_{1z}} - \frac{\mu_2}{\mu_1}\frac{k_{1z}}{k_{2z}}\right]\sin k_{2z}\ell, \tag{10.18}$$

$$M_{22}^{(s)} = \cos k_{2z}\ell - \frac{i}{2}\left[\frac{\mu_1}{\mu_2}\frac{k_{2z}}{k_{1z}} + \frac{\mu_2}{\mu_1}\frac{k_{1z}}{k_{2z}}\right]\sin k_{2z}\ell, \tag{10.19}$$

$$M_{21}^{(s)} = -\frac{i}{2}\left[\frac{\mu_1}{\mu_2}\frac{k_{2z}}{k_{1z}} - \frac{\mu_2}{\mu_1}\frac{k_{1z}}{k_{2z}}\right]\sin k_{2z}\ell. \tag{10.20}$$

Note that, in this case,

$$\det \mathbf{M}^{(s)} = 1. \tag{10.21}$$

The transmission amplitude reads

$$t_s = \frac{1}{M_{22}}$$

(10.22)

and the reflection amplitude is

$$r_s = -\frac{M_{12}}{M_{22}}.$$

(10.23)

Similar formulas could be derived also for the TM-polarized wave.

Note that, the transmission coefficient, given by equation (9.29), reduces to the simple expression

$$T = |t_s|^2,$$

(10.24)

equivalent to the transmission coefficient for a quantum particle described in chapter 1. The origin of this simplicity lies in the equivalence of the left and right media. The incoming and outgoing waves propagate in the same medium.

We remind the reader that all the above formulas are valid also for the case when the wave vector k_{2z} *inside* the slab is complex. However, k_{1z} should be real in equation (10.24). We find imaginary values of k_{1z} in the case of bound states inside the slab. In this particular case, the transmission coefficient is zero.

10.2 Dielectric Slab Embedded in Vacuum

Consider now the more specific problem of transmission of an electromagnetic wave through a dielectric slab embedded in vacuum. The slab thickness is ℓ. We assume that the permittivity ε_2 is real and define the permeability $\mu_2 = 1$. This means

$$\mu_1 = \mu_2 = \mu_3 = 1$$

(10.25)

and

$$\varepsilon_1 = \varepsilon_3.$$

(10.26)

In this case, the elements of the transfer matrix, given by equations (10.17)–(10.20), further reduce to

$$M_{11}^{(s)} = \cos k_{2z}\ell + \frac{i}{2}\left[\frac{k_{2z}}{k_{1z}} + \frac{k_{1z}}{k_{2z}}\right]\sin k_{2z}\ell,$$

(10.27)

$$M_{12}^{(s)} = \frac{i}{2}\left[\frac{k_{2z}}{k_{1z}} - \frac{k_{1z}}{k_{2z}}\right]\sin k_{2z}\ell,$$

(10.28)

$$M_{22}^{(s)} = \cos k_{2z}\ell - \frac{i}{2}\left[\frac{k_{2z}}{k_{1z}} + \frac{k_{1z}}{k_{2z}}\right]\sin k_{2z}\ell,$$

(10.29)

$$M_{21}^{(s)} = -\frac{i}{2}\left[\frac{k_{2z}}{k_{1z}} - \frac{k_{1z}}{k_{2z}}\right]\sin k_{2z}\ell.$$

(10.30)

Note that, in the case when both k_{1z} and k_{2z} are real, we have $M_{22} = M_{11}^*$ and $M_{21} = M_{12}^*$. The formulas obtained are formally equivalent to the transfer matrix for the rectangular potential, derived in chapter 2. To understand the analogy between these two problems in detail, we write the dispersion relations for the electromagnetic wave outside and inside the slab:

$$\frac{\omega^2}{c^2}\varepsilon_1 = k_x^2 + k_{1z}^2 \quad \text{and} \quad \frac{\omega^2}{c^2}\varepsilon_2 = k_x^2 + k_{2z}^2. \tag{10.31}$$

Subtracting the second equation from the first one gives us the equation

$$k_{1z}^2 - k_{2z}^2 = \frac{\omega^2}{c^2}[\varepsilon_1 - \varepsilon_2], \tag{10.32}$$

which is formally equivalent to the relationship between the wave vectors of a quantum particle,

$$k^2 - k'^2 = \frac{2mV_0}{\hbar}, \tag{10.33}$$

which follows from equations (2.7) and (2.8).

From the two equations (10.32) and (10.33), we see that the difference between permittivities, $\varepsilon_1 - \varepsilon_2$, plays the role of the potential step in the quantum problem. When the permittivity of the slab is larger than the permittivity of the embedding medium, we have a problem analogous to the problem of a quantum particle in a potential well ($V_0 < 0$). By this analogy, we expect to observe resonant transmission for special frequencies, as well as a bound state, which propagates inside the slab but decays exponentially in the embedding medium. In the opposite case, $\varepsilon_2 < \varepsilon_1$, the problem is analogous to the problem of a quantum particle propagating through a potential barrier. Again, we might have resonant transmission for propagating solutions and tunneling of the electromagnetic wave through the slab.

To complete the analogy between the two phenomena, we also need to determine the quantity that corresponds to the energy of a quantum particle. To do so, we rewrite equations (2.4) and (2.5) in the form

$$k^2 = \frac{2mE}{\hbar^2}, \qquad k'^2 = \frac{2m(E - V_0)}{\hbar^2}, \tag{10.34}$$

and compare them with the dispersion relations equation (10.31):

$$k_{1z}^2 = \frac{\omega^2}{c^2}\varepsilon_1 - k_x^2, \qquad k_{2z}^2 = \frac{\omega^2}{c^2}\varepsilon_2 - k_x^2. \tag{10.35}$$

We see that the energy $2mE/\hbar^2$ is equivalent to $\omega^2\varepsilon_1/c^2 - k_x^2$.

In contrast to the electron, which propagates only in a one-dimensional system, an electromagnetic wave has two degrees of freedom—it propagates not only in the direction perpendicular to the interface of the slab, but also in the direction parallel to the slab. This additional degree of freedom makes the problem slightly more complicated. On the other hand, propagation in the x-direction is a necessary condition for the observation of bound states, since bound states in quantum mechanics have negative energy. We do not have negative frequency in the electromagnetic problem, but we can have a situation

Table 10.1. Correspondence between the various parameters used in the analysis of the rectangular potential barrier and those used in the analysis of the transmission through the dielectric slab.

Rectangular Potential	Dielectric Slab
$\dfrac{2mE}{\hbar^2}$	$\dfrac{\omega^2}{c^2}\varepsilon_1 - k_x^2$
$\dfrac{2mV_0}{\hbar^2}$	$\dfrac{\omega^2}{c^2}(\varepsilon_1 - \varepsilon_2)$
$\dfrac{E}{V_0}$	$\dfrac{[\omega^2/c^2\varepsilon_1] - k_x^2}{[\omega^2/c^2](\varepsilon_1 - \varepsilon_2)}$
β^2	$\dfrac{\omega^2\ell^2}{4c^2}|\varepsilon_2 - \varepsilon_1|$

in which $k_x^2 > \omega^2/c^2$. Therefore, bound states cannot be observed without simultaneous propagation of the wave along the boundary.

Another difference between a quantum particle and an electromagnetic wave is that the electromagnetic wave possesses two polarizations. The analogy between the two phenomena works only for the TE polarization. For the TM polarization, the transfer matrix is of the form

$$\tilde{M}_{11}^{(p)} = \cos k_{2z}\ell + \frac{i}{2}\left[\frac{\varepsilon_1\,k_{2z}}{\varepsilon_2\,k_{1z}} + \frac{\varepsilon_2\,k_{1z}}{\varepsilon_1\,k_{2z}}\right]\sin k_{2z}\ell, \tag{10.36}$$

$$\tilde{M}_{12}^{(p)} = \frac{i}{2}\left[\frac{\varepsilon_1\,k_{2z}}{\varepsilon_2\,k_{1z}} - \frac{\varepsilon_2\,k_{1z}}{\varepsilon_1\,k_{2z}}\right]\sin k_{2z}\ell, \tag{10.37}$$

$$\tilde{M}_{22}^{(p)} = \cos k_{2z}\ell - \frac{i}{2}\left[\frac{\varepsilon_1\,k_{2z}}{\varepsilon_2\,k_{1z}} + \frac{\varepsilon_2\,k_{1z}}{\varepsilon_1\,k_{2z}}\right]\sin k_{2z}\ell, \tag{10.38}$$

$$\tilde{M}_{12}^{(p)} = -\frac{i}{2}\left[\frac{\varepsilon_1\,k_{2z}}{\varepsilon_2\,k_{1z}} - \frac{\varepsilon_2\,k_{1z}}{\varepsilon_1\,k_{2z}}\right]\sin k_{2z}\ell. \tag{10.39}$$

Again, for real wave vectors k_{1z} and k_{2z} we have $\tilde{M}_{22}^{(p)} = [\tilde{M}_{11}^{(p)}]^*$ and $\tilde{M}_{21}^{(p)} = [\tilde{M}_{12}^{(p)}]^*$. The explicit dependence of the transfer matrix $\tilde{\mathbf{M}}^{(p)}$ on the permittivity brings some quantitative differences between the transmission of TE and TM modes. Nevertheless, the physical principles of the propagation are the same. In this section, we will discuss mostly the TE polarization and leave the derivation of the formulas for the TM polarization to the reader.

10.2.1 Fabry-Pérot Resonances

In the case when the propagation of electromagnetic waves is possible in both media, we study how the transmission depends on the frequency of incoming waves. For the TE mode, we express the transmission coefficient $T_s = |t_s|^2 = |M_{22}^{(s)}|^{-2}$ in the form

$$T_s = \frac{1}{1 + \dfrac{1}{4}\left[\dfrac{k_{2z}}{k_{1z}} - \dfrac{k_{1z}}{k_{2z}}\right]^2 \sin^2 k_{2z}\ell}. \tag{10.40}$$

In the same way as for the case of rectangular barriers, we see that the transmission $T = 1$ when $\sin k_{2z}\ell = 0$. This means that the wave vector k_{2z} takes the values

$$k_{2z}\ell = m\pi, \quad m = 1, 2, \ldots. \tag{10.41}$$

Since $k_{2z} = k_2 \cos\theta_2$ and $k_2 = \omega\sqrt{\varepsilon_2}/c$, the condition (10.41) can be written as

$$\omega\frac{\ell}{c}\sqrt{\varepsilon_2}\cos\theta_2 = m\pi. \tag{10.42}$$

We remind the reader that θ_2 is the angle of propagation inside the slab. θ_2 is related to θ_1 by Snell's law as

$$\sin\theta_1 = \frac{\sqrt{\varepsilon_2}}{\sqrt{\varepsilon_1}}\sin\theta_2. \tag{10.43}$$

For a fixed incident angle θ_1, we find resonant transmission if the frequency of the incident wave is

$$\sqrt{\varepsilon_1}\omega_m\frac{\ell}{c} = m\frac{\pi}{\sqrt{\varepsilon_2/\varepsilon_1 - \sin^2\theta_1}}. \tag{10.44}$$

To understand the physical meaning of the condition (10.41), we recall that the field inside the slab consists of waves propagating to the right and other waves propagating to the left. The resonant transmission appears when all waves propagating to the right have the same phase. For a given z, the right-going waves have the phase $\phi_{n+1} = k_{2z}(2n\ell + z)$ where n counts how many times the wave was reflected from the right boundary (figure 10.2). The requirement that *all* these waves must be in phase means that $\phi_{n+1} - \phi_1 = 2\pi m$ (m integer) for all n. This is satisfied only when $k_{2z}\ell = m\pi$. For the special case of normal incidence, $\theta_1 = \theta_2 = 0$, we recover from equation (10.41) the condition that the wavelength $\lambda = 2\ell/m$, known already from our studies of the quantum propagation through a rectangular barrier.

The transmission coefficient as a function of frequency is plotted in figure 10.3. We plotted also the transmission coefficient for the TM-polarized wave. When compared with figure 2.2, we see that in the present case the transmission coefficient is a *periodic function of the frequency*. The period is

$$\Delta\omega = \frac{\pi c/(\ell\sqrt{\varepsilon_1})}{\sqrt{\varepsilon_2/\varepsilon_1 - \sin^2\theta_1}}. \tag{10.45}$$

This difference between the electromagnetic and quantum problems is due to the different dispersion relations for a quantum particle and for an electromagnetic wave. While the energy of a quantum particle is a quadratic function of the wave vector, $E \propto k^2$, the frequency of the electromagnetic wave is a linear function of the wave vector, $\omega = kc/n$.

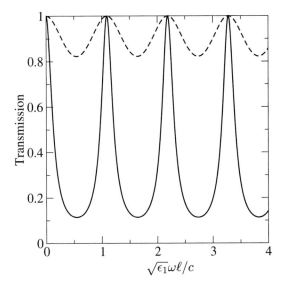

FIGURE 10.3. Transmission coefficient of a TE electromagnetic wave propagating through a dielectric slab with $\varepsilon_2 = 9$ and thickness ℓ. The frequency of the incoming wave is measured in dimensionless units $\omega\ell/c$. The angle of incidence is $\theta_1 = \pi/3$. The first resonant maximum $T = 1$ is observed for the frequencies $\omega = \omega_{\pi/3} = (\pi c/\ell)/[\sqrt{\varepsilon_2 - \sin^2 \theta_1}]$. Other maxima are observed for frequencies $m \times \omega_{\pi/3}$, $m = 1, 2, \ldots$, in agreement with equation (10.41). The dashed line is the transmission coefficient for the TM-polarized wave, calculated in problem 10.2 [equation (10.85)]. $\varepsilon_1 = 1$.

Figure 10.4 shows how the transmission coefficient depends on the incident angle θ_1 for three different values of the incident wave's frequencies. For a sufficiently high frequency of the incident electromagnetic wave, we again see the series of resonant maxima. Their positions can be estimated from the resonant condition, $k_{2z}\ell = m\pi$, given by equation (10.41).

Note that Fabry-Pérot oscillations exist also in a dielectric slab, embedded in a medium with *higher* permittivity. In this case, however, we are restricted to the incident angles $\theta_1 < \theta_{\mathrm{crit}}$, since for higher incident angles there is no transmission in the slab. We show here the transmission coefficient for the TE-polarized wave only. As shown in figure 10.5, the transmission coefficient oscillates as a function of frequency with a period given by equation (10.45). In figure 10.6, we show that the transmission coefficient oscillates as a function of incident angle θ_1. For $\theta_1 > \theta_{\mathrm{crit}}$ another transport regime, tunneling, takes place. We will analyze this phenomenon in section 10.2.3.

10.2.2 Bound States in the Slab

If the permittivity of the slab, ε_2, is larger than the permittivity of the embedding medium, we expect, from the analogy with the potential well, that the system possesses solutions that decay exponentially on both sides of the slab, but propagate inside the slab. We call such solutions *bound states*.

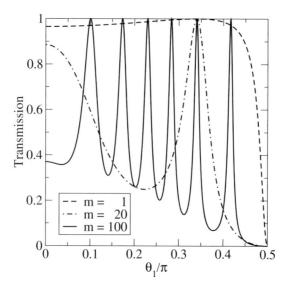

FIGURE 10.4. Transmission coefficient of a TE electromagnetic wave for a constant incident frequency through a dielectric slab as a function of incident angle θ_1. The permittivities of the slab are $\varepsilon_2 = 9$ and $\varepsilon_1 = 1$. The frequency of the incident wave is chosen as $m \times \omega_{\pi/3}$, where $\omega = \omega_{\pi/3}$ is the frequency at which the first resonant maximum appears if the incident angle $\theta_1 = \pi/3$ (figure 10.3). Three different frequencies were used, $m = 1$, 20, and 100.

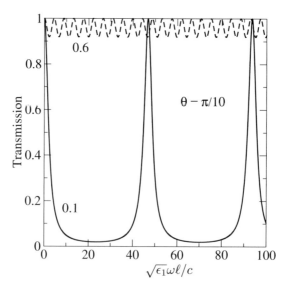

FIGURE 10.5. Transmission coefficient for a dielectric slab with dielectric permittivity ε_2, embedded in a material with higher permittivity. The ratio $\varepsilon_2/\varepsilon_1 = 0.6$ (dashed line) and 0.1 (solid line). The angle of incidence is $\theta_1 = \pi/10$. We see the Fabry-Pérot oscillations for frequencies given by equation (10.44).

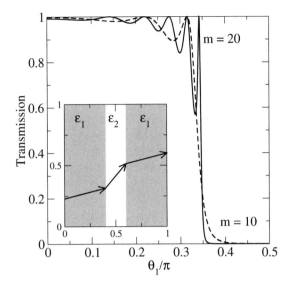

FIGURE 10.6. Transmission coefficient for a dielectric slab with dielectric permittivity ε_2 embedded in a material with higher permittivity $\varepsilon_1 > \varepsilon_2$. We see Fabry-Pérot oscillations for angles, $\theta_1 < \theta_{\text{crit}}$, and tunneling through the slab for $\theta_1 > \theta_{\text{crit}}$. The frequency of the incoming wave is $\sqrt{\varepsilon_1}\omega\ell/c = 10$ and 20. The ratio $\varepsilon_2/\varepsilon_1 = 0.8$ and $\varepsilon_1 = 1$.

We will consider only the TE-polarized wave, where the vector of the electric field $\vec{E}$ is parallel to the interface. We will find that in this case the bound electromagnetic states inside the slab are fully equivalent to the bounded states of electrons in a rectangular potential well.

To obtain an electric field decreasing exponentially outside the slab, we assume that

$$k_{1z} = i\kappa_1, \quad \kappa_1 > 0, \tag{10.46}$$

and obtain the following dispersion relations:

$$\omega^2 = \frac{c^2}{\varepsilon_1} \left[k_x^2 - \kappa_1^2 \right] \tag{10.47}$$

and

$$\omega^2 = \frac{c^2}{\varepsilon_2} \left[k_x^2 + k_{2z}^2 \right]. \tag{10.48}$$

Note that the wave vector k_{2z} inside the slab is *real*.

The bound state must follow the equation

$$M_{22}(i\kappa_1, k_{2z}) = 0. \tag{10.49}$$

Equation (10.49) can be written as follows:

$$\cos k_{2z}\ell - \frac{i}{2}\sin k_{2z}\ell \left[\frac{i\kappa_1}{k_{2z}} + \frac{k_{2z}}{i\kappa_1} \right] = 0. \tag{10.50}$$

This leads to the equation

$$\cot k_{2z}\ell = -\frac{1}{2}\left[\frac{\kappa_1}{k_{2z}} - \frac{k_{2z}}{\kappa_1}\right]. \tag{10.51}$$

Using the relation $\cot 2x = [\cot x - \tan x]/2$ we find two branches of possible solutions. We have either

$$\tan\frac{k_{2z}\ell}{2} = \frac{\kappa_1}{k_{2z}} \tag{10.52}$$

or

$$\tan\frac{k_{2z}\ell}{2} = -\frac{k_{2z}}{\kappa_1}. \tag{10.53}$$

Equations (10.52) and (10.53) are equivalent to the equation for bound states of the potential well given by equation (2.73). Indeed, by defining β and χ as

$$\beta^2 = \frac{\omega^2\ell^2}{4c^2}(\varepsilon_2 - \varepsilon_1) \tag{10.54}$$

and

$$\chi = \frac{k_{2z}\ell}{2}, \tag{10.55}$$

we express

$$\kappa_1 = \sqrt{4\beta^2 - k_{2z}^2}, \tag{10.56}$$

and obtain the following equations for k_{2z}:

$$\tan\chi = \frac{1}{\chi}\sqrt{\beta^2 - \chi^2} \tag{10.57}$$

for the first branch, and

$$\tan\chi = -\frac{\chi}{\sqrt{\beta^2 - \chi^2}} \tag{10.58}$$

for the second branch. The graphical solution of the above equations is given in figure 2.11 of chapter 2 and is not repeated here.

From the known values of the z-component, k_{2z}, of the wave vector, we calculate k_x given by equation (10.48). The dispersion relation $\omega = \omega(k_x)$ for waves propagating *inside* the slab is shown in figure 10.7. We see that there is always at least one solution which corresponds to $k_x\ell < \pi/2$. With increasing frequency, more branches appear because the number of possible solutions of equations (10.57) and (10.58) increases. The $(m+1)$th branch appears when β reaches the value $\pi/2 \times m$. From equation (10.54), we see that this happens when

$$\frac{\omega\ell}{c} = \frac{\omega_{m+1}\ell}{c} = \frac{\pi(m+1)}{\sqrt{\varepsilon_2 - \varepsilon_1}}. \tag{10.59}$$

The branch exists only on the right of the dispersion relation in vacuum, $\omega = kc$.

Figure 10.8 shows the spatial dependence of the electric field inside the slab for two bound states. An electromagnetic wave propagates along the x-direction, and is bound to

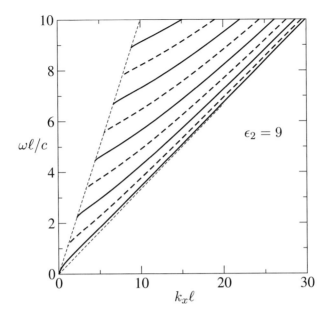

FIGURE 10.7. Dispersion relations for electromagnetic waves traveling in the x-direction inside a dielectric slab with permittivity $\varepsilon_2 = 9$, embedded in vacuum ($\varepsilon_1 \equiv 1$). Solid (dashed) lines correspond to the solutions of equations (10.57) and (10.58), respectively. The thin dashed lines are the dispersion relations for a dielectric slab, $\sqrt{\varepsilon_2}\,\omega\ell/c = k\ell$ and the vacuum, $\sqrt{\varepsilon_1}\omega\ell/c = k\ell$. Each branch exists only between these two lines, since we are interested only in solutions with $\sqrt{\varepsilon_1}\omega/c < k_x < \sqrt{\varepsilon_2}\omega/c$.

the region of the slab in the z-direction. The energy of the wave consists of the energy of oscillations in both perpendicular directions. Since the sum of these energies is the same, k_{1x} is smaller for higher bound states. Indeed, we see that the wavelength of the propagation along the x-direction is larger for higher bound states.

10.2.3 Tunneling through a Dielectric Slab

Consider now a dielectric slab with dielectric constant

$$\varepsilon_2 < \varepsilon_1. \tag{10.60}$$

In section 9.5, we learned that an electromagnetic wave, incident from a medium with permittivity ε_1, is totally reflected on a single interface, if the angle of incidence θ_1 is larger than θ_{crit}. From Snell's law we found the critical angle,

$$\sin \theta_{\text{crit}} = \frac{n_2}{n_1} = \sqrt{\frac{\varepsilon_2}{\varepsilon_1}}. \tag{10.61}$$

We will now investigate how an electromagnetic wave propagates through a slab of dielectric material ε_2, embedded into two dielectric layers with permittivity $\varepsilon_1 > \varepsilon_2$. One can intuitively argue that, even when $\theta_1 > \theta_{\text{crit}}$, some small portion of the electromagnetic

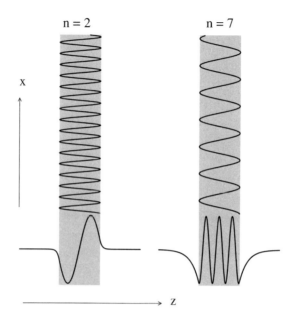

FIGURE 10.8. Spatial dependence of bound electric field inside a dielectric slab. The permittivity of the slab is $\varepsilon_2 = 9$. The frequency $\omega\ell/c = \sqrt{50} \approx 7.07$ and $\varepsilon_1 = 1$. We see in figure 10.7 that for this frequency we have 7 modes, which correspond to 7 solutions of equations (10.57) and (10.58). Two solutions are shown, left for $n = 2$ and right for $n = 7$. The lower figures show the z dependence of the field and the upper figures show the x dependence, $\cos k_x x$. Note that, for this particular choice of the frequency, the parameter $\beta = 10$, so that the z dependence of the field is identical with the form of the bound states of the wave function of the quantum particle, shown in figure 2.12.

wave can transfer through the slab. This intuition is supported by our experience with tunneling of quantum particles through a rectangular barrier (see section 2.3).

If the slab thickness ℓ is not too large, we indeed see that some portion of the energy tunnels through the slab and gives rise to a plane wave, propagating in the embedding medium on the opposite site of the slab. To get a quantitative description of the tunneling, we will again use our formulas (10.27) and (10.28) for the transfer matrix and calculate both the transmission and the reflection coefficients.

The transmission amplitude is given again by the M_{22} element of the transfer matrix:

$$t_s^{-1} = M_{22} = \cosh \kappa_2 \ell + \frac{i}{2} \left[\frac{\kappa_2}{k_{1z}} - \frac{k_{1z}}{\kappa_2} \right] \sinh \kappa_2 \ell, \tag{10.62}$$

where we use that $k_{2z} = i\kappa_2$. From equaitons (9.119) and (9.122), we have

$$\frac{\omega^2}{c^2} [\varepsilon_1 - \varepsilon_2] = k_{1z}^2 + \kappa_2^2. \tag{10.63}$$

Equations (10.62) and (10.63) are formally equivalent to equations (2.48) and (2.45) for the tunneling of a quantum particle through a rectangular potential barrier. Since the media left and right of the slab are the same, we obtain the transmission coefficient directly from

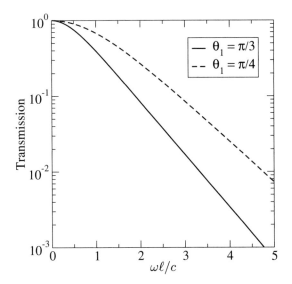

FIGURE 10.9. Transmission of an electromagnetic wave through a dielectric slab with $\varepsilon_2 = 1$ as a function of the frequency of the incoming wave for two different incident angles. Note the logarithmic scale on the vertical axis. The transmission decreases exponentially when the frequency ω increases, since the exponential decrease of the evanescent wave is faster for higher frequencies, in agreement with equaiton (10.65). The permittivity of the embedded medium is $\varepsilon_1 = 9$.

the transmission amplitude,

$$T_s = |t_s|^2 = \frac{1}{1 + \frac{1}{4}\left[\frac{\kappa_2}{k_{1z}} + \frac{k_{1z}}{\kappa_2}\right]^2 \sinh^2 \kappa_2 \ell}. \tag{10.64}$$

Equation (10.64) is exactly equivalent to equation (2.48), which describes the transmission coefficient through a rectangular barrier for a quantum particle.

From equation (10.64) one sees that T decreases exponentially, if the width of the slab $\ell \gg \kappa_2^{-1}$. For a given slab width, the transmission through the slab strongly depends on the incident angle. From equation (10.63) with $k_{1z} = \omega/c\sqrt{\varepsilon_1}\cos\theta_1$, one can show that

$$\kappa_2 = \frac{2\pi}{\lambda_1}\sqrt{\sin^2\theta_1 - \frac{\varepsilon_2}{\varepsilon_1}}. \tag{10.65}$$

Note that an increase of θ_1 causes an increase of κ_2 and, consequently, a decrease of the transmission coefficient.

Figure 10.9 shows the transmission as a function of the frequency for an incident electromagnetic wave with a given incidence angle.

The formula for the transmission coefficient deserves a more detailed analysis. If some portion of the electromagnetic wave tunnels through the slab and propagates on its opposite side, then also some portion of the energy should "tunnel" through the slab. We learned in section 9.6.1 that evanescent waves do not transfer any energy. However, in the present problem, we have a slab of air embedded into two dielectrics. In this case, we do not have a single evanescent wave, but a superposition of two evanescent waves decaying in opposite directions,

$$E_y = \left[E_y^+ e^{-\kappa z} + E_y^- e^{+\kappa(z-2\ell)}\right] e^{i k_x x}, \tag{10.66}$$

where E_y is the y-component of the electric field, and E_y^+ (E_y^-) denotes the electromagnetic wave propagating to the right (left), respectively. In equation (10.66), z is restricted to $0 \le z \le \ell$. The second wave, E_y^-, represents the evanescent wave reflected from the right boundary of the slab. We assume that both E_y^+ and E_y^- are of order unity.

Now, we calculate the Poynting vector, given by equation (B.60),

$$S_z = \frac{c}{8\pi} \, \text{Re} \left[E_y H_x^*\right], \tag{10.67}$$

With the help of Maxwell's equation (B.17), $\vec{k} \times \vec{E} = +(\mu\omega/c) \, \vec{H}$, we express the magnetic fields inside the slab in terms of an electric field. We obtain

$$S_z = \frac{c^2}{8\pi\mu\omega} \, \text{Re} \left[\left(E_y^+ e^{-\kappa z} + E_y^- e^{\kappa(z-2\ell)}\right)\left(i\kappa E_y^+ e^{-\kappa z} - i\kappa E_y^- e^{\kappa(z-2\ell)}\right)^*\right], \tag{10.68}$$

which gives

$$S_z = \frac{c^2\kappa}{4\pi\mu\omega} \, \text{Re}[E_y^+ E_y^{-*}]e^{-2\kappa\ell}. \tag{10.69}$$

Expression (10.69) should be compared with formula (2.63) for the current density of a quantum particle tunneling through a rectangular barrier. We see that a small portion of energy (of order of $e^{-2\kappa\ell}$) is indeed transferred through the slab. This is the energy of the wave that propagates in the dielectric material on the opposite side of the slab.

For completeness, we present also an expression for the x-component of the Poynting vector, S_x [85]. It represents the energy that propagates along an interface. Since both fields E^+ and $E^- \sim e^{ik_x}$,

$$\begin{aligned} S_x(z) &= \frac{c^2}{8\pi\mu\omega} \, \text{Re} \left[k_x(E_y^+ e^{-\kappa z} + E_y^- e^{\kappa(z-2\ell)})(E_y^{+*} e^{-\kappa z} + E_y^{-*} e^{\kappa(z-2\ell)})\right] \\ &= \frac{c^2}{8\pi\mu\omega} k_x \left[|E_y^+|^2 e^{-2\kappa z} + |E_y^-|^2 e^{\kappa(z-2\ell)} + 2\text{Re}(E_y^+ E_y^{-*})e^{-2\kappa\ell}\right]. \end{aligned} \tag{10.70}$$

As expected, S_x decreases exponentially with the distance from the left interface. At the right interface, $z = \ell$, we have

$$S_x(z = \ell) = \frac{c^2}{8\pi\mu\omega} k_x \left[|E_y^+|^2 + |E_y^-|^2 + 2\text{Re}(E_y^+ E_y^{-*})\right] e^{-2\kappa\ell}. \tag{10.71}$$

10.3 Transmission through a Metallic Slab

Transmission through a metallic slab of thickness ℓ can be again obtained from the formula $T = |M_{22}|^{-2}$, where the matrix element M_{22} is given by equation (10.19). However, in contrast to dielectrics, here the wave vector inside the metallic slab, k_{2z}, is complex.

We remind the reader that the permittivity of a metal is given by the Drude formula (see appendix B.7 for details)

$$\varepsilon_m = 1 - \frac{\omega_p^2}{\omega^2 + i\gamma\omega}. \tag{10.72}$$

We will discuss only the high-frequency limit, $\omega \gg \gamma$. In this case, the real and imaginary parts of the permittivity can be written as

$$\varepsilon_r \approx 1 - \frac{\omega_p^2}{\omega^2} \quad \text{and} \quad \varepsilon_i \approx \frac{\omega_p^2 \gamma}{\omega^3}, \tag{10.73}$$

and $\varepsilon_i/|\varepsilon_r| = \gamma/\omega \ll 1$, so that the metallic permittivity is mostly real and negative. Consider only the case of normal incidence, $k_x = 0$. Then we have

$$k_{2z} = \frac{\omega}{c}\sqrt{\varepsilon_m}. \tag{10.74}$$

It possesses a large imaginary part $i\kappa$ and a small real part k_{2r}. Explicit formulas for the wave vector were derived in section 9.4.

The transmission and reflection coefficients are

$$T = \frac{1}{|M_{22}|^2} \quad \text{and} \quad R = \frac{|M_{12}|^2}{|M_{22}|^2}, \tag{10.75}$$

where the elements of the transfer matrix $\mathbf{M}$ are given by equations (10.17)–(10.20). Since the wave vector k_2 is complex, we have $|M_{22}|^2 + |M_{12}|^2 < 1$ so that also $T + R < 1$. The difference $A = 1 - T - R$ is the absorption. It represents the amount of energy absorbed inside the metallic slab.

The transmission coefficient as a function of the frequency is shown in figure 10.10. We see that below the plasma frequency the transmission is small and the reflection coefficient is close to 1, in agreement with our estimations in section 9.4. Only when the frequency approaches the plasma frequency ω_p, does the transmission start to increase. For $\omega > \omega_p$, the metal behaves as a lossy dielectric and we also see typical Fabry-Pérot oscillations with resonant transmission smaller than 1 because of the absorption.

We want to estimate how the transmission of the electromagnetic wave depends on the thickness of the slab. To simplify our analysis, we neglect the real part k_{2r} of the wave vector in the metal with respect to the imaginary part κ. This is a good approximation, since $k_{2r} \approx (\gamma/2\omega)\kappa \ll \kappa$, as shown in equation (9.88) in section 9.4.2. With this approximation we express an approximate formula for the transmission coefficients as follows:

$$T \approx \frac{1}{1 + \frac{1}{4}\left[\dfrac{\kappa}{k_{1z}} + \dfrac{k_{1z}}{\kappa}\right]^2 \sinh^2 \kappa\ell}. \tag{10.76}$$

From the dispersion relation given by equation (10.74) and the relation for the metallic permittivity, eq (10.72), we also see that

$$\frac{|k_2|}{|k_1|} = \frac{\kappa}{k_{1z}} = \sqrt{|\varepsilon_m|} \gg 1. \tag{10.77}$$

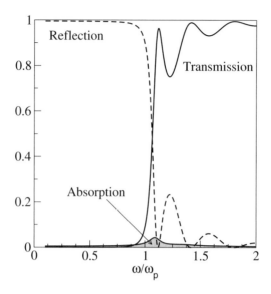

FIGURE 10.10. Transmission (solid line), reflection (dashed line), and absorption (shaded area) of an electromagnetic wave through a metallic slab as a function of frequency. The thickness of the slab is $\ell = \pi c / \omega_p$ and $\gamma = 0.0025\omega_p$.

From equation (10.76) we know that the transmission is close to 1, only when the second term in the denominator is small. Since $\sinh \kappa \ell$ is small only when $\kappa \ell \ll 1$, the metallic slab transmits electromagnetic waves only for very small slab thickness ℓ. To estimate the thickness for which the slab remains transparent, we consider the absolute value of the expression

$$\frac{1}{4}\left[\frac{\kappa}{k_{1z}} + \frac{k_{1z}}{\kappa}\right]^2 \sinh^2 \kappa \ell. \tag{10.78}$$

For small $\kappa \ell$, we expand $\sinh \kappa \ell \approx \kappa \ell$. We can also neglect the term k_{1z}/κ, which is much smaller than κ / k_{1z}, due to the relation (10.77). We end up with the estimation that the metallic slab is transparent only when

$$\left(\frac{1}{2}\frac{\kappa^2}{k_{1z}}\ell\right)^2 \ll 1 \tag{10.79}$$

or

$$\ell \ll \frac{2c}{\omega|\varepsilon_m|}. \tag{10.80}$$

For $\omega \gg \gamma$ we know that $|\varepsilon_m| \approx \omega_p^2/\omega^2$ and the skin depth is $\delta = c/\omega_p$. Inserting these relations into equaiton (10.80), we finally obtain the relation

$$\ell \ll \frac{2\omega}{\omega_p}\delta. \tag{10.81}$$

So when the thickness of the slab ℓ is much less than the skin depth δ, the metallic slab is transparent to the incident wave.

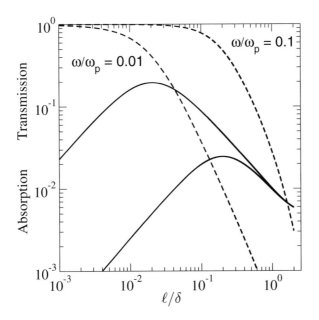

FIGURE 10.11. Transmission (dashed lines) and absorption (solid lines) of an electromagnetic wave as a function of the thickness of a metallic slab for two different frequencies of the electromagnetic waves. $\delta = c/\omega_p$ is the skin depth. The metallic permittivity is given by equation (10.72) with $\gamma = 0.0025\omega_p$. Note the logarithmic scale on both axes.

In figure 10.11 we show the transmission and absorption of the electromagnetic waves propagating through a metallic slab. The numerical data confirm our result given by equation (10.81). Although the skin depth is approximately the same for both frequencies $\omega = 0.1\omega_p$ and $\omega = 0.01\omega_p$, the thickness for which the metallic slab is transparent, is ten times smaller for the lower frequency.

10.4 Problems

Problems with Solutions

Problem 10.1
Derive the transfer matrix $\mathbf{M}^{(p)}$ for the TM-polarized wave propagating through a slab embedded in a homogeneous medium.

Solution. The matrices $\mathbf{M}^{(p)}$ and $\tilde{\mathbf{M}}^{(p)}$ are defined as

$$\begin{pmatrix} E_3^+ \\ E_3^- \end{pmatrix} = \mathbf{M}^{(p)} \begin{pmatrix} E_1^+ \\ E_1^- \end{pmatrix} \quad \text{and} \quad \begin{pmatrix} H_3^+ \\ H_3^- \end{pmatrix} = \tilde{\mathbf{M}}^{(p)} \begin{pmatrix} H_1^+ \\ H_1^- \end{pmatrix}. \tag{10.82}$$

Since $\varepsilon_3 = \varepsilon_1$ and $\mu_3 = \mu_1$, we obtain from equation (B.25) that

$$\frac{E_3}{E_1} = \frac{H_3}{H_1}. \tag{10.83}$$

Therefore we have

$$\mathbf{M}^{(p)} \equiv \tilde{\mathbf{M}}^{(p)}. \tag{10.84}$$

Note that equation (10.84) holds only if the incoming medium is identical with the outgoing medium. An explicit expression for both matrices is given by equations (10.36)–(10.39).

Problems without Solutions

Problem 10.2

From the general formula for the element $\tilde{M}_{22}^{(p)}$ of the transfer matrix $\tilde{\mathbf{M}}^{(p)}$, given by equation (10.13), derive

$$T_p = \frac{1}{1 + \dfrac{1}{4}\left[\dfrac{k_{2z}}{\varepsilon_2 k_{1z}} - \dfrac{\varepsilon_2 k_{1z}}{k_{2z}}\right]^2 \sin^2 k_{2z}\ell}. \tag{10.85}$$

Problem 10.3

Derive the equations for the TM-polarized bound states of the dielectric slab:

$$\tan k_{2z}\ell = \begin{cases} \dfrac{\varepsilon_2}{\varepsilon_1}\dfrac{\kappa_1}{k_{2z}}, \\[2ex] -\dfrac{\varepsilon_1}{\varepsilon_2}\dfrac{k_{2z}}{\kappa_1}. \end{cases} \tag{10.86}$$

Solve equation (10.86) and plot the dispersion relations for waves propagating inside the dielectric slab. Use $\varepsilon_2 = 9$ for the permittivity of the slab.

11 Surface Waves

In this chapter we study an interesting phenomenon, namely, the excitation of surface waves. We will see that for an appropriate choice of the electromagnetic parameters an interface between two media can support the excitation of surface waves [2,58]. Surface waves can propagate along the interface and decay exponentially as a function of the distance from the surface, shown in figure 11.1. This phenomenon has no analogy in quantum physics.

We analyze first the surface waves on a single interface between two media. We find that surface waves propagate only along the interface separating two media with opposite signs of either permittivity or permeability. Two interfaces are examined in detail. The first one is an interface between vacuum and metal. In this case, the electric permittivity changes sign across the interface [58]. Such an interface supports only the excitation of TM-polarized surface waves. In the second example, we examine an interface between a vacuum and a left-handed material [136]. In this case, both the permittivity and the permeability change sign at the interface, so that both TE and TM surface waves can be excited. For both cases (TE and TM polarization), we derive the dispersion relation between the wave vector and the frequency of the surface wave.

Then we discuss the excitation of surface waves on the two opposite interfaces of the slabs [58,137]. As a particular example, we consider a slab of a left-handed material. We again derive the dispersion relations for the two surface waves and study how the two surface waves interact with each other.

Finally, we discuss an experimental configuration which enables us to observe the excitation of surface waves [135,137]. As a particular example, we examine the excitation of the TE-polarized surface waves at the interface of vacuum and a left-handed material.

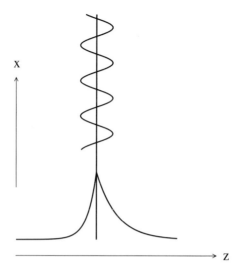

FIGURE 11.1. A surface wave propagates along the boundary between two media (the x-direction in our case), and decreases exponentially on both sides of the interface. Note that the exponential decrease depends upon the properties of the media, such as ε and μ, so that it can be different on both sides of the interface.

11.1 Surface Waves at the Interface between Two Media

Since a surface wave propagates along the interface, its wave vector is real only in the direction of the interface. In our geometry, this means that the wave vector $\vec{k}$ of the surface wave has a real component k_x along the x-direction and an imaginary component along the z-direction,

$$k_{1z} = i\kappa_1 \quad \text{and} \quad k_{2z} = i\kappa_2. \tag{11.1}$$

We assume that both κ_1 and κ_2 are real and positive. The values of κ_1 and κ_2 determine the exponential decrease of the amplitude of the electric field from the boundary between the two media. In general, the two media have different electromagnetic properties, such as ε and μ, so that κ_1 and κ_2 also differ from each other. We assume that the interface is located at $z = 0$. Thus, the electric field in the first medium can be written as

$$E_1 = E_1^+ e^{-\kappa_1 z + i k_x x} + E_1^- e^{+\kappa_1 z + i k_x x}, \quad z < 0, \tag{11.2}$$

and the electric field in the second medium can be written as

$$\vec{E}_2 = E_2^+ e^{-\kappa_2 z + i k_x x} + E_2^- e^{+\kappa_2 z + i k_x x}, \quad z > 0. \tag{11.3}$$

To avoid any divergences of the electromagnetic fields for $z \to \pm\infty$, we require that

$$E_1^+ \equiv 0 \tag{11.4}$$

and

$$E_2^- \equiv 0. \tag{11.5}$$

Otherwise, the electric field would increase exponentially with increasing distance from the interface, which is not allowed. The electric fields E_1 and E_2 are related by the transfer matrix $\mathbf{M}^{(s)}$ derived in chapter 9. Using the known form of the transfer matrix $\mathbf{M}^{(s)}$ and the two conditions given by equations (11.4) and (11.5), we can find the eigenfrequencies of the surface state from the known form of the transfer matrix $\mathbf{M}^{(s)}$ in the same way as we found bound states of the one-dimensional potential in quantum mechanics (chapter 2) or of the dielectric slab (section 10.2.2). Inserting equations (11.4) and (11.5) into equation (9.23), we calculate that

$$E_2^+ = M_{12}^{(s)} E_1^-$$ (11.6)

and

$$E_2^- = M_{22}^{(s)} E_1^-.$$ (11.7)

Since $E_2^- = 0$ but $E_1^- \neq 0$, the last equation is satisfied only when the M_{22} element of the transfer matrix is equal to zero,

$$\mathbf{M}_{22}^{(s)}(i\kappa_1, i\kappa_2) = 0.$$ (11.8)

In the same way, we obtain the following condition for the excitation of the TM surface wave:

$$\tilde{\mathbf{M}}_{22}^{(p)}(i\kappa_1, i\kappa_2) = 0.$$ (11.9)

From equations (11.8) and (11.9) we have the dispersion relations of surface waves. For instance, for the s-polarization, we obtain from equation (9.22) that

$$1 + \frac{\mu_2}{\mu_1}\frac{\kappa_1}{\kappa_2} = 0.$$ (11.10)

This can be written in a more symmetric form,

$$\frac{\kappa_1}{\mu_1} + \frac{\kappa_2}{\mu_2} = 0, \quad \text{TE polarization.}$$ (11.11)

In the same way, we can obtain for the TM polarization that

$$\frac{\kappa_1}{\varepsilon_1} + \frac{\kappa_2}{\varepsilon_2} = 0, \quad \text{TM polarization.}$$ (11.12)

In order for an interface to support surface waves, it needs to satisfy equation (11.11) or equation (11.12) for the TE and TM waves, respectively. Notice that for regular interfaces of two dielectrics with positive dielectric constants no surface wave exists. One of the materials must have either a negative permittivity ε (like a metal) or negative permeability μ (like a ferromagnetic resonance material or left-handed material).

The two parameters κ_1 and κ_2 are related by the dispersion relations of the electromagnetic waves inside the two media, which, in our case, can be written as follows:

$$\omega^2 = \frac{c^2}{n_1^2}\left[k_x^2 - \kappa_1^2\right] = \frac{c^2}{n_2^2}\left[k_x^2 - \kappa_2^2\right].$$ (11.13)

For a given frequency dependence of the permittivity and permeability, we easily find the relation $\omega = \omega(k_x)$, which determines the allowed frequencies of surface waves. We express κ_1 and κ_2 from relations (11.13) and insert them into equation (11.11). We obtain that

$$\frac{\kappa_1}{\kappa_2} = \sqrt{\frac{k_x^2 - \omega^2 \varepsilon_1 \mu_1/c^2}{k_x^2 - \omega^2 \varepsilon_2 \mu_2/c^2}} = -\frac{\mu_1}{\mu_2}. \tag{11.14}$$

Equation (11.14) can be rewritten in the following form:

$$k_x^2 = \frac{\omega^2}{c^2} \varepsilon_1 \mu_1 \frac{1 - \dfrac{\varepsilon_2 \mu_1}{\varepsilon_1 \mu_2}}{1 - \dfrac{\mu_1^2}{\mu_2^2}}, \tag{11.15}$$

which should be solved for a given dispersion relation $\varepsilon_1 = \varepsilon_1(\omega)$, $\mu_1 = \mu_1(\omega)$ and $\varepsilon_2 = \varepsilon_2(\omega)$, $\mu_2 = \mu_2(\omega)$.

In a similar way, for the TM polarization, we calculate that

$$k_x^2 = \frac{\omega^2}{c^2} \varepsilon_1 \mu_1 \frac{1 - \dfrac{\varepsilon_1 \mu_2}{\varepsilon_2 \mu_1}}{1 - \dfrac{\varepsilon_1^2}{\varepsilon_2^2}}. \tag{11.16}$$

11.1.1 Surface Waves at an Interface between a Vacuum and a Metal

The simplest example of a surface wave is the TM-polarized surface wave at the interface between a vacuum and a metal. In the frequency region $\gamma \ll \omega < \omega_p$, the metal can be assumed to be a lossless medium with negative permittivity given by

$$\varepsilon_m(\omega) = 1 - \frac{\omega_p^2}{\omega^2}. \tag{11.17}$$

Therefore, we expect that an interface between a metal and a vacuum can support excitation of the TM-polarized surface waves. Inserting $\varepsilon_1 = \mu_1 = \mu_2 = 1$ and $\varepsilon_2 = \varepsilon_m(\omega)$ given by equation (11.17), in the dispersion relation (11.16) we obtain

$$\frac{k_x}{k_p} = \frac{\omega}{\omega_p} \sqrt{\frac{\omega^2 - \omega_p^2}{2\omega^2 - \omega_p^2}}, \tag{11.18}$$

where $k_p = \omega_p/c$. In order to have k_x real, the expression inside the square root must be positive. For a metal with $\varepsilon(\omega) < 0$, we have that $\omega < \omega_p$, which means that the numerator of equation (11.18) is negative. So the denominator $2\omega^2 - \omega_p^2$ must be negative too. This gives

$$\omega < \omega_2 = \frac{\omega_p}{\sqrt{2}}. \tag{11.19}$$

Note that ω_2 is exactly the frequency for which the permittivity of a metal equals -1,

$$\varepsilon(\omega_2) = -1. \tag{11.20}$$

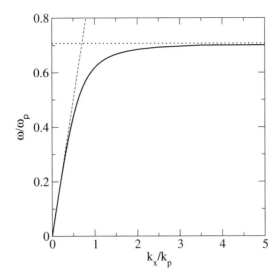

FIGURE 11.2. Dispersion relation for a surface wave at an interface of lossless metal and vacuum. For small frequencies, $\omega \approx ck_x$. In the limit of $k_x \to \infty$, the frequency of the surface wave saturates to $\omega_p/\sqrt{2}$, shown by the dotted horizontal line. The dashed line is the dispersion relation in vacuum, $\omega = ck$. The frequency ω_p is the plasma frequency of the metal and $k_p = \omega_p/c$.

The dispersion relation (11.18) is shown in figure 11.2. Notice that we always have a surface wave with the p (TM) polarization for $\omega < \omega_p/\sqrt{2}$. Note that the surface wave exists only for

$$k_x > \omega/c. \tag{11.21}$$

If we had $k_x < \omega/c$ (this region is called a light cone), we would have from the dispersion relation $\omega^2/c^2 = k_x^2 + k_{1z}^2$ that k_{1z} is real, so that we would have propagating solutions in vacuum. The condition (11.21) must always be satisfied in order to have solutions that decay exponentially on the vacuum side of the interface. Note that, in the metal, each solution decays exponentially with distance from the interface, since there are no propagating waves for $\omega < \omega_p$.

11.1.2 Surface Waves at an Interface between Vacuum and a Left-Handed Medium

Another interface that supports surface waves is that of a vacuum and a left-handed medium. Since both the permittivity and the permeability are negative in the left-handed material, we can have both TE- and TM-polarized surface waves. The dispersion relation for these surface waves can easily be obtained from the general formulas given by equations (11.15) and (11.16). We put $\varepsilon_1 = \mu_1 = 1$ for the first medium (vacuum), and use the following relations for the permittivity and the permeability of the left-handed medium on the right side of the interface:

$$\varepsilon_2 = \varepsilon_2(\omega) = 1 - \frac{\omega_p^2}{\omega^2}, \quad \mu_2 = \mu_2(\omega) = 1 - \frac{F\omega^2}{\omega^2 - \omega_0^2}. \tag{11.22}$$

As a physical case of a left-handed medium, we choose the plasma frequency $\nu_p = \omega_p/2\pi = 10\,\text{GHz}$, the resonant magnetic frequency $\nu_0 = \omega_0/2\pi = 4\,\text{GHz}$, and the filling factor $F = 0.56$.

As discussed before, surface waves exist if the permittivity or the permeability changes its sign across the interface. This means that we can expect a TE-polarized surface mode in the frequency interval where the permeability $\mu_2 < 0$. From equation (11.22) we see that the left-handed medium possesses negative permittivity in the frequency interval between ω_0 and $\omega_0/\sqrt{1-F}$. These two lines are shown as dotted horizontal lines in figure 11.3.

Similarly, TM-polarized surface modes exist in the frequency interval where permittivity is negative. In our case, this means that $0 < \omega < \omega_p$.

There are also other requirements for surface waves to be excited. Since the electromagnetic waves must decrease exponentially on both sides of the interface, the x-component k_x of the wave vector must be

$$k_x > \frac{\omega}{c}. \tag{11.23}$$

In this way, the z-component of the wave vector is imaginary in vacuum. For surface waves inside the left-handed region, the following condition must be satisfied:

$$k_x^2 > \varepsilon_2 \mu_2 \frac{\omega^2}{c^2} \tag{11.24}$$

to obtain an exponential decrease of the electromagnetic wave in the left-handed medium. Outside the left-handed frequency region, we have $\mu_2 > 0$ and $\varepsilon_2 < 0$, so that no wave propagation is possible. A plot of the relation (11.24) is shown in figure 11.3 by the solid line at the border between the two shaded areas.

To calculate the dispersion curve $\omega = \omega(k_x)$ for surface waves on the interface between a vacuum and a left-handed medium, we insert the permittivity and the permeability given by equation (11.22) into the dispersion relations of the surface modes, given by equations (11.15) and (11.16). We obtain

$$\frac{k_x}{k_p} = \frac{\omega}{\omega_p}\sqrt{\mu_2 \frac{\mu_2 - \varepsilon_2}{\mu_2^2 - 1}} \tag{11.25}$$

for the TE-polarized surface waves, and

$$\frac{k_x}{k_p} = \frac{\omega}{\omega_p}\sqrt{\varepsilon_2 \frac{\varepsilon_2 - \mu_2}{\varepsilon_2^2 - 1}} \tag{11.26}$$

for the TM-polarized surface waves.

From equation (11.25) we see that k_x is real only when

$$\mu_2(\omega) > -1 \quad \text{and} \quad \mu_2 > \varepsilon_2 \tag{11.27}$$

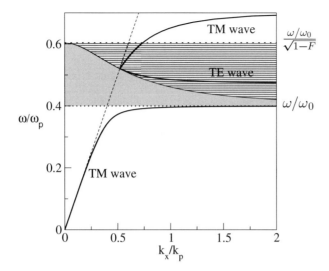

FIGURE 11.3. Dispersion relation of surface waves at the interface between a vacuum and a left-handed medium. The left-handed medium has a frequency dependence of the permittivity and the permeability given by equations (11.22). The effective permittivity ε_2 is negative for $\omega < \omega_p$. The two shaded areas show the frequency interval, $\omega_0 < \omega < \omega_0/\sqrt{1-F}$, in which the second medium is left handed. The solid line between the two shaded areas describes the relation $k_x/k_p = (\omega/\omega_p)\sqrt{\varepsilon_2(\omega)\mu_2(\omega)}$, where $k_p = \omega_p/c$. Surface waves exist only in the region right of this line. The dashed line is the dispersion relation for the vacuum, $\omega = k_x c$. Again, from equation (11.23) it follows that excitation of surface waves is possible only for parameters to the right of this line. We have one TE-polarized surface wave inside the left-handed band. Two TM-polarized solutions originate from the TM-polarized surface wave known from metals (figure 11.2) [136].

or

$$\mu_2(\omega) < -1 \quad \text{and} \quad \mu_2 < \varepsilon_2. \tag{11.28}$$

After some careful examination, we find that only the first condition can be satisfied for our choice of parameters of the left-handed medium. Then, the TE surface wave possesses the frequency band

$$\omega_1 < \omega < \omega', \tag{11.29}$$

where

$$\omega_1 = \sqrt{\frac{2}{2-F}}\,\omega_0 \tag{11.30}$$

is the frequency for which $\mu_2(\omega_1) = -1$, and ω' is the frequency for which $\varepsilon_2(\omega')\mu_2(\omega') = 1$. Note that, ω' is the highest frequency for which both conditions (11.23) and (11.24) can be satisfied simultaneously. From figure 11.3 we see that ω' is given by

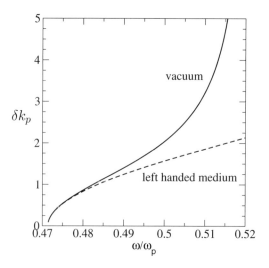

FIGURE 11.4. Dimensionless length δk_p $(k_p = \omega_p/c)$ to which the TE surface wave penetrates into the vacuum (solid line) and left-handed medium (dashed line). The width of the surface wave is of the order of $\delta \sim 2/k_p = \lambda_p/\pi$ $(\lambda_p = 2\pi c/\omega_p)$ on the side of the left-handed medium, and depends only weakly on the frequency. On the vacuum side, the width of the surface waves increases considerably when the frequency approaches the upper band edge. The frequency band of the TE waves is $0.4714 \le \omega/\omega_p \le 0.52$.

the crossover frequency of the two lines,

$$k_x^2 = \frac{\omega^2}{c^2} = \frac{\omega^2}{c^2}\mu(\omega')\varepsilon(\omega'). \tag{11.31}$$

The two TM-polarized surface waves originate from the metallic surface mode shown in figure 11.2 (note that the frequency dependence of the permittivity of the left-handed medium is the same as that of a metal). The lower branch starts at $\omega = 0$. For small frequencies, $\omega \ll \omega_0$, this branch is similar to the TM mode of a metal. However, when ω approaches ω_0, which is the lower band edge of the left-handed medium, the dispersion relation for the surface mode bends and approaches ω_0 in the limit of $k_x \to \infty$. Above the frequency ω_0, there are no surface waves because the refractive index $n_2 = \sqrt{\varepsilon_2\mu_2}$ is real, so that the left-handed medium allows propagation of plane waves. The upper branch starts at frequency ω'. It is necessary to satisfy the condition (11.24) to have an exponentially decaying solution in the left-handed medium. Thus, the upper TM-polarized surface wave frequency band spans between ω' and ω_2 defined by the relation $\varepsilon(\omega_2) = -1$. Only part of this band lies in the left-handed frequency region.

Inserting k_x given by equation (11.15) into equation (11.13), we also find the dependency of the parameters κ_1 and κ_2, which determine the exponential decrease of the surface wave along the z-direction, on the frequency ω (see figure 11.4). These results give us a quantitative estimation of the distance from the interface, where the electromagnetic field of the surface mode is still detectable. Notice that the surface waves decay much faster in the left-handed medium than in the vacuum.

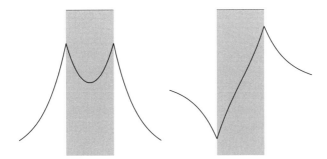

FIGURE 11.5. Two surface modes excited on both interfaces of the slab.
Left: symmetric excitation; right: antisymmetric excitation.

11.2 Surface Modes on a Slab

We learned in the previous section that the interface between two media might support surface waves. Surface waves decrease exponentially at both sides of the interface. Their decay length might be different in the two media. Now we want to investigate the existence of similar solutions for the case of a slab. Here, two surface waves are excited, one for each interface, as schematically shown in figure 11.5. Because of the overlap of the fields which decay from both interfaces inside the slab, the two surface waves interact with each other. This interaction influences the dispersion relation for the surface modes.

Note that, the excitation of the two coupled surface waves, shown in figure 11.5, differs qualitatively from the excitation of the bound state on a slab of dielectric material, discussed in section 10.2.2. The surface wave always decays exponentially on both sides of the interfaces. Therefore, in the case of two surface waves excited on two opposite sides of the slab, we do not have any propagating waves inside the slab. In other words, surface waves are excited in a frequency region where no wave propagation inside the slab is allowed. Contrary to this situation, in the case of bound states in a dielectric slab, the electromagnetic waves always propagate *inside the slab*, as is shown in figure 10.8. Thus, bound states exist only in the frequency region which allows plane wave propagation inside the slab, and surface waves might be excited only in the frequency region, where no propagation solution inside the slab exists.

Excitation of surface waves on the two interfaces of the slab is a different physical phenomena than the amplification of the evanescent waves in the left-handed medium, discussed in chapter 16. We have learned in the present chapter that surface waves exist only in the frequency region where no propagation is allowed inside the material. Also, for a given value of the wave vector, we have only one frequency for which the surface wave can be excited. Thus, although the surface waves transfer the evanescent waves through the slab shown in figure 11.5, this transmission is possible only for specific values of the parameters (k_x, ω). Contrary to this, we will see in chapter 16 that the amplification of the evanescent waves in the left-handed slab takes place for any frequency from the left-handed frequency band. This is very important for the construction of the perfect lens, since we want left-handed slab to transmit both evanescent and propagating waves simultaneously.

We will derive a dispersion relation for the surface modes. For simplicity, we assume that the first and the third medium are identical. This means that $\varepsilon_3 = \varepsilon_1$ and $\mu_3 = \mu_1$.

In analogy with the case of a single interface discussed in the previous section, all possible eigenfrequencies of the waves localized at the surface slab can be obtained from the requirements for the wave vectors $k_{1z} = i\kappa_1$ and $k_{2z} = i\kappa_2$. The M_{22} element of the transfer matrix must be zero:

$$M_{22}(i\kappa_1, i\kappa_2) = 0, \quad \kappa_1, \kappa_2 > 0. \tag{11.32}$$

Inserting the explicit form of the element M_{22} of the transfer matrix for the slab, given by equation (10.19), into equation (11.32), we obtain

$$\cosh \kappa_2 \ell + \frac{1}{2} \left[z + \frac{1}{z} \right] \sinh \kappa_2 \ell = 0, \tag{11.33}$$

where we substitute

$$z = \frac{\mu_2}{\mu_1} \frac{\kappa_1}{\kappa_2} \tag{11.34}$$

for the TE mode. Similarly, for the TM mode we obtain

$$z = \frac{\varepsilon_2}{\varepsilon_1} \frac{\kappa_1}{\kappa_2}. \tag{11.35}$$

Equation (11.33) has two solutions

$$\frac{\mu_2}{\mu_1} = \begin{cases} -\dfrac{\kappa_2}{\kappa_1} \tanh \kappa_2 \ell/2, \\ -\dfrac{\kappa_2}{\kappa_1} \coth \kappa_2 \ell/2 \end{cases} \tag{11.36}$$

for the TE mode, and another two solutions

$$\frac{\varepsilon_2}{\varepsilon_1} = \begin{cases} -\dfrac{\kappa_2}{\kappa_1} \tanh \kappa_2 \ell/2, \\ -\dfrac{\kappa_2}{\kappa_1} \coth \kappa_2 \ell/2 \end{cases} \tag{11.37}$$

for the TM mode.

The relations (11.36) and (11.37) represent the general form of the dispersion relation for the surface waves excited on both sides of the interface. For a given material, we must substitute the dispersion relations $\mu_2(\omega)$ and $\varepsilon_2(\omega)$, and the relation (11.13) between the frequency ω and the wave vectors. The solution of the given set of equations is similar to that for the surface mode on a single interface. Indeed, the only difference between equations (11.11) and (11.36) is in the factor of $\tanh \kappa_2 \ell$ or $\coth \kappa_2 \ell$ in (11.36). Since both $\tanh x$ and $\coth x$ converge to 1 when $x \to \infty$, we obtain in the limit of

$$\kappa_2 \ell \to \infty \tag{11.38}$$

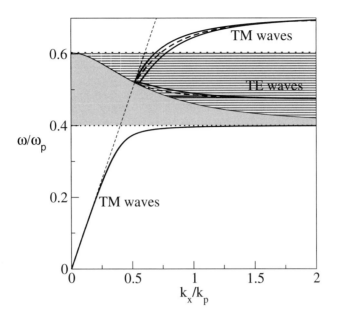

FIGURE 11.6. Spectrum of surface waves for a slab of a left-handed material. The thickness of the slab is $k_0\ell = 1.5$. The dashed lines show the spectrum of surface waves for a single interface. Notice that both TE and TM surface waves are split into two modes. This split is almost invisible for the lower TM mode. $k_0 = \omega_0/c$ and ω_0 is the resonant magnetic frequency of the left-handed medium, defined in equation (11.22) [137].

that two surface modes, located on the opposite sides of the slab, have the same frequency. This is easy to understand, since in the limit (11.38) there is no overlap of the electromagnetic fields decaying from both sides of the interface. When the slab becomes narrower, so that the fields inside the slab overlap, the two surface modes start to interact. This interaction removes the degeneracy, and the frequencies of the two resulting modes follow the dispersion relations (11.36).

To obtain the relationship between the frequency ω and the wave vector k_x of the surface waves, we again use the dispersion relation given by equation (11.13). However, in this case the relation $k_x = k_x(\omega)$ cannot be expressed by an analytical function because of the nontrivial κ dependence of the r.h.s. of equations (11.36) and (11.37). We can find the dispersion relation only numerically.

11.2.1 Slab of a Left-Handed Material

We will demonstrate the existence of surface modes on a slab of a left-handed material. The frequency dependence of the permittivity and permeability are given by equations (11.22).

Figure 11.6 shows the solution of equations (11.36) and (11.37) for the left-handed slab of thickness $k_0\ell = 1.5$, where $k_0 = \omega_0/c$. Since equations (11.36) and (11.37) are transcendental, we must solve them numerically. We see that for finite ℓ each single-surface mode splits into two modes. Only in the limit $\ell \to \infty$ do these two modes coincide.

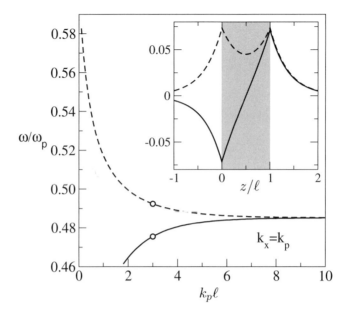

FIGURE 11.7. Eigenfrequencies of the TE surface wave for a left-handed slab of thickness ℓ. For definiteness, we fix $k_x = k_p$. The two branches are due to the interaction of surface waves on the opposite sides of the slab. For $\ell \to \infty$ the interaction is negligible and we recover the eigenfrequency of a surface wave at a single interface. The inset shows the z dependence of the electric field of the two surface waves for a slab of thickness $k_p \ell = 3$. The frequencies of these surface waves, $\omega = 0.4755 \omega_p$ and $\omega = 0.4924 \omega_p$, are marked by small circles.

To show how the splitting of the surface modes depends on the thickness of the slab, we calculate how the frequency of the two TE-polarized surface waves depend on the slab thickness ℓ. We fix the wave vector $k_x = 2.5 k_0 = k_p$. For the case of a single interface, discussed in section 11.1.2, this value of k_x corresponds to the frequency of the TE wave $\omega = 0.4852 \omega_p$. This is also the eigenfrequency of two surface waves on opposite sides of the slab, when the thickness of the slab, ℓ, is such that $\kappa_2 \ell \gg 1$. When ℓ decreases, two surface waves start to interact. This interaction changes both frequencies of surface modes, as is shown in figure 11.7. We see that, when ℓ decreases below a certain limit, $k_p \ell < 2$, the lower TE mode no longer exists. In analogy with the case of the bound states of two attractive δ-function potential wells, we conclude that this mode is an antisymmetric mode. As shown in the inset of figure 11.7, we indeed have bonding and antibonding of surface states. The dashed line corresponds to the bonding (symmetric) solution and the solid line corresponds to the antibonding (antisymmetric) solution.

The above analysis explains why the split of two lower TM surface modes is almost invisible in figure 11.6. Note that, the permittivity of the left-handed medium $|\varepsilon(\omega)| \gg 1$, when $\omega \ll \omega_p$. Therefore, the relation $\kappa \ell \ll 1$ is satisfied already for a much smaller thickness of the vacuum slab. As a consequence, the two TM surface modes do not "see" each other for the slab thickness $k_0 \ell = 1.5$, used in the figure 11.6.

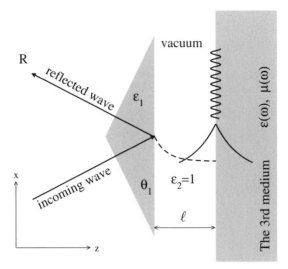

FIGURE 11.8. Experimental setup for the observation of surface waves. A dielectric wedge is used to produce the evanescent waves in the vacuum slab. Electromagnetic waves are totally reflected from the interface of the dielectric wedge with a vacuum slab. There are only evanescent waves in the vacuum slab. The reflection coefficient $R = 1$ for total internal reflection. If the third medium comes close to the wedge and supports surface modes, it might be possible to excite these surface modes, and the reflection coefficient R will be smaller than 1 at the frequency that excites the surface waves.

11.3 Experimental Observation of Surface Waves

We know already that the surface waves decrease exponentially on both sides of the interface. Therefore, the surface waves could be excited by evanescent waves approaching the surface from the outside. The easiest way to generate such evanescent waves is to use the phenomenon of total internal reflection.

Consider a possible experimental setup for observing surface waves, shown in figure 11.8. Electromagnetic waves propagate inside the wedge with a permittivity $\varepsilon_1 > 1$ (we consider $\varepsilon_1 = 9$ in this case). The wedge angle is $\theta_1 > \theta_{crit}$, so that we have total reflection on the interface between the wedge and the vacuum. Then there are no propagating waves in the space between the wedge and the third material. However, there are evanescent waves, decaying from the interface between the dielectric and the vacuum, as shown in figure 11.8. If the interface between the vacuum and the third material supports the excitation of surface waves, we can, by a change of the frequency of the incoming electromagnetic wave coming from the wedge, find the eigenfrequencies of the surface modes.

In this experiment, one measures the reflection coefficient R. If no surface mode is excited on the interface between the vacuum and the third medium, then all the energy of the incoming wave is reflected back because of the total internal reflection, and $R = 1$.

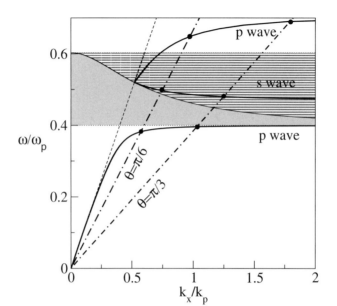

FIGURE 11.9. Searching for the dispersion relation of surface waves. Dot-dashed lines are dispersion relations of incoming waves, $\omega/c = k_x/(\sqrt{\varepsilon_1} \sin \theta_1)$, for two incident angles, $\theta_1 = \pi/6$ and $\theta_1 = \pi/3$. The full lines show dispersion relationships of surface waves at the interface between vacuum and a left-handed medium (figure 11.3). A cross-section of the dispersion line of incoming waves with the dispersion curve of surface waves determines the frequency of incoming wave that excites a surface wave. Of course, a TM-polarized incoming wave can excite only a TM surface wave, and the same holds for the TE polarization.

However, if both the frequency ω and the x-component of the wave vector,

$$k_x = \frac{\omega}{c} \sqrt{\varepsilon_1} \sin \theta_1, \tag{11.39}$$

of an incident wave coincide with the frequency and the wave vector of the surface wave, then the evanescent waves inside the vacuum slab will excite the surface waves at the interface between the vacuum and the third medium. This excitation costs energy. The only source of energy is the incoming wave. This means that part of the energy of the incoming wave is transferred through the vacuum slab. Consequently, the reflection coefficient R is smaller than 1 for this specific value of the frequency.

Figure 11.9 shows how to find the frequency of the surface wave. If we fix the incident angle and change the frequency, we move in the (k_x, ω) plane along the dot-dashed line. The cross section of the dot-dashed line with the dispersion curve of the surface waves gives us the frequency and the wave vector of the surface wave.

We can obtain results for the above experimental setup with the help of the transfer matrix for the slab derived in chapter 10. To simplify this experiment, consider the vacuum slab ($\varepsilon_2 = \mu_2 = 1$) sandwiched, from the left side, by a semi-infinite dielectric with permittivity $\varepsilon_1 = 9$ and permeability $\mu_1 = 1$, and from the right side by a semi-infinite left-handed material with permittivity $\varepsilon(\omega)$ and $\mu(\omega)$ given by equations (11.22).

An electromagnetic wave propagates in the dielectric medium and strikes the interface dielectric-vacuum with an incident angle $\theta_1 > \theta_{\text{crit}}$. In this way we will have total internal reflection. Then, in the vacuum we only have waves that decrease exponentially along the z-direction.

We can calculate the reflection coefficient

$$R = |r_s|^2, \qquad r_s = \frac{M_{21}^{(s)}}{M_{22}^{(s)}} \tag{11.40}$$

from the elements of the transfer matrix $\mathbf{M}^{(s)}$, derived in section 10.1. The general form for the elements of the transfer matrix, M_{22} and M_{21}, for the case of a slab located between two different media are given by equations (10.6) and (10.7), respectively. For our special case, these matrix elements are given as follows:

$$M_{22}^{(s)} = \frac{1}{2}\left[1 - i\mu_3 \frac{k_{1z}}{\kappa_3}\right]\cosh k_{2z}\ell + \frac{1}{2}\left[\mu_3 \frac{\kappa_2}{\kappa_3} - i\frac{k_{1z}}{\kappa_2}\right]\sinh k_{2z}\ell \tag{11.41}$$

and

$$M_{21}^{(s)} = \frac{1}{2}\left[1 + i\mu_3 \frac{k_{1z}}{\kappa_3}\right]\cosh k_{2z}\ell + \frac{1}{2}\left[\mu_3 \frac{\kappa_2}{\kappa_3} + i\frac{k_{1z}}{\kappa_2}\right]\sinh k_{2z}\ell, \tag{11.42}$$

with $\mu_3 = \mu(\omega)$ given by equation (11.22). The components of the wave vectors are related by the dispersion relations for materials 1, 2, and 3:

$$
\begin{aligned}
\frac{\omega^2}{c^2}\varepsilon_1 &= k_x^2 + k_{1z}^2, \\
\frac{\omega^2}{c^2} &= k_x^2 - \kappa_2^2, \\
\frac{\omega^2}{c^2}\varepsilon(\omega)\mu(\omega) &= k_x^2 - \kappa_3^2.
\end{aligned} \tag{11.43}
$$

For a fixed incident angle θ_1, we look for the value of the frequency ω of the incident electromagnetic wave that satisfies the condition for the existence of surface waves given by equation (11.11),

$$\kappa_2(\omega, k_x) + \frac{\kappa_3(\omega, k_x)}{\mu(\omega)} = 0. \tag{11.44}$$

However, the reflection coefficient R given by equation (11.40) is always 1 when the permittivity and permeability of the left-handed slab, $\varepsilon(\omega)$ and $\mu(\omega)$, are real. Indeed, we see from equations (11.41) and (11.42) that $M_{21}^{(s)} = [M_{22}^{(s)}]^*$. Hence, the absolute value of their ratio is always 1. This result is consistent with our definition of the transmission and reflection coefficients. We remind the reader that $R = 1 - T$ in lossless materials, and that the transmission coefficient T is given by the portion of the energy of the incoming wave transmitted form the dielectric through the vacuum slab and propagating in the third medium along the z-direction. In our case, no energy propagates in the z-direction, independently of whether the surface wave is excited or not. Consequently, $T = 0$ and $R = 1$. To avoid this difficulty and to detect the excitation of the surface waves, we must introduce a small imaginary part into the permittivity $\varepsilon(\omega)$ and permeability $\mu(\omega)$ of the left-handed material. This means that we use the dispersion relation given by

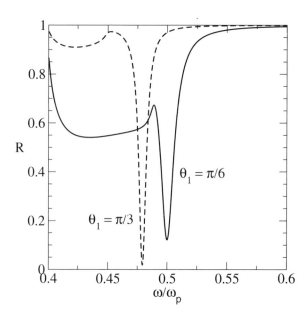

FIGURE 11.10. Frequency dependence of the reflection coefficient R for two different values of the incident angle θ_1. A decrease in the reflection coefficient means that the incident energy is transferred, through the vacuum slab, to the excitation of the surface wave. The position of the minimum in $R(\omega)$ agrees with the frequency of the TE surface wave marked in figure 11.9. For instance, in the case of $\theta = \pi/3$, we see that the frequency of the surface wave is $0.476\omega_p$. The width of the vacuum slab is $\ell = \lambda_p/4 = \pi c/(2\omega_p)$. To observe the dip in the frequency dependence of the reflection coefficient, we add a small imaginary part to both the permittivity and the permeability of the left-handed medium.

equations (16.31) and (16.53) with the loss term $\gamma = 0.003\omega_p$. Then, the energy of the surface wave is absorbed in the left-handed medium when a surface wave is excited. The absorption of the energy causes a decrease of the reflection coefficient.

Figure 11.10 shows the frequency dependence of the reflection coefficient R, for two different incident angles θ_1. We indeed see a sharp dip in the reflection coefficient when the frequency of the incident wave coincides with the frequency of surface wave.

Figure 11.11 shows how the frequency dip of the reflection coefficient depends on the width ℓ of the vacuum slab. Intuitively, it is clear that, if the distance between the dielectric and the left-handed medium is very large, then no surface waves can be excited. This is confirmed by our data. For instance, for the frequency $\omega = 0.476$, which is the position of the sharp dip in $R(\omega)$ observed in figure 11.10 for $\theta_1 = \pi/3$, we have from figure 11.4 that $\kappa_2/k_p = 0.6$. In a vacuum slab of thickness $\ell = 2\pi/k_p$, the evanescent wave decays by a factor of $e^{-2\pi/0.6} = e^{-10.5} = 3 \times 10^{-5}$. Thus, the electromagnetic field is too weak to excite a surface wave on the other side of the vacuum slab. Indeed, we see in figure 11.11 that the reflection coefficient is constant and equal to 1 for any frequency. When the width of the vacuum slab decreases by a factor of 4, $\ell = 2\pi c/(4k_p)$, then the evanescent wave decreases by only a factor of $e^{-10.5/4} \approx 0.07$. We expect that the electromagnetic

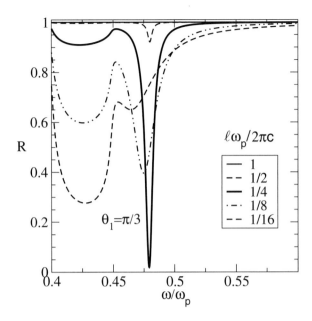

FIGURE 11.11. Frequency dependence of the reflection coefficient for two different incident angles and for different widths of the vacuum slab.

field is strong enough to excite a surface wave. Indeed, we observe a dip in the frequency dependence of the reflection coefficient. This is clearly seen in figure 11.11 as a solid line.

Note, however, that the above analysis depends also on the value of the damping factor γ. For smaller values of γ, the resonant dips become narrower and we observe dips also for a larger distance ℓ between the dielectric and the left-handed medium.

When the thickness of the slab further decreases, the surface waves start to interact with the dielectric medium. This changes the eigenfrequency of the surface waves. In figure 11.11 we see that the positions of the reflection dips change when the thickness of the vacuum slab decreases. An even further decrease in the vacuum thickness changes the physics completely. The vacuum slab becomes too narrow and we find that propagating waves exist in left-handed media when the frequency $\omega \sim 0.425\omega_p$. Note that, in this frequency region, the refractive index of the left-handed medium is real and negative but large in absolute value. Therefore, we have tunneling of the decayed electromagnetic wave from the dielectric medium into the left-handed material when the vacuum slab is very narrow.

11.4 Problems

Problem 11.1

Find the bound states of the slab of a left-handed material. First, use the left-handed material with a frequency-independent permittivity $\varepsilon = -9$ and permeability $\mu = -1$.

Then consider a realistic frequency-dependent permittivity and permeability, given by equations (11.22).

The first problem is, in fact, identical with that of the bound states of a dielectric slab, analyzed in section 10.2.2. The second problem is more complicated, since the frequency and the wave vector of the bound states lie in the interval in which the left-handed medium allows propagating solutions. This is the lower shaded region in figure 11.3. Since the electromagnetic wave must decay in vacuum, solutions are possible only for $k_x > \omega/c$, which is the region on the right of the dashed line in figure 11.3.

Problem 11.2
Find numerically the dispersion relations for two surface modes excited on opposite sides of thin metallic slab with negative permittivity $\varepsilon = 1 - \omega_p^2/\omega^2$. Estimate the thickness of the metallic slab at which two surface modes interact with each other.

Problem 11.3
Consider now an opposite configuration in which a thin dielectric slab is contained between two semi-infinite metallic media. Find the dispersion relation for the surface waves excited at two interfaces between the dielectric and metal [58].

Problem 11.4
Consider again the experimental setup shown in figure 11.8 with a slab of finite thickness ℓ_{LHM}, instead of a semi-infinite left-handed medium. Incident evanescent waves excite surface waves on the left interface of the slab. If the thickness of the left-handed slab, ℓ_{LHM}, is small, then we expect that the electromagnetic wave, decaying from the left interface of the slab, can excite a surface wave also at the second interface. When ℓ_{LHM} increases to infinity, no surface wave is excited on the second interface. Give a qualitative estimate of ℓ_{LHM} for which the second surface wave can be excited.

12 Resonant Tunneling through Double-Layer Structures

In chapter 10 we learned that the transmission through a dielectric slab is determined by the relation of the slab thickness to the wavelength of the electromagnetic wave inside the slab. In particular, the transmission coefficient is close to 1 when the wavelength of the electromagnetic wave in the z-direction is proportional to even multiples of the slab thickness.

Now, we will study the transmission of the electromagnetic wave through a system of two slabs, embedded in homogeneous material, as shown in figure 12.1 [39]. We concentrate on the case when there is no wave propagation in the layer a, because the z-component of the wave vector inside the layer a, k_{az}, is complex. This corresponds either to tunneling of the electromagnetic wave through a dielectric layer a, discussed in section 10.2.3, or to transmission through a thin metallic layer, discussed in section 10.3. Although the transmission through a single layer a is small, because of the exponential decrease of the electric and magnetic fields inside the slab a, we will show that the system of two layers of the same material a, separated by a distance ℓ_b, can, at certain frequencies, exhibit resonant transmission $T = 1$. These resonant frequencies are determined by the distance ℓ_b between the two layers a. Resonant transmission through a double-layered structure is the basis of the *Fabry-Pérot resonator*.

12.1 Transmission through Two Dielectric Layers

The calculation of the transmission coefficient of a double-dielectric-layer system is easy. We use the transfer matrix $\mathbf{M}^{(s)}$ or $\mathbf{M}^{(p)}$ for a slab embedded in a homogeneous medium,

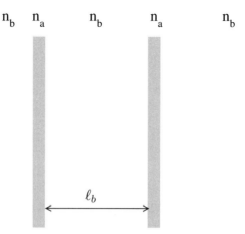

FIGURE 12.1. Experimental setup for transmission experiments through two thin layers a.

given by equations (10.17)–(10.20), and add a vacuum layer of thickness ℓ_b,

$$\mathbf{M} = \begin{pmatrix} e^{ik\ell_b} & 0 \\ 0 & e^{-ik\ell_b} \end{pmatrix} \mathbf{M}^{(s,p)}. \tag{12.1}$$

For a TE wave, we obtain the result that the ratio of the transmission amplitudes for a single slab a embedded in the medium b is

$$\frac{r}{t} = M_{12}^{(s)} e^{ik_{bz}\ell_b} = \frac{i}{2}\left[\frac{\mu_b}{\mu_a}\frac{k_{az}}{k_{bz}} - \frac{\mu_a}{\mu_b}\frac{k_{bz}}{k_{az}}\right]\sin k_{az}\ell_a e^{ik_{bz}\ell_b}, \tag{12.2}$$

where we use equation (10.18) for the matrix element $M_{12}^{(s)}$ of the transfer matrix. The transmission coefficient for a single slab a is given by

$$T_a = \frac{1}{1 + |M_{12}|^2}. \tag{12.3}$$

Now we apply Chebyshev's formula for the transmission through N barriers, given by equation (1.106),

$$T_N = \frac{1}{1 + |M_{12}|^2\,\dfrac{\sin^2 Nq\ell}{\sin^2 q\ell}}. \tag{12.4}$$

In our case, $N = 2$, so that $\sin Nq\ell/\sin q\ell = 2\cos q\ell$, and the transmission through the two slabs is given by the expression

$$T = \frac{1}{1 + 4|M_{12}|^2\cos^2 q\ell}, \tag{12.5}$$

where

$$\cos q\ell = \cos k_{az}\ell_a \cos k_{bz}\ell_b - \frac{1}{2}\left[\frac{\mu_b}{\mu_a}\frac{k_{az}}{k_{bz}} + \frac{\mu_a}{\mu_b}\frac{k_{bz}}{k_{az}}\right]\sin k_{az}\ell_a \sin k_{bz}\ell_b \tag{12.6}$$

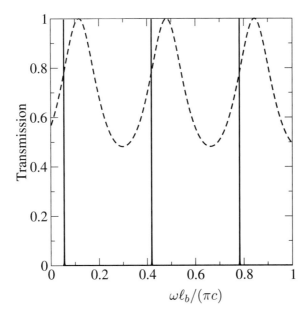

FIGURE 12.2. Transmission of electromagnetic waves through the double-layer structure shown in figure 12.1. The refractive indices of layers a and b are $n_a = 1$, and $n_b = 3$, respectively. The angle of propagation in medium b is $\theta_1 = \sin^{-1} 0.4$, so that there is total reflection at the interface between b and a. Nevertheless, the transmission through the two layers is $T = 1$ for distances ℓ_b for which $\cos q\ell$, given by equation (12.6), is zero. Note that the transmission is a periodic function of $k_{bz}\ell_b$, with period given by equation (12.8), where $k_{bz} = \omega n_b \cos \theta_b / c = 2.7497\omega/c$ in this particular case. Solid line depicts the transmission for $\ell_a = 5c/\omega$; dashed line is for $\ell_a = 0.5c/\omega$.

and

$$\ell = \ell_a + \ell_b. \tag{12.7}$$

We see that resonant transmission through a single layer, $T_a = 1$, is possible only when $M_{12} = 0$, but resonant transmission through a double-slab structure appears also at frequencies for which $\cos q\ell = 0$. Thus, the system of two layers can possess resonant transmission $T = 1$ also in the case when no resonant transmission through the single barrier is possible.

Consider now the case when two dielectric layers are embedded in a medium with a higher refractive index, $n_b > n_a$. The propagation direction of the electromagnetic wave in the dielectric b is such that waves are totally reflected at the interface between the dielectric b and the vacuum slab a: $\theta_b > \theta_{\text{crit}} = \sin^{-1}(n_a/n_b)$. Then T_a is always less than 1, since there is only tunneling through the layer a, as discussed in section 10.2.3. However, a system of two vacuum layers a can exhibit resonant transmission.

Figure 12.2 shows the transmission coefficient through a system of two vacuum layers a embedded in a dielectric medium with refractive index $n_b = 3$. The direction of propagation in a dielectric medium is given by the angle $\theta_b = \sin^{-1} 0.4$.

Since $\theta_b > \theta_{\mathrm{crit}} = \sin^{-1}(n_a/n_b)$, we have total internal reflection at the interface between the dielectric and vacuum. However, if $q\ell = (2m+1)\pi/2$, then resonant transmission appears. Note that T is a periodic function of $k_{bz}\ell_b$. Therefore, if $\cos q\ell = 0$ for a given distance ℓ_b, then it is also zero for the distance $\ell_b' = \ell_b + \pi/k_{bz}$. Thus, the transmission coefficient is a periodic function of ℓ_b with period

$$\Delta \ell_b = \frac{\pi}{k_{bz}} = \frac{\pi c}{\omega \cos \theta_b}, \tag{12.8}$$

as is clearly seen in figure 12.2.

Of course, the transmission strongly depends on the thickness of the vacuum layer a. We consider two cases in figure 12.2. If layer a is thin, $\ell_a = 0.5c/\omega$, then the transmission coefficient is high for any distance ℓ_b. This is because the tunneling transmission through layer a is high (see section 10.2.3 for details). However, for a much thicker layer $\ell_a = 5c/\omega$, there is almost no transmission through a single barrier. In the double-layered system, transmission is possible only for very specific distances ℓ_b between two layers a, as shown in figure 12.2.

12.2 Transmission through Two Metallic Layers

A similar analysis can be done for the case of two thin metallic layers. In this case, we consider only normal incidence of the electromagnetic wave. We remind the reader that the metallic layer a has a complex permittivity, given by the Drude formula,

$$\varepsilon(\omega) = 1 - \frac{\omega_p^2}{\omega^2 + i\omega\gamma}. \tag{12.9}$$

We will discuss only the high-frequency limit, $\gamma \ll \omega < \omega_p$.

First, we neglect the loss term. If $\gamma = 0$, then the metallic permittivity is real and negative. The electromagnetic wave decreases in the metallic layer as $E(z) \sim e^{-z/\delta}$, where $\delta = c/\omega_p$ is the skin depth. Nevertheless, the transmission through two metallic layers can be resonant. We calculate the transmission coefficient for the system of two metallic layers in the same way as for the case of two dielectric layers. Using equation (12.6), we obtain the transmission coefficient as a function of the distance ℓ_b between layers. Results are shown in figure 12.3 for three values of the thickness of the metallic slab, ℓ_a.

If the absorption inside the metal is considered, then we cannot expect resonant transmission to appear. The transmission coefficient is $T < 1$. Nevertheless, we see in figure 12.4, that for certain distances ℓ_b the transmission through two metallic layers is rather high, much higher than that through the single metallic slab.

Note that in in this case we cannot calculate the transmission from the Chebyshev formula, given by equations (12.5) and (12.6), because $R + T < 1$. We remind the reader that the Chebyshev formula was derived in section 1.4.5, using the assumption $|r|^2 + |t|^2 = 1$ [see equation (1.102)]. Therefore, we have to calculate the product of transfer

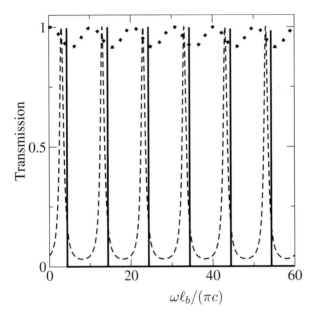

FIGURE 12.3. Transmission through a system of two *lossless* metallic layers of thickness ℓ_a. The thickness is 0.03, 0.3, and 3 $\times\delta$ ($\delta = c/\omega_p$ is the skin depth) depicted by dotted, dashed, and solid lines, respectively. The frequency $\omega = 0.1\omega_p$ and $\gamma \equiv 0$. The transmission is a periodic function of the distance ℓ_b, with period equal to $\pi c/\omega$, which is half of the wavelength of the propagating electromagnetic wave.

matrices,

$$M_{\text{double}} = M_a \begin{pmatrix} e^{ik_b\ell_b} & 0 \\ 0 & e^{-ik_b\ell_b} \end{pmatrix} M_a, \tag{12.10}$$

where M_a is the transfer matrix for a homogeneous metallic slab a embedded in the medium b, given by equations (10.17)–(10.20). The transmission is then given as

$$T = \frac{1}{|M_{22}|^2}. \tag{12.11}$$

An alternative Chebyshev formula for periodic systems with absorption is given in problem 12.2.

Figure 12.4 shows the transmission as a function of the distance between two metallic layers for three values of the thickness of the metallic layers. We see that the transmission is much larger than for the case of a single metallic slab, shown in figure 10.11. For instance, if $\ell_a/\delta = 0.3$, we have that the transmission through a single slab is $T_a = 0.3$, but the transmission through the double-layer system can be as large as 0.85. Transmission can be rather high even for the case when the thickness of the metallic layer is $\ell_a = 3\delta$, as is shown in the lower panel of figure 12.4.

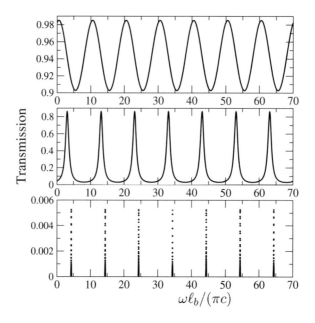

Figure 12.4. Transmission coefficient *versus* distance between two metallic layers, ℓ_b. Note the different scales on the horizontal axis. The frequency $\omega = 0.1\omega_p$ and $\gamma = 0.0025\omega_p$. The thickness of the metallic layer, ℓ_a, is 0.03 (top), 0.3 (middle), and 3 (bottom) $\times\delta$ ($\delta = c/\omega_p$ is the skin depth). The transmission is a periodic function of the distance ℓ_b, with period equal to $\pi c/\omega$, which is half of the wavelength of the propagating electromagnetic wave.

12.3 Problems

Problem 12.1

Calculate the transmission coefficient for single lossless metallic layers of thickness $\ell_a = 0.03\delta$ and $\ell_a = 3\delta$ ($\delta = c/\omega_p$ is the skin depth). The frequency of the electromagnetic wave is $\omega = 0.1\omega_p$. The metallic permittivity is given by the Drude formula equation (12.9) with $\gamma = 0$. Compare your results with the transmission through a system of two metallic layers, shown in figure 12.3.

Problem 12.2

Using the general expression $T = |M_{22}|^{-2}$ and Chebyshev's identity, given by equation (1.99), derive an alternative form of Chebyshev's formula for the transmission coefficient through a periodic system of N samples with absorption,

$$T = \left| \frac{\sin q\ell}{M_{22}\sin Nq\ell - \sin(N-1)q\ell} \right|^2, \tag{12.12}$$

where M_{22} is an element of the transfer matrix of one period and $q\ell$ is given by equation (12.6).

13 Layered Electromagnetic Medium: Photonic Crystals

Previously, in chapters 9 and 10, we analyzed the transmission of electromagnetic waves through a single interface and through a thin slab of width ℓ. In chapter 12 we learned that the transmission through two slabs of the same material leads to resonant transmission, even if resonant transmission through one layer is not possible. This indicates that more complicated structures could have new transmission properties not observable in single components of the structure. To investigate this problem in more detail, in this chapter we will apply the transfer matrix formalism to analysis of the transmission coefficient through layered *periodic* media, which consist of an infinite repetition of two layers a and b [39]. Such a periodic medium represents the simplest model for a one-dimensional *photonic crystal*. We show that the spectrum of the periodic layered structure consists of bands and gaps, and give their simple physical interpretation. Photonic crystals represent a new class of materials intensively studied both theoretically and experimentally [21, 22, 27, 33, 35, 36, 39, 42, 50, 53, 61, 64, 72, 102, 139, 164, 165]. While the one-dimensional case, discussed in this chapter, can be analyzed analytically using the transfer matrix method, most known results for two- and three-dimensional structures must be obtained numerically, using the generalized transfer matrix method [118, 122, 123, 160] or time domain methods [38].

We discuss both the TE and TM polarizations of propagating electromagnetic waves. For the non-normal direction of propagation, the polarization strongly influences the structure of the bands of allowed frequencies. In particular, the TM-polarized wave propagating in the direction given by Brewster's angle possesses no transmission gap.

We will also discuss the coupling of the layered medium with the surrounding environment, and calculate the transmission of the electromagnetic wave through a

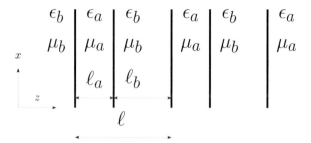

FIGURE 13.1. Infinite periodic layered structure.

finite periodic medium. As the number of layers increases, we clearly see the development of bands and gaps through studies of the transmission coefficient. We also discuss finite layered media, which possess *omnidirectional* gaps [162]. This means that such a medium behaves in a given frequency interval as a perfect mirror, reflecting electromagnetic waves of both polarizations and all incident angles.

We also analyze layered media that contain dispersive materials, such as metals and left-handed materials. For a periodic arrangement of very thin metallic layers, we find a very simple band structure. We find no propagation for frequencies less than a cutoff frequency, which is inversely proportional to the distance between the thin metallic layers. We also examine layered materials with one of the layers being a left-handed material and the other either dielectric or vacuum. In this case a new type of gap appears, which depends only on the average value of the index of refraction [94]. Unlike gaps observed in dielectric photonic crystals, this new gap does not change its position if we change the widths of the layers.

Finally, we discuss a simplified model of the layered medium, the Kronig-Penney model, with infinitesimally narrow layers *a* possessing an infinite refractive index [97].

13.1 Photonic Crystals: Infinite Periodic Layered Medium

The simplest layered medium is a periodic structure consisting of an infinite periodic arrangement of two layers *a* and *b*. Layer *a* has a thickness ℓ_a, permittivity ε_a, and permeability μ_a; and layer *b* has parameters ℓ_b, ε_b, and μ_b, as shown in figure 13.1. We want to find the band structure of such a periodic medium.

The simplest way to find the band structure is to determine the transfer matrix **M** of one unit cell, and calculate its eigenvalues. Propagating solutions correspond to eigenvalues $\lambda_{1,2} = e^{\pm iq\ell}$ with real momentum q.

The transfer matrix of one unit cell, **M**, can be obtained with the help of the transfer matrix for a slab embedded in a homogeneous medium. This transfer matrix was derived in section 10.1.1 for the case of TE polarization. To obtain the transfer matrix of the unit cell, we need to multiply the transfer matrix of the slab *b*, $\mathbf{M}_{\text{slab}}$, by the translational matrix given by

$$\mathbf{M}^{(s,p)} = \begin{pmatrix} e^{ik_{bz}\ell_b} & 0 \\ 0 & e^{-ik_{bz}\ell_b} \end{pmatrix} \mathbf{M}^{(s,p)}_{\text{slab}}, \tag{13.1}$$

where the matrix $\mathbf{M}^{(s,p)}_{\text{slab}}$ is the transfer matrix of slab a embedded in the medium b, given by equations (10.17)–(10.20) for the TE- or TM-polarized wave. Note that, since the medium on both sides of the slab b is the same, the differences between matrices $\mathbf{M}$ and $\tilde{\mathbf{M}}$ for the electric and magnetic fields are not relevant (see problem 10.1).

After some simple algebra, we obtain

$$\Gamma^{(s,p)} = \text{Tr}\, \mathbf{M}^{(s,p)} = 2\cos\phi_a \cos\phi_b - \left(z_{ab}^{(s,p)} + \frac{1}{z_{ab}^{(s,p)}} \right) \sin\phi_a \sin\phi_b, \tag{13.2}$$

where we introduced

$$\phi_a = \ell_a k_{az}, \quad \phi_b = \ell_b k_{bz}, \tag{13.3}$$

and

$$z_{ab}^{(s)} = \frac{\mu_b}{\mu_a} \frac{k_{az}}{k_{bz}} \tag{13.4}$$

for the case of the TE-polarized wave, and

$$z_{ab}^{(p)} = \frac{\varepsilon_b}{\varepsilon_a} \frac{k_{az}}{k_{bz}} \tag{13.5}$$

for the case of the TM-polarized wave.

We remind the reader that the dispersion relationships for the electromagnetic wave inside the material a and b are

$$\frac{\omega^2}{c^2}\varepsilon_a\mu_a = k_x^2 + k_{az}^2 \quad \text{and} \quad \frac{\omega^2}{c^2}\varepsilon_b\mu_b = k_x^2 + k_{bz}^2. \tag{13.6}$$

Here, k_x is the x-component of the wave vector, conserved inside the structure. k_x determines the direction of the propagation of the electromagnetic wave inside the medium.

Since absorption destroys any propagation in infinite systems, we do not consider systems with absorption in this chapter. That means that all the electromagnetic parameters, the permittivities and permeabilities, are supposed to be real. Then, $\text{Tr}\,\mathbf{M}^{(s,p)}$ is real, too, and we can use the theorem from section 1.4.2, that the absolute value of the trace of the transfer matrix $|\text{Tr}\,\mathbf{M}|$ for the propagating solutions is less than 2. Thus, propagating solutions exist in the periodic medium only when

$$|\Gamma| < 2. \tag{13.7}$$

Then, $\text{Tr}\,\mathbf{M} = 2\cos q\ell$, where $\ell = \ell_a + \ell_b$ is the spatial period of the layered structure, and q is a *real* wave vector inside the structure. Our aim is to find all the allowed frequencies and to calculate the dispersion relation $\omega = \omega(q)$. First, we investigate in detail the case when both materials a and b are dielectrics.

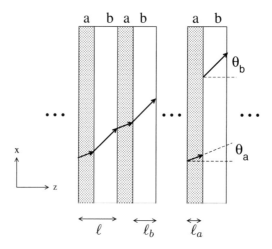

FIGURE 13.2. Structure of an infinite periodic layered medium. Two layers, a and b, have thicknesses ℓ_a and ℓ_b, respectively, and refractive indices $n_a > n_b$. The spatial period of the structure is $\ell = \ell_a + \ell_b$. Also shown are the angles θ_a and θ_b of propagation in layers a and b, respectively.

13.2 Periodic Arrangement of Dielectric Layers

For dielectric layers, we have $\mu_a = \mu_b = 1$ and $\varepsilon_a^2 = n_a$ and $\varepsilon_b^2 = n_b$, where n_a and n_b are the refractive indices of the materials, a and b, respectively. Then we write $k_{az} = k_a \cos\theta_a$ and $k_{bz} = k_b \cos\theta_b$. The two angles of propagation, θ_a and θ_b, are related to each other by Snell's law, $n_a \sin\theta_a = n_b \sin\theta_b$. We consider the case $n_a > n_b$.

In the case of dielectric layers a and b, the expressions for z_{ab} are given by

$$z_{ab}^{(s)} = \frac{n_b \cos\theta_a}{n_a \cos\theta_b} \quad \text{and} \quad z_{ab}^{(p)} = \frac{n_a \cos\theta_a}{n_b \cos\theta_b}. \tag{13.8}$$

13.2.1 Symmetries

Before the analysis of the dispersive relation, we will discuss the parameters of the model. At present, we have six parameters: the frequency ω, two lengths ℓ_a and ℓ_b, two refractive indices n_a and n_b, and, finally, the direction of propagation, given by the x-component of the wave vector (figure 13.2). We remind the reader that the x-component of the wave vector is conserved in the structure so that $k_{ax} = k_{bx} = k_x$.

Fortunately, we can use some symmetries that are obeyed by equation (13.2), and reduce the number of free parameters to four. First, note that equation (13.2) does not change if we make the following transformations:

$$n_a \rightarrow a n_a, \quad n_b \rightarrow a n_b, \quad \text{and} \quad \omega \rightarrow \omega/a. \tag{13.9}$$

This means that, the transmission properties of the dielectric layered medium are given only by the ratio of the refractive indices, n_a/n_b. Two layered media, which differ only in the refractive indices n_a and n_b in the first structure and n_a' and n_b' in the second structure, have the same band structure (after rescaling of the frequency), if $n_a/n_b = n_a'/n_b'$. Without loss of generality, we can consider $n_b = 1$. For other values of the refractive index n_b, we rescale the frequency according to equation (13.9).

Second, note that equation (13.2) does not change if we rescale the lengths ℓ_a and ℓ_b and the frequency ω, as follows:

$$\ell_a \to a\ell_a, \quad \ell_b \to a\ell_b, \quad \text{and} \quad \omega \to \omega/a. \tag{13.10}$$

We want to emphasize that the above symmetries can be applied only in the case when the materials a and b are not dispersive. This means that the permittivity of both materials does not depend on the frequency.

With the help of the above symmetries, we see that the transmission properties of a given structure are given only by two parameters: the ratio n_a/n_b of the refractive indices and the ratio ℓ_a/ℓ_b of the layer widths. The two other parameters, the frequency ω and the x-component of the wave vector, k_x, determine the energy and the direction of propagation of the electromagnetic wave.

13.2.2 Quarter Stack

It is also convenient to consider the frequency with respect to the reference frequency ω_0 defined by the relation

$$k_0 n_b \ell_b = \frac{\pi}{2}, \quad k_0 = \frac{\omega_0}{c}. \tag{13.11}$$

With the help of ω_0, we have $k_b \ell_b = (\omega/ck_0)k_0 n_b \ell_b = (\omega/\omega_0) \times \pi/2$, so that

$$\phi_b = k_{bz}\ell_b = k_b\ell_b \cos\theta_b = \frac{\pi}{2}\frac{\omega}{\omega_0}\cos\theta_b \tag{13.12}$$

and

$$\phi_a = k_{az}\ell_a = \left(\frac{n_a\ell_a}{n_b\ell_b}\right) k_b\ell_b \cos\theta_a = \frac{\pi}{2}\frac{\omega}{\omega_0}\left(\frac{n_a\ell_a}{n_b\ell_b}\right)\cos\theta_a. \tag{13.13}$$

We will first study a special case of a layered medium, where

$$\frac{n_b\ell_b}{n_a\ell_a} = 1. \tag{13.14}$$

Such a structure is called a *quarter stack* layered medium. Consider an electromagnetic wave with frequency ω_0. Then, from equation (13.11) we have

$$k_b\ell_b = 2\pi\frac{\ell_b}{\lambda_b} = \frac{\pi}{2}. \tag{13.15}$$

so that

$$\ell_b = \frac{\lambda_b}{4}. \tag{13.16}$$

Also, with the use of equation (13.14) we obtain

$$\ell_a = \frac{\lambda_a}{4}. \tag{13.17}$$

We obtain that, in the case of normal incidence ($\theta_a = \theta_b = 0$), the width of each slab is exactly equal to one-quarter of the wavelength of the electromagnetic wave propagating in a given medium with frequency ω_0.

13.3 Band Structure of Photonic Crystals

Now we can analyze the band structure of layered media. The simplest way to obtain the bands and gaps is to calculate the wave vector $q\ell$ for many different values of the parameters k_x and ω. If $q\ell$ is real, then a given pair (k_x, ω) belongs to the energy band. If $q\ell$ is imaginary, then our parameters belong to the frequency gap.

But, before using this technique, let us try to find some simple arguments for the existence of gaps and bands. First, note that, if $\phi_a = m_a\pi$ but $\phi_b \neq m_b\pi$ (m_a and m_b are integers), then $|\Gamma| \leq 2$. Therefore, this special case corresponds to a propagating solution. Next, note that $|\Gamma| = 2$, if both $\phi_a = m_a\pi$ and $\phi_b = m_b\pi$. Thus, we expect that the corresponding parameters (k_x, ω) determine the band edges of the transmission bands.

Then consider the situation when $\phi_a \neq m\pi$, $\phi_b \neq m\pi$, but their sum

$$\phi_a + \phi_b = m\pi, \quad m = 1, 2, \ldots . \tag{13.18}$$

Then, $\phi_b = m\pi - \phi_a$, $\cos\phi_b = (-1)^m \cos\phi_a$ and $\sin\phi_b = (-1)^{m+1} \sin\phi_a$. Inserting this information into equation (13.2), we immediately obtain

$$|\Gamma| = 2\left[1 + \frac{1}{2}\left(\frac{1}{z_{ab}} - z_{ab}\right)^2 \sin^2\phi_a\right] > 2. \tag{13.19}$$

So we see that the layered structure possesses a gap for frequencies that satisfy the condition (13.18).

13.3.1 The Limit of $n_a \gg n_b$

The analysis of the band structure becomes very simple in the limit of $n_a \gg n_b$. In this limit, the term $z_{ab} + z_{ab}^{-1}$ becomes very large. Since propagating solutions exist only when the absolute value of the right-hand side of equation (13.2) is smaller than 2,

$$\left|2\cos\phi_a \cos\phi_b - \left(z_{ab}^{s,p} + \frac{1}{z_{ab}^{s,p}}\right)\sin\phi_a \sin\phi_b\right| < 2, \tag{13.20}$$

we obtain the result that propagating solutions exist in the limit of $n_a/n_b = \infty$, only if

$$\sin\phi_a = 0 \tag{13.21}$$

or

$$\sin \phi_b = 0. \tag{13.22}$$

The condition (13.21) is satisfied if $\phi_a = (\pi/2)(\omega/\omega_0) \cos \theta_a = m\pi$, where m is an integer. This gives us the allowed frequency

$$\frac{\omega}{\omega_0} = \frac{2m}{\cos \theta_a} \approx 2m. \tag{13.23}$$

The last relationship follows from Snell's law, $n_a \sin \theta_a = n_b \sin \theta_b$, which tells us that $\theta_a \sim n_b/n_a \ll 1$ so that $\cos \theta_a \approx 1$. Thus, we obtain that transmission is possible only if $\omega = 2m\omega_0$. This transmission band does not depend on the direction of propagation.

The second condition for the existence of bands, given by equation (13.22), is satisfied if

$$\frac{\pi}{2} \frac{\omega}{\omega_0} \cos \theta_b = m\pi, \tag{13.24}$$

where m is an integer. This means that

$$k_{bz}/k_0 = 2m. \tag{13.25}$$

Using the dispersion relation for slab b written in the form $n_b^2 \omega^2 = k_{bz}^2 + k_x^2$, and the relation $\omega_0^2/c^2 = k_0^2$, we obtain

$$n_b \frac{\omega}{\omega_0} = \sqrt{(2m)^2 + (k_x/k_0)^2}. \tag{13.26}$$

Note that, the two relations (13.21) and (13.22) represent the condition (10.41) for the Fabry-Pérot resonances in the slab a and b, respectively. Thus, in the limit of $n_a \gg n_b$, all transmission bands in photonic crystals can be explained as a consequence of the Fabry-Pérot resonances either in slab a or in slab b.

The band structure of the quarter stack with $n_a = 20$ and $n_b = 1$ is shown in figure 13.3. We see that in figure 13.3 that the bands of allowed frequencies are indeed very narrow when $n_a/n_b = 20$. The positions of bands agree very well with our predictions given by equations (13.23) and (13.26).

13.3.2 Band Structure: General Case

The band structure of the quarter stack with $n_b = 1$ and $n_a = 2$ is shown in figure 13.4. We see that the spectrum of the allowed energies depends strongly on the direction of propagation of the electromagnetic wave inside the layered medium. The band structure is given as a function of two parameters, the frequency ω/ω_0 and the x component of the wave vectors, k_x/k_0, which define the direction of propagation. To find bands and gaps for a given direction of propagation, we must draw a line from the origin in the direction $\omega = k_x/\sin \theta_b$. A few such lines are drawn in figure 13.4. The thin dashed line is for the case of normal incidence ($k_x = 0$), the dotted line is for the case of Brewster's angle θ_{Bp}, and the thick solid and dashed lines determine the light cone $\omega = ck/n_b$ ($\omega = ck/n_a$)

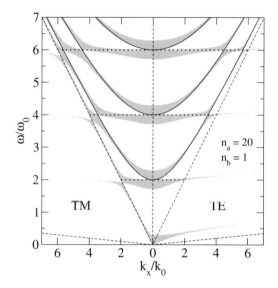

FIGURE 13.3. Band structure of a quarter stack with $n_a/n_b = 20$. Dashed lines are dispersion relations $\omega = ck/n_a$ and $\omega = ck/n_b$ for the materials a and b, respectively. In the limit of $n_a/n_b \to \infty$, the transmission bands reduce to $\omega = 2m\omega_0$ and $\omega/\omega_0 = \sqrt{(2m)^2 - k_{bx}^2/k_0^2}$, shown by dotted and solid lines, respectively.

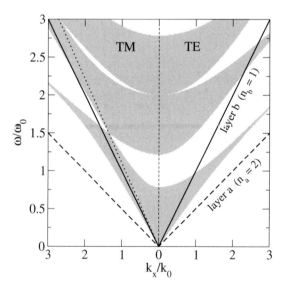

FIGURE 13.4. Band structure of a photonic crystal with $n_a = 2$ and $n_b = 1$. The solid (dashed) lines are dispersion relations for the material b (a), respectively. The dotted line is the relation $k_x = \omega \sin\theta_{Bp}$, where θ_{Bp} is Brewster's angle, $\tan\theta_{Bp} = n_b/n_a$ for the TM-polarized electromagnetic wave. Note that, the TM wave is totally transmitted through each interface so that there is only a right-going wave for this particular propagation direction.

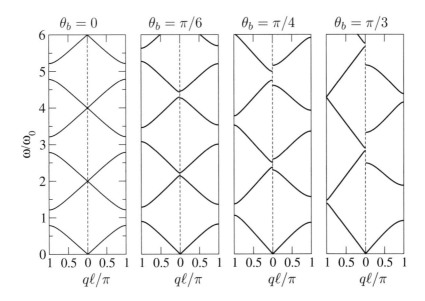

FIGURE 13.5. Frequency versus wave vector $q\ell$ of an electromagnetic wave propagating inside an infinite layered medium with $\ell_a n_a = \ell_b n_b$ and $n_a = 2$, $n_b = 1$. The propagation direction is given by the angle θ_b. The left and right sides of each panel show the band structure for the TM and TE modes, respectively. Note that, the two polarizations possess different band structures for non-normal propagation direction given by the angle θ_b.

for the materials b and a, respectively. We remind the reader that the light cone of the material with refractive index n is defined by the region between the lines $k \equiv 0$ and $k = \omega n/c$. It is the region of parameters that allow propagation of electromagnetic waves.

For normal incidence, $k_x = 0$, our problem reduces to the problem of finding the energy bands of a one-dimensional periodic arrangement of rectangular wells, as discussed in section 4.7.2. We remind the reader that we have observed similar band structures already in previous chapters, when we studied an infinite Kronig-Penney model of periodic δ-function potential wells (chapter 4) and also the tight binding model with various spatial periodicities (chapter 5). We have learned that the band structure is a consequence of the interference of waves, which are transmitted and reflected inside the periodic structure.

Note that, in the case of normal incidence, we can find the band structure analytically. The solution is given in the solved problem 13.1.

Nonzero values of k_x correspond to nonzero angles of propagation θ_a and θ_b. We see that the structure of the bands and gaps depends on these angles, which determine the propagation of the electromagnetic wave in the structure. To see the band structure for a given direction of propagation θ_b, in figure 13.4 we simply draw the line $\omega = k_x/(n_b \sin \theta_b)$. When the frequency increases, we move along this line and cross either a band or a gap of the structure. The band structure for four directions of propagation is shown in figure 13.5.

For $\theta_b = 0$, which is the case of propagation perpendicular to the interface between the layers, the band structure of both polarizations is the same. However, for nonzero angle θ_b, the two polarization modes, TE and TM, possess different band structures. Contrary to the TE mode, where the gaps become broader as θ_b increases, the gaps become *narrower* for the TM mode, and finally disappear when θ_b equals the Brewster angle θ_{Bp}. We remind the reader that the TM wave incident to the interface with incidence angle θ_{Bp} is totally transmitted through the interface. Note that, if θ_b is the Brewster angle for the interface $b \to a$, then θ_a is the Brewster angle for the interface $a \to b$. Therefore, the wave propagating in the direction of the Brewster angle is totally transmitted through the structure. In figure 13.4 we show the wave propagation in the direction of the Brewster angle by a dotted line. Indeed, all gaps close for this direction of propagation.

Note also that the layered structure allows propagation of the electromagnetic wave outside the light cone of the medium b. Indeed, we see that the bands span also the region between the two light cones, which is the region where $\theta_a > \theta_{crit}$ (we remind the reader that $n_a > n_b$). For this direction of propagation, θ_b and k_{bz} are imaginary, because $k_x > \omega$. In layer a, which has a larger index of refraction, the electromagnetic wave still propagates. These propagating states become very rare in the limit $n_a \gg n_b$ (see figure 13.3), because the electromagnetic wave decays exponentially inside each slab b,

$$\vec{E}_b = \vec{E}_b^+ e^{-\kappa_b z} + \vec{E}_b^- e^{-\kappa_b(\ell_b - z)}, \tag{13.27}$$

where z is measured from the left interface of the slab b and κ_b was derived in section 9.6,

$$\kappa_b = \frac{\omega^2}{c^2} \left[\varepsilon_a \sin^2 \theta_a - \varepsilon_b \right], \tag{13.28}$$

as given by equation (9.123). We see that κ_b increases with frequency, so tunneling through slab b becomes more difficult for higher frequencies.

In the limit of $n_a/n_b \to 1$, the layered medium becomes similar to a homogeneous material. Gaps becomes narrower and disappear for $n_a = n_b$.

13.4 Coupling to a Finite Photonic Crystal

Each photonic crystal has a finite size. The electromagnetic wave propagates in the embedding medium (mostly air), strikes the interface between air and the layered structure, then propagates inside the layered structure, and finally leaves the structure through the interface on the opposite side. The coupling between the layered structure, i.e., the photonic crystal, and the embedding medium must be considered when we are interested in the transmission coefficient of a *finite* layered structure (figure 13.6). The transfer matrix of this process can be written as

$$\mathbf{M} = \mathbf{M}_{slab}^{0b} \mathbf{M}_{hom}^{b} \left[\mathbf{M}_{slab}^{bb} \right]^{N-1} \mathbf{M}_{int}^{b0}, \tag{13.29}$$

where the transfer matrix $\mathbf{M}_{int}^{b0}$ is the transfer matrix for the interface between the air and the first slab b on the right side of the layered structure, $\mathbf{M}_{slab}^{bb}$ is the transfer matrix for traveling from the right edge of the slab b to the right edge of the next slab b, given by equation (13.1), $\mathbf{M}_{hom}^{b}$ is the diagonal transfer matrix for propagation inside the

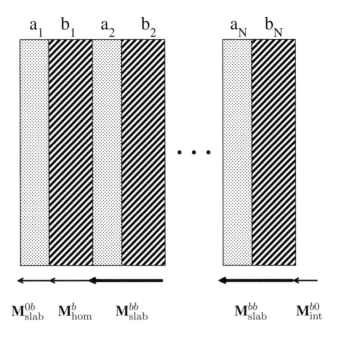

FIGURE 13.6. Calculation of the transfer matrix for a layered medium embedded in a vacuum. Arrows shows the distance the wave travels due to the corresponding transfer matrix.

homogeneous slab b, and finally, the transfer matrix $\mathbf{M}_{\text{slab}}^{0b}$ is the transfer matrix from the right edge of the last slab b to air, derived in section 10.1. The calculation of the transfer matrix is shown in figure 13.6.

Another method of calculation of the transfer matrix of a periodic system of finite length is described in problem 13.3.

We will first study the simpler problem when the refractive index n_b equals the refractive index of the embedding medium. In this case, the transfer matrix, given by equation (13.29), simplifies considerably. Then we discuss also the general case, when n_b differs from the refractive index of the embedding medium. This case is very important, since it enables us to obtain a layered medium that, in a given frequency interval, reflects all electromagnetic waves for both polarizations and for all incident angles.

13.4.1 The Transmission Coefficient for a Finite Layered Medium

In the special case when the layer b is also a vacuum, the transfer matrix (13.29) reduces to

$$\mathbf{M} = \left[\mathbf{M}_{\text{slab}}^{bb}\right]^{N}. \tag{13.30}$$

In this particular case, we can easily calculate the transmission with the help of Chebyshev's formula. We find the M_{12} element of the transfer matrix $\mathbf{M}$ given by

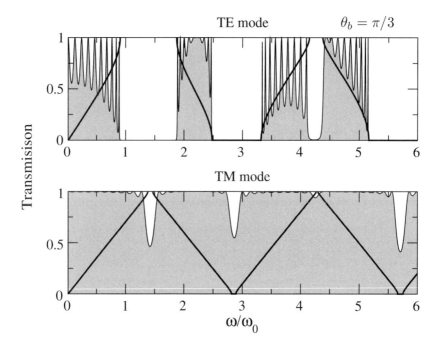

FIGURE 13.7. Transmission coefficient versus frequency for a layered dielectric structure of $N = 10$ periods for the incident angle $\theta_b = \pi/3$. The index of refraction is $n_a = 2$ and $n_b = 1$ for layers a and b, respectively. The top (bottom) panel shows the transmission for the TE (TM) mode. The solid line shows the wave vector $q\ell$, which is real inside the transmission band and imaginary in the gaps.

equation (13.1),

$$M_{12} = \frac{1}{2} e^{ik_{bz}\ell_b} \left[z_{ab} - \frac{1}{z_{ab}} \right] \sin \phi_a,$$ (13.31)

and obtain the transmission coefficient T_N for the N layers, given by

$$T_N = \frac{1}{1 + |M_{12}|^2 \dfrac{\sin^2 Nq\ell}{\sin q\ell}},$$ (13.32)

where $q\ell$ is given by equation (13.2) and $\ell = \ell_a + \ell_b$. In figure 13.7 we show the transmission coefficient for an electromagnetic wave propagating through a system of $N = 10$ layers a with refractive index $n_a = 2$, separated by a vacuum slab with thickness $\ell_b = 2\ell_a$. We see that the positions of the transmission bands and gaps agree well with the gaps obtained for the infinite layered medium, shown in figures 13.4 and 13.5. To show this equivalence between the finite and the infinite media more clearly, we plot in figure 13.7 also the wave vector $q\ell$. We see that the frequency intervals in which the transmission is nonzero coincide with the intervals in which $q\ell$ is real. Some doubts

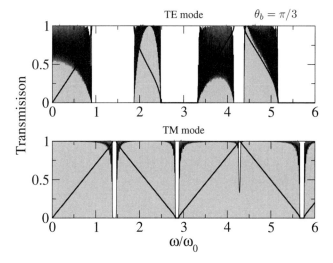

FIGURE 13.8. Transmission coefficient versus frequency for a layered dielectric structure of $N = 50$ periods for an incident angle $\theta_b = \pi/3$. The index of refraction is $n_a = 2$ and $n_b = 1$ for layers a and b, respectively. The top (bottom) panel shows the transmission for the TE (TM) mode. The solid line shows the wave vector $q\ell$, which is real inside the transmission band and imaginary in the gaps.

about the validity of this statement appear for the TM wave, where the gaps are not so pronounced for the $N = 10$ layers and $q\ell$ is also real inside the gaps. However, if we calculate the transmission through a system of $N = 50$ layers, we see in figure 13.8 that in this case the gaps are well developed also for the TM mode and $q\ell$ is real only inside the gaps.

Thus, we see that equation (13.32) is a very important formula for obtaining the transmission coefficient for any number of layers and for any incident angle. We need only calculate the M_{12} element of the transfer matrix for one unit cell, which is proportional to $|r/t|^2$. However, we must keep in mind that for systems with absorption, we need to calculate the transmission coefficient from the relationship $T = |M_{22}|^{-2}$, since $|M_{12}|^2 + |M_{22}|^2 < 1$ (see problem 12.2 for details).

13.4.2 Band Gap for Oblique Angles

We have seen already that an infinite layered medium does not possess an *omnidirectional gap*. This means that, for each frequency, there is at least one allowed propagation direction, given by Brewster's angle. Now we want to answer the following question: is there a region of frequencies in which a *finite* layered medium reflects back all the electromagnetic waves incident from an embedding medium, independent of the incident angle?

The answer is simple [162]. Yes, it is easy to create such a layered medium. We can obtain an omnidirectional gap of the finite layered medium as follows. If we rescale

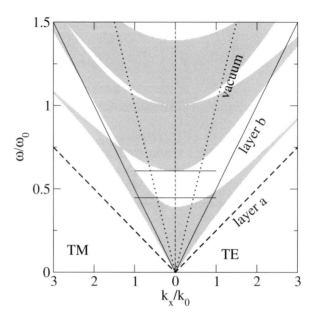

FIGURE 13.9. Band structure of a quarter stack composed of layers a and b with refractive indices $n_a = 4$ and $n_b = 2$. Dotted, solid, and dashed lines show the region of propagating solutions in vacuum, in layer b, and in layer a, respectively. We see that the structure possesses a full gap (marked by two horizontal solid lines) in the frequency region $0.447 < \omega/\omega_0 < 0.608$. That means that any wave incident on the surface of the one-dimensional photonic crystal from vacuum is reflected back.

both the refractive indices n_a and n_b by a factor of a, we obtain a layered medium with the same band structure, but with frequency rescaled by a factor of a^{-1}. We show in figure 13.9 the band structure of the layered medium with refractive indices $n_a = 4$ and $n_b = 2$. Indeed, the band structure is the same as that shown in figure 13.4 for the case $n_a = 2$ and $n_b = 1$. The only difference is that the frequency is rescaled by a factor of 2. However, the dispersion relation of the electromagnetic wave in vacuum is now different from the dispersion relation of waves propagating in layer b. The light cone of the layer b is broader. We see in figure 13.9 that there is an interval of frequencies where no propagation of a wave incident from the vacuum is possible. This means that the electromagnetic waves incoming from the vacuum are always totally reflected, for every incident angle.

To show that our layered medium indeed possesses the omnidirectional gap, we calculate the transmission coefficient for the electromagnetic wave incident from a vacuum to a periodic layered medium composed of $N = 10$ alternating layers a and b with refractive indices $n_a = 4$ and $n_b = 2$. Since the refractive index of the embedding medium is 1, we must use the general formula for the transfer matrix for the layered periodic medium embedded in the homogeneous material given by equation (13.29). Our data for the transmission, shown in figure 13.10, confirm that there is no transmission in

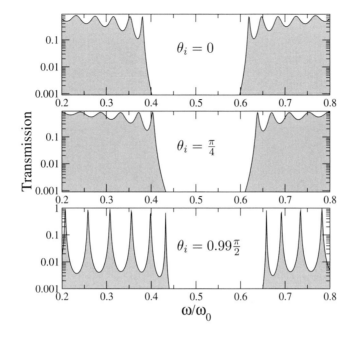

FIGURE 13.10. Transmission coefficient for $N = 10$ layers of a quarter stack with $n_a = 4$ and $n_b = 2$ for three incident angles. The band structure of an infinite quarter stack was shown in figure 13.9. Note that a gap exists for all incident angles. Therefore, we have an omni-directional gap with a one-dimensional photonic crystal [162].

the frequency interval $0.44 < \omega < 0.6$. These results agree with results of our analysis of the band structure shown in figure 13.9.

13.5 Layered Dispersive Media

If layer a is a dispersive medium, such as a metal [73, 113, 144] or left-handed medium, then the analysis of the band structure is more difficult than it was for the case of a layered dielectric medium. In particular, we do not have any scaling relation between the frequency and the refractive index.

We will study two dispersive layered materials. The first case is that layer a is made from a metal. We show that the layered medium behaves as a medium with reduced plasma frequency. The second example is a layered medium with left-handed slabs. This medium possesses an additional gap in the spectrum. In contrast to other band gaps, this new gap does not change its position when the thickness of all layers change.

We remind the reader that we do not consider absorption. It is clear that absorption destroys any propagation in infinite systems. Nevertheless, for a system of finite length, transmission is possible also when absorption is considered. We expect that small absorption does not change the position of bands and gaps considerably, it only reduces the value of the transmission of propagating wave (problem 13.4).

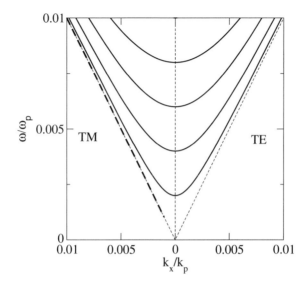

FIGURE 13.11. Band structure of a layered medium in which layer a is a metal. The metallic layer has thickness $\ell_a = \lambda_p/4$, where $\lambda_p = 2\pi c/\omega_p$. The vacuum layer b has thickness $\ell_b = 1000\ell_a$. Thick dashed line is the dispersion relation of the surface waves. For very small frequencies, the permittivity of the metal $|\varepsilon_a| \to \infty$ and surface waves on the opposite sides of the metallic layer do not overlap. This is the reason that the surface wave branch ends up at a nonzero frequency. A typical value of the plasma frequency is $\nu_p = 2000\,\text{THz}$. Using this value one obtains $\lambda_p = 150\,\text{nm}$, $\ell_a = 37.5\,\text{nm}$, and $\ell_b = 37.5\,\mu\text{m}$. The lowest frequency that allows a propagating solution in the metallic layered structure is given by $\omega'_p = \omega_p/500$.

13.5.1 Metallic Layered Medium

If the layer a is a metal with permittivity

$$\varepsilon(\omega) = 1 - \omega_p^2/\omega^2, \tag{13.33}$$

then its thickness must be very small in order to keep the electromagnetic wave propagating through the metallic layer. An interesting property of the layered metallic structure is that it enables propagation for frequencies much lower than the plasma frequency of metal. To show this, we plot in figure 13.11 the band structure of a layered system with the thickness of the metallic slab, $\ell_a = \lambda_p/4 = \pi c/(2\omega_p)$ (ω_p is the plasma frequency of the metal), and the thickness of the vacuum slab, $\ell_b = 1000\ell_a$. We see that the effective plasma frequency of this system is much lower, and decreases as the inverse of the spatial period of the system. This is easy to understand. Consider first the case of normal incidence. Then, propagating solutions are possible only when the distance ℓ_b is an integer number of half the wavelength, i.e.,

$$\ell_b = m\frac{\lambda_b}{2}, \quad m = 1, 2, \ldots, \tag{13.34}$$

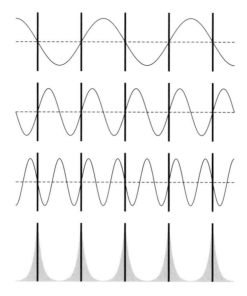

FIGURE 13.12. Propagating solutions of a layered medium composed of thin metallic slabs separated by vacuum. The thickness of the vacuum slab, $\ell_b = 1000\ell_a$. The three upper panels show propagating solutions. Note that, the waves have nodes exactly at the positions of the metallic slabs. The lower panel shows a propagating solution composed of surface modes from metallic slabs.

or $k_b\ell_b = m\pi$. Then we obtain the propagation solutions given by

$$\frac{\omega_m n_b}{c} = \frac{m\pi}{\ell_b}. \tag{13.35}$$

Now, inserting $\ell_b = N\ell_a$ ($N = 1000$ in the case shown in figure 13.11), and $\ell_a = \lambda_p/4 = \pi c/(2\omega_p)$, we obtain that $\ell_b = (N/2)\pi c/\omega_p$ and

$$\omega_m = \frac{2m}{Nn_b}\omega_p. \tag{13.36}$$

In the particular case of $n_b = 1$, we obtained that the first allowed solution ($m = 1$) is given by $\omega_1 = (2/N)\omega_p$. This frequency plays the role of a new plasma frequency ω_p'.

We remind the reader that we discussed a similar problem in chapter 4, when we studied the Kronig-Penney model with high values of the parameter β. In the present case, the permittivity of the metal is very high, so that a thin metallic layer plays the role of a thin high potential barrier.

In order to have propagating solutions for non-normal incidence, we need to satisfy the relation $k_{bz}\ell_b = m\pi$. Therefore,

$$\omega^2 = (c^2 k_{bz}^2 + c^2 k_x^2)/n_b^2 = \omega_m^2 + k_x^2 c^2/n_b^2. \tag{13.37}$$

Solutions given by equation (13.37) are shown in figure 13.11 as solid lines. The spatial dependence of the three lowest propagating solutions through the layered system of metallic slabs is shown in figure 13.12.

The thick dashed line shown in figure 13.12 close to the light cone $\omega = ck_x$ for the TM-polarized wave is the surface wave at the interface of the metallic layers. The corresponding propagating solution for the layered system is schematically shown in the lower panel of figure 13.12. We see that this band is created by the waves localized in the z-direction. Such bands are studied in chapter 5, where we discuss tight binding models. Note that, the width of the band depends upon the overlap of the surface waves along the z-direction. This overlap depends on the thickness of the metallic slab, ℓ_a, and on the frequency of the electromagnetic wave, ω. For small values of ω, the metallic permittivity ε_a is large in absolute value, and negative. This decreases the overlap of the surface waves at two neighboring interfaces, [(see equation (9.123)]. This is a reason that the surface wave band terminates at a nonzero frequency in figure 13.11. Also, if we add a small imaginary part into the dispersion relation for the metallic permittivity, given by equation (13.33), the surface mode disappears at low frequencies. In contrast to other modes shown in figure 13.11, this mode requires the electric field to be localized inside the metal, where it is absorbed.

To conclude, we see that, apart from the surface mode, there are no propagating solutions for frequencies $\omega < \omega_1$. Reduction of the plasma frequency of metallic structures enables us to engineer so-called *metamaterials*.

13.5.2 Left-Handed Layered Media

Of special interest is the layered medium in which layer b is a vacuum and layer a is a left-handed medium. Such a structure possesses new gaps in the frequency region where the average refractive index is zero,

$$\bar{n} = \frac{n_a \ell_a + n_b \ell_b}{\ell_a + \ell_b}. \tag{13.38}$$

To understand the origin of this gap, we remind the reader that a wave propagating in a left-handed medium possesses a *negative* phase, $\phi_a < 0$. If $\bar{n} = 0$, we have

$$\phi_a + \phi_b = 0, \tag{13.39}$$

[94] which is an additional condition for the existence of a gap, given by equation (13.18). Note that, the requirement (13.39) is never satisfied in a dielectric layered medium, where both ϕ_a and ϕ_b are positive. In contrast to gaps in dielectric photonic crystals, which scale according to equation (13.10), the $\bar{n} = 0$ gap does not change its position, because the condition given by equation (13.39) does not change when we rescale both the widths ℓ_a and ℓ_b by the same multiplicative factor.

In figure 13.13 we show the band structure for two layered media composed of slabs of vacuum and left-handed material. Both layers have the same thickness,

$$\ell_a = \ell_b = \ell. \tag{13.40}$$

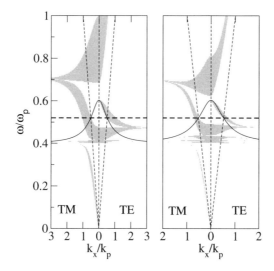

FIGURE 13.13. Band structure of two layered media in which layer a is a left-handed medium and layer b is a vacuum. The thickness of the layers is $\ell_a = \ell_b = \lambda_p/4$ ($= \lambda_p/2$) for the left (right) panel. Note that, the position of the $\bar{n} = 0$ gap is the same for both structures (shown by horizontal dashed line). The thin solid line is the relation $k_x/k_p = \omega/\omega_p \sqrt{\varepsilon(\omega)\mu(\omega)}$.

The left-handed medium has a permittivity and permeability given by equations (16.31) and (16.53), respectively:

$$\varepsilon(\omega) = 1 - \frac{\omega_p^2}{\omega^2} \tag{13.41}$$

and

$$\mu(\omega) = 1 - \frac{F\omega^2}{\omega^2 - \omega_0^2}. \tag{13.42}$$

The resonant frequency $\omega_0 = 0.4\omega_p$ and $F = 0.56$. The two structures shown in figure 13.13 differ only in the thickness of the layers. In the left panel, we have $\ell = \lambda_p/4$, where $\lambda_p = 2\pi c/\omega_p$, and in the right panel, we have $\ell = \lambda_p/2$.

We see in figure 13.13 that both structures possess the $\bar{n} = 0$ gap around the frequency $\omega = 0.52\omega_p$, where the refractive index of the left-handed media is close to -1. This gap does not move when the thickness of both layers is scaled by the same factor.

In addition to the $\bar{n} = 0$ gap, the band structure of the left-handed layered medium deserves some comments. We see that for $\omega < \omega_0$ there are no propagating solutions for the TE mode. For the TM mode, however, we clearly see the surface mode branch, which starts at $\omega = 0$. This branch becomes less pronounced when the thickness of the layers increases (right panel of figure 13.13), and disappears for larger ℓ_a. For layered structures with $\ell = 5\lambda_p/4$ and $5\lambda_p/2$ (shown in figure 13.14), there are already no propagating solutions for $\omega < \omega_0$. The reason for this is obvious; the overlap of surface states of neighboring interfaces is too small for larger layer thicknesses. We remind the reader of the definition of the tight binding model (chapter 5). The transmission

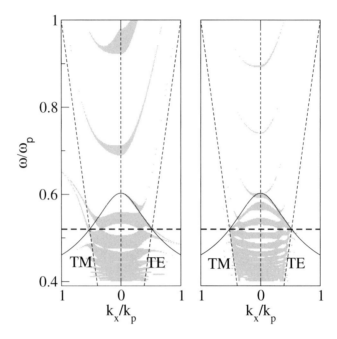

FIGURE 13.14. Band structure of two layered media in which layer a is a left-handed medium and layer b is a vacuum. The thickness of layers is $\ell_a = \ell_b = 5\lambda_p/4$ ($= 10\lambda_p/4$) for the left (right) panel. Note that, since there are no propagating states for $\omega < \omega_0$, the scale of the horizontal axis is shifted with respect to that of figure 13.13. Note also, that the position of the $\bar{n} = 0$ gap (marked as horizontal dashed line) is the same for both structures. The thin solid line is the relationship $k_x/k_p = \omega/\omega_p\sqrt{\varepsilon(\omega)\mu(\omega)}$.

bands in figure 13.13 originate from the surface states, localized at the interfaces between neighboring layers. This is exactly the same as in the tight binding band discussed in chapter 5, where the bands originate from the overlap of the electron's wave function, localized around neighboring atoms. Clearly, when the thickness of the layers increases, the overlap of neighboring surface waves decreases and the corresponding bands disappear.

Above the resonant frequency, we see in figure 13.13 a transmission band, which terminates due to the existence of the $\bar{n} = 0$ gap. Two propagating branches outside the vacuum light cone are again due to the surface modes. They can be identified with the surface modes shown in figure 11.3.

The upper bands, in the frequency region where $\mu > 0$ and $\varepsilon < 0$, have the same origin as the bands in metallic structures discussed in section 13.5.1 (see figure 13.11). To see this mode clearly, we show in figure 13.14 the band structure for $\ell = 5\lambda_p/4$ and $\ell = 10\lambda_p/2$. Note that, the last structure exhibits also a series of Bragg gaps in the region where both ε and μ are negative. These gaps are not visible in figure 13.13, since the conditions $\phi_a = \ell_a k_{az} = m\pi$ cannot be satisfied for small layer thickness ℓ_a. Note also the negative curvature of the Bragg bands, $\omega \sim -k^2$, inside the left-handed frequency region.

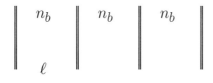

FIGURE 13.15. Structure of an infinite periodic structure with infinite-simally narrow slabs a.

13.6 Kronig-Penney Model of a Photonic Crystal

The analysis of the band structure of a layered medium simplifies considerably in the limit of $n_a \to \infty$ and $\ell_a \to 0$ in such a way that the product

$$n_a^2 \ell_a = \varepsilon_a \ell_a = U \tag{13.43}$$

is constant (figure 13.15). From Snell's law $n_a \sin \theta_a = n_b \sin \theta_b$, we immediately see that $\sin \theta_a \sim n_a^{-1}$, so that $\theta_a = 0$ and $\cos \theta_a = 1$ in the limit of $n_a \to \infty$. Since $n_a^2 \ell_a = U = $ const, we have that the product $n_a \ell_a \sim U/n_a \to 0$. We also have that $\ell = \ell_b$. Under these circumstances, $\cos[k_a \ell_a \cos \theta_a] = \cos[(\omega/c) n_a \ell_a \cos \theta_a] \to \cos 0 = 1$ and $\sin[k_a \ell_a \cos \theta_a] \approx (\omega/c) n_a \ell_a \ll 1$. Note that, in equation (13.2) we have the product

$$\frac{n_a}{n_b} \frac{\cos \theta_b}{\cos \theta_a} \sin k_a \ell_a \cos \theta_a, \tag{13.44}$$

which converges to a finite number in the limiting procedure we consider. The limit reads

$$\frac{\omega \cos \theta_b}{c n_b} U. \tag{13.45}$$

Equation (13.2) then reduces to

$$\cos q\ell = \cos k\ell - \frac{a_{s,p}}{2} k\ell \sin k\ell, \tag{13.46}$$

where

$$k\ell = k_b \ell \cos \theta_b = \frac{\omega}{c} n_b \ell \cos \theta_b \tag{13.47}$$

and

$$a_s = \frac{U \cos \theta_b}{c^2 \ell} \quad \text{and} \quad a_p = \frac{U}{c^2 \ell \cos \theta_b}, \tag{13.48}$$

for the TE- and TM-polarized waves, respectively.

The relationships obtained are very similar to those studied in the chapter 4. The main difference between the present model and the Kronig-Penney model is that here the parameter a also depends on the energy of the propagating wave. Nevertheless, we can apply the formalism developed in chapter 4, and find the band structure of the model (13.46). For instance, from the plot of the right-hand side of equation (13.46) shown in figure 13.16, we immediately find the positions of bands and gaps in the frequency spectrum.

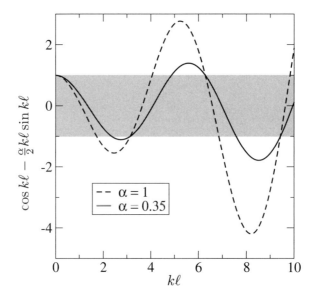

FIGURE 13.16. $\cos q\ell = \cos k\ell - (a/2)k\ell \sin k\ell$ as a function of $k\ell$, given by equation (13.46), for two values of the parameter a. The shaded area highlights the propagating solutions, when $|\cos q\ell| < 1$. Note that, an increase of a causes broadening of the band gaps. Band edges starts at $k\ell = m\pi$, $m = 0, 1, \ldots$.

In order to calculate the transmission coefficient through a finite periodic arrangement of δ-function layers, we follow section 3.4. We define the transfer matrix

$$\mathbf{M}_1 = \begin{pmatrix} e^{ik\ell}\left[1 - \dfrac{a}{2i}k\ell\right] & -\dfrac{a}{2i}k\ell e^{ik\ell} \\ \dfrac{a}{2i}k\ell e^{-ik\ell} & e^{-ik\ell}\left[1 + \dfrac{a}{2i}k\ell\right] \end{pmatrix} \tag{13.49}$$

of one segment consisting of one δ-function layer and free space of length ℓ. In contrast to the transfer matrix for one δ function quantum potential, given by equation (3.42), the transfer matrix $\mathbf{M}_1$ in equation (13.49) possesses a wave vector $k\ell$ in the numerator of each matrix element. This follows from the relationship between the parameters β and a: $\beta = -a(k\ell)^2$, obtained by a comparison of equation (3.46) with equation (13.46).

The transmission through N δ-function layers can be calculated with the use of Chebyshev's formula,

$$T_N = |t_N|^2 = \frac{1}{1 + \left|\dfrac{r_1^2}{t_1^2}\right| \dfrac{\sin^2 Nq\ell}{\sin^2 q\ell}} = \frac{1}{1 + \dfrac{\beta^2 k^2 \ell^2}{4} \dfrac{\sin^2 Nq\ell}{\sin^2 q\ell}}, \tag{13.50}$$

where $q\ell$ is given by equation (13.46).

In figure 13.17 we plot the transmission coefficient, given by equation (13.50), for $N = 45$ δ-function layers and two values of a. By comparing these results with the data for the transmission of a quantum particle through N δ-function potential wells, studied in chapter 3, we see that the first transmission band always starts at $ka = 0$. Thus, the lower band edges are given by the relation $ka = m\pi$ with $m = 0, 1, \ldots$.

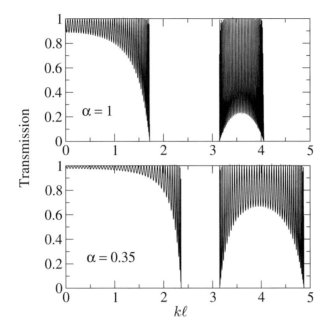

FIGURE 13.17. Transmission coefficient for a system of $N = 45$ δ-function layers with $a = 1$ (upper panel) and 0.35 (lower panel). Transmission bands and gaps are clearly visible. Their position agrees with the estimation given by the solution of equation $|\cos q\ell| = 2$ (figure 13.16).

We will use the present results in chapter 15, which deals with periodic nonlinear dielectric systems.

13.7 Problems

Problems with Solutions

Problem 13.1
Find the band structure of the quarter stack one-dimensional photonic crystal for the case of normal incidence.

Solution. For the quarter stack, we have

$$n_a \ell_a = n_b \ell_b = \frac{\lambda_0}{4}, \tag{13.51}$$

where $\lambda_0 = 2\pi\omega_0/c$. For the case of *normal incidence*, we have

$$\phi_a = \phi_b = \frac{\pi}{2}\frac{\omega}{\omega_0} \tag{13.52}$$

and $z_{ab} = n_a/n_b = n$.

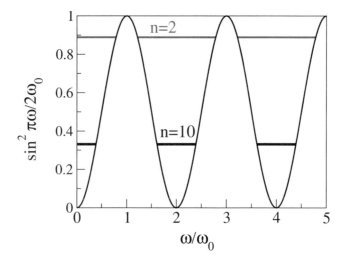

FIGURE 13.18. Graphical solution of equations (13.54) and (13.55). Horizontal lines show the intervals of the allowed energies for $n = 2$ and 10. Note that, the allowed intervals are the same for n and for n^{-1}, as they should be.

Substituting in equation (13.2), we obtain a simple expression for $\Gamma = \text{Tr } \mathbf{M}$,

$$\Gamma = 2 - \frac{(n+1)^2}{n} \sin^2 \frac{\pi\omega}{2\omega_0}. \tag{13.53}$$

Propagating solutions satisfy the condition $|\Gamma| < 2$. It is easy to see that the frequency region of propagating solutions is given by

$$0 < \frac{\omega}{\omega_0} \leq \Delta,$$

$$\tag{13.54}$$

$$2m - 1 + \frac{\Delta}{2} < \frac{\omega}{\omega_0} < 2m + 1 - \frac{\Delta}{2}, \quad m = 1, 2, \ldots,$$

where the parameter Δ is given by

$$\Delta = 2 \frac{2}{\pi} \sin^{-1} \sqrt{\frac{4n}{(n+1)^2}} \tag{13.55}$$

and determines the width of the gap. Note that, all gaps have the same width for normal incidence propagation.

In figure 13.18 we show the allowed frequency bands for $n = 2$ and 10. We see that gaps close only for $n = 1$, which is the case of a homogeneous medium. In the opposite limit, $n \to \infty$, only waves with the frequency

$$\omega = 2m\omega_0 \tag{13.56}$$

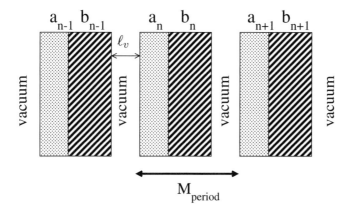

FIGURE 13.19. Layered structure with a thin layer of vacuum added between each layer b_{n-1} and a_n. The transfer matrix, **M**, relates electromagnetic fields left of the interface of the layer a_{n+1} with electromagnetic field left of the layer a_n.

propagate. Note that the limit $n_a/n_b \to \infty$ corresponds to the limit $n_a \to \infty$, since $n_b = 1$. Since the product $n_b \ell_b$ is constant, due to the condition (13.51), the width of the layer b, $\ell_b \to 0$. The allowed frequencies given by equation (13.56) correspond to waves that have zeros at the ends of layers a.

Problems without Solutions

Problem 13.2

Consider a more general case of a dielectric layered medium where the ratio

$$\frac{n_a}{n_b} \frac{\ell_a}{\ell_b} \neq 1. \tag{13.57}$$

Give a qualitative analysis of the position of bands and gaps in the frequency spectrum of such a medium.

Problem 13.3

An alternative expression for the transfer matrix of a layered medium embedded in vacuum can be found as follows. Consider first a medium with an additional vacuum slab between each layer b_{n-1} and a_n (see figure 13.19). The thickness of the vacuum slab is ℓ_h.

Calculate the transfer matrix $\mathbf{M}_{\text{period}}$ for one segment of this material, as shown in figure 13.19. Then, express the transfer matrix of the structure consisting of N segments as

$$\mathbf{M}_N = \mathbf{M}_{\text{period}}^N. \tag{13.58}$$

Show that in the limit of $\ell_v \to 0$ we obtain the required transfer matrix for the layered medium discussed in figure 13.6.

Use the transfer matrix $\mathbf{M}$ given by equation (13.58) for the calculation of the transmission coefficient of N alternating layers a and b embedded in vacuum.

Problem 13.4

Until now, we have not considered absorption. It is clear that no propagating solutions exist in an infinite layered medium with absorption. Finite periodic systems, however, can have propagating solutions. Calculate the band structure of a layered dielectric quarter stack consisting of $N = 10$ layers of material a with refractive index $n_a = 2 + 0.01i$ and $n_b = 1$. Compare your results with the data shown in figure 13.7.

14 Effective Parameters

Up to now, we have analyzed the transmission of electromagnetic waves in homogeneous media. The only inhomogeneities were given by the interfaces between two homogeneous materials. We assumed that the distance between two adjacent interfaces is larger than, or at least comparable to, the wavelength of the propagating electromagnetic wave.

Now we will analyze structures that possess inhomogeneities much smaller than the wavelength. A simple example of such a structure is the layered medium shown in figure 13.2, where the thickness of each slab, ℓ_a and ℓ_b, is much smaller than λ. In such a case, the propagating electromagnetic wave does not see the details of the sample. It propagates as if the medium is homogeneous, with an *effective* permittivity ε_{eff} and an effective permeability μ_{eff}.

Another example of inhomogeneous media are the so-called metamaterials, which consist of periodic arrangement of small unit cells. Each unit is created from various materials. An example of the unit cell of the simplest left-handed material is shown in figure 16.12. If the size of the unit cell is much smaller than the wavelength of the electromagnetic wave, we can characterize the electromagnetic response of such a metamaterial by effective permittivity and permeability.

Our aim in this chapter is to calculate the effective parameters. First, we introduce the effective theory for a layered medium. Then, we solve the inversion problem for a single slab of inhomogeneous material. In the last section, we apply the theory developed to a simple layered medium of alternating layers with negative permittivity and permeability.

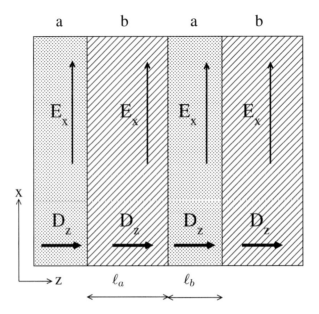

FIGURE 14.1. A layered medium. Shown are the preserved components of the electric field, namely, electric intensity E parallel to the interface, and electric displacement D perpendicular to the interface.

14.1 Effective Parameters of a Layered Medium

Consider again the same one-dimensional layered medium as in chapter 13 (figure 14.1), but now assume that the wavelength of the propagating field is much larger than the period of the layered medium. Then we can consider the medium as a homogeneous material, which can be characterized by effective electromagnetic parameters. Since the structure is not isotropic, we must distinguish the effective parameters for different orientations of the electric and magnetic fields.

Consider first the electric field E_x parallel to the interface between two layers, as shown in figure 14.1. From the boundary conditions at the interface between two media given by equation (B.68), we know that E_x is conserved along the interface, possessing the same value in both layers a and b. Therefore, if we calculate the average electric field, we obtain

$$E_x^{av} = E_x^a = E_x^b. \tag{14.1}$$

Now we calculate the electric displacement D_x. Using the permittivities ε_a and ε_b, we calculate the electric displacement $D_x^a = \varepsilon_a E_x^a$ and $D_x^b = \varepsilon_b E_x^b$ in layers a and b, respectively. Then the average value of the displacement D_x^{av} is given by

$$D_x^{av} = \frac{\ell_a D_x^a + \ell_b D_x^b}{\ell_a + \ell_b} = \frac{\ell_a \varepsilon_a + \ell_b \varepsilon_b}{\ell_a + \ell_b} E_x^{av}. \tag{14.2}$$

We define the effective permittivity $\varepsilon_{eff}^{\parallel}$ by the relation

$$D_x^{av} = \varepsilon_{eff}^{\parallel} E_x^{av}. \tag{14.3}$$

Inserting in equation (14.3), the expressions for E_x^{av} and D_x^{av} given by equations (14.1) and (14.2), respectively, we obtain the result that the effective permittivity $\varepsilon_{\mathrm{eff}}^{\parallel}$ for the direction parallel to the interface between two layers is

$$\varepsilon_{\mathrm{eff}}^{\parallel} = \frac{\ell_a \varepsilon_a + \ell_b \varepsilon_b}{\ell_a + \ell_b}, \tag{14.4}$$

which is just the arithmetic average of the permittivity.

Now, consider the case when the electric field is perpendicular to the interfaces between the two layers. In this case, from equation (B.67) we have that the vector of the electric displacement D is conserved across the interface,

$$D_z^a = D_z^b = D_z^{\mathrm{av}}. \tag{14.5}$$

We calculate the average value of the electric field E_z^{av}, given by

$$E_z^{\mathrm{av}} = \frac{\ell_a E_z^a + \ell_b E_z^b}{\ell_a + \ell_b}. \tag{14.6}$$

Using the relations $D_z^a = \varepsilon_a E_z^a$ and $D_z^b = \varepsilon_b E_z^b$, we obtain that

$$E_z^{\mathrm{av}} = \frac{\ell_a \dfrac{D_z^a}{\varepsilon_a} + \ell_b \dfrac{D_z^b}{\varepsilon_b}}{\ell_a + \ell_b}$$

$$= \frac{\dfrac{\ell_a}{\varepsilon_a} + \dfrac{\ell_b}{\varepsilon_b}}{\ell_a + \ell_b} D_z^{\mathrm{av}}. \tag{14.7}$$

This relation defines the effective permittivity $\varepsilon_{\mathrm{eff}}^{\perp}$ for the direction perpendicular to the interfaces,

$$E_z^{\mathrm{av}} = \frac{1}{\varepsilon_{\mathrm{eff}}^{\perp}} D_z^{\mathrm{av}}. \tag{14.8}$$

Comparing equation (14.8) with equation (14.7), we obtain that the effective permittivity in the direction perpendicular to the interface is

$$\varepsilon_{\mathrm{eff}}^{\perp} = \varepsilon_a \varepsilon_b \frac{\ell_a + \ell_b}{\ell_a \varepsilon_b + \ell_b \varepsilon_a}. \tag{14.9}$$

We conclude that the effective permittivity of the layered medium is a diagonal tensor, with a diagonal component given by $\varepsilon_{\mathrm{eff}}^{\parallel}$ and $\varepsilon_{\mathrm{eff}}^{\perp}$,

$$\varepsilon_{\mathrm{eff}} = \begin{pmatrix} \varepsilon_{\mathrm{eff}}^{\parallel} & 0 & 0 \\ 0 & e_{\mathrm{eff}}^{\parallel} & 0 \\ 0 & 0 & \varepsilon_{\mathrm{eff}}^{\perp} \end{pmatrix}. \tag{14.10}$$

The two components of the tensor that correspond to the orientation of the electric field in two directions parallel to the interface (x and y in the present case) are equal to each other.

In the general case of arbitrarily distributed nonhomogeneities, it can be shown (see [18] for details) that $\varepsilon_{\mathrm{eff}}^{\parallel}$ and $\varepsilon_{\mathrm{eff}}^{\perp}$ represent the upper and lower bounds of the effective permittivity:

$$\varepsilon_{\text{eff}}^{\perp} \leq \varepsilon_{\text{eff}} \leq \varepsilon_{\text{eff}}^{\parallel}. \tag{14.11}$$

In a similar way, we calculate the tensor of the magnetic permeability. Using the continuity requirement of H_x and B_z given by equations (B.69) and (B.67), respectively, we calculate the tensor of the effective magnetic permeability as follows:

$$\mu_{\text{eff}} = \begin{pmatrix} \varepsilon_{\text{eff}}^{\parallel} & 0 & 0 \\ 0 & \mu_{\text{eff}}^{\parallel} & 0 \\ 0 & 0 & \mu_{\text{eff}}^{\perp} \end{pmatrix}, \tag{14.12}$$

where

$$\mu_{\text{eff}}^{\parallel} = \frac{\ell_a \mu_a + \ell_b \mu_b}{\ell_a + \ell_b} \tag{14.13}$$

and

$$\mu_{\text{eff}}^{\perp} = \mu_a \mu_b \frac{\ell_a + \ell_b}{\ell_a \mu_b + \ell_b \mu_a}. \tag{14.14}$$

If the electromagnetic wave propagates in an effectively anisotropic medium, then the response of the medium depends on the propagation direction of the electromagnetic wave. This response is simple in the following two cases.

For the case of normal incidence (the wave propagates along the z-direction), both electric and magnetic fields are parallel to the interface. Therefore, we find that the layered medium behaves as a homogeneous medium with effective permittivity $\varepsilon_{\text{eff}}^{\parallel}$ and effective permeability $\mu_{\text{eff}}^{\parallel}$, given by equations (14.4) and (14.13), respectively. The wave equation for the electric field is

$$\frac{\partial^2 E_x}{\partial z^2} = \frac{\varepsilon_{\text{eff}}^{\parallel} \mu_{\text{eff}}^{\parallel}}{c^2} \frac{\partial^2 E_x}{\partial t^2}. \tag{14.15}$$

If the wave propagates in the direction parallel to the interface (for instance along the x-axis), the two polarizations TE and TM "see" a different effective media. The wave polarized with electric field parallel to the interface (i.e., oriented along the y-axis) propagates in an effective medium with permittivity $\varepsilon_{\text{eff}}^{\parallel}$ and permeability $\mu_{\text{eff}}^{\perp}$. The other polarization has a magnetic field parallel to the interface and an electric field perpendicular to the interface, so this wave propagates in a medium with permittivity $\varepsilon_{\text{eff}}^{\perp}$ and permeability $\mu_{\text{eff}}^{\parallel}$. The wave equation for the electric field E_y is

$$\frac{\partial^2 E_y}{\partial x^2} = \frac{\varepsilon_{\text{eff}}^{\parallel} \mu_{\text{eff}}^{\perp}}{c^2} \frac{\partial^2 E_y}{\partial t^2}, \tag{14.16}$$

and for the electric field E_z is

$$\frac{\partial^2 E_z}{\partial x^2} = \frac{\varepsilon_{\text{eff}}^{\perp} \mu_{\text{eff}}^{\parallel}}{c^2} \frac{\partial^2 E_z}{\partial t^2}. \tag{14.17}$$

Note that the three wave equations determine three different phase velocities.

A general description of propagation of electromagnetic waves in anisotropic media is beyond the scope of the present book. We refer the readers to [2, 20, 26, 39].

14.2 Retrieval Procedure

In chapter 10 we calculated the transmission and reflection amplitudes for a homogeneous slab. We found that the transmission properties of the slab are completely determined by its permittivity and permeability, and the slab thickness ℓ. Now we want to solve the inverse problem. Suppose we know the transmission and reflection amplitudes t and r for an electromagnetic wave propagating through a slab of thickness ℓ, embedded in a vacuum. From these data, can we determine the permittivity and permeability of the slab?

This problem is of special interest in the case when the slab consists of small components (much smaller than the wavelength of the propagating wave), which themselves have very different electromagnetic properties. In many applications, we can consider such a slab to be a homogeneous medium. As an example of such a slab, consider the left-handed medium that will be described in chapter 16. Such a medium consists of a periodic array of so-called split-ring resonators and very thin metallic wires. From a macroscopic point of view, the medium is built from "atoms" of finite size. The effective parameters are given by the electromagnetic response of each atom. For instance, the negative permittivity of left-handed media is due to the resonant behavior of the split-ring resonator. Thus, the left-handed medium has negative permeability, although it contains no magnetic material.

The solution of the problem seems to be straightforward. What we need is to express the transmission and reflection *amplitudes* in terms of the permittivity and permeability of the slab, and then to invert these relationships to obtain the permittivity and permeability in terms of the transmission and reflection amplitudes [143, 146]. The problem is similar to that solved in section 2.6 for the case of a quantum particle propagating through a rectangular potential.

The first step can be done immediately with the help of the transfer matrix. We recall that the transmission amplitude t is given by

$$t = \frac{1}{M_{22}},\tag{14.18}$$

and the matrix element M_{12} determines the ratio r/t of the reflection and transmission amplitudes,

$$\frac{r}{t} = M_{12}.\tag{14.19}$$

Using the explicit form of the transfer matrix given for the case of the TE-polarized wave by equations (10.17)–(10.20), we obtain

$$t = \frac{1}{\cos k_z\ell - \frac{i}{2}\left(z + \frac{1}{z}\right)\sin k_z\ell}\tag{14.20}$$

and

$$\frac{r}{t} = -\frac{i}{2}\left(z - \frac{1}{z}\right)\sin k_z\ell.\tag{14.21}$$

In equations (14.20) and (14.21) we use the parameter

$$\gamma^{-1} = \frac{\mu_1}{\mu_2} \frac{k_z}{k_{0z}}, \tag{14.22}$$

with k (k_0) being the wave vector inside (outside) the slab.

For simplicity, we will discuss only the case of normal incidence. Then the wave vector inside the slab, $k = k_z = \omega n/c$, and $z = \sqrt{\mu/\varepsilon}$. Using the relation $k = n\omega/c$, we obtain from equations (14.20) and (14.21) the following expressions for t and r:

$$t = \frac{1}{M_{22}} = \frac{1}{\cos \dfrac{n\omega\ell}{c} - \dfrac{i}{2}\left(z + \dfrac{1}{z}\right)\sin \dfrac{n\omega\ell}{c}} \tag{14.23}$$

and

$$\frac{r}{t} = M_{12} = -\frac{i}{2}\left(z - \frac{1}{z}\right)\sin \frac{n\omega\ell}{c}. \tag{14.24}$$

The last two equations can be solved for the refractive index n and the impedance z as follows:

$$\cos n\omega\ell/c = \frac{1}{2t}\left[1 - r^2 + t^2\right] \tag{14.25}$$

and

$$z = \pm\sqrt{\frac{(1+r)^2 - t^2}{(1-r)^2 - t^2}}. \tag{14.26}$$

From known values of n and z, we obtain ε and μ,

$$\mu = nz \quad \text{and} \quad \varepsilon = \frac{n}{z}. \tag{14.27}$$

The problem of finding the permittivity and permeability seems to be solved with equations (14.25) and (14.26). However, note that equations (14.25) and (14.26) have multiple solutions. Therefore, we need additional information, which enables us to determine which solution is correct. For the impedance z, we use the physical requirement derived in Appendix B.4 that the real part of the impedance, z_r, must be positive. This requirement determines the sign in front of the square root on the r.h.s. of equation (14.26).

The problem is slightly more difficult for the index of refraction. With the help of the relationship $e^{ix} + e^{-ix} = 2\cos x$, we rewrite equation (14.25) as follows:

$$e^{in\omega\ell/c} = X \pm i\sqrt{1 - X^2}, \quad X = \frac{1}{2t}\left[1 - r^2 + t^2\right]. \tag{14.28}$$

To find the true physical solution of equation (14.25), we must consider both real and imaginary parts of the refractive index, $n = n_r + in_i$. Then equation (14.28) reads

$$e^{-n_i\omega\ell/c}\left[\cos \frac{n_r\omega\ell}{c} + i\sin \frac{n_r\omega\ell}{c}\right] = X \pm i\sqrt{1 - X^2}. \tag{14.29}$$

Since the imaginary part of the refractive index, n_i, must be positive (see Appendix B.6 for details), both sides of equation (14.29) must be less than 1. This determines the sign on

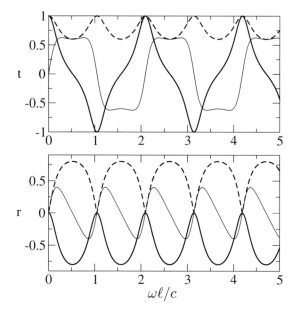

FIGURE 14.2. Transmission (t, upper panel) and reflection (r, lower panel) amplitudes of a homogeneous dielectric slab *versus* the slab thickness ℓ. The thick solid line is the real part, the thin solid line is the imaginary part, and the dashed line is the absolute value of the transmission and reflection amplitudes. The permittivity of the slab is $\varepsilon = 9$ and the permeability is $\mu = 1$.

the r.h.s of equation (14.29). Then, by comparing the real and imaginary parts of equation (14.29), we can determine the real part of the refractive index, n_r.

However, we are still not ready, because equation (14.29) has multiple solutions for n_r. Indeed, if n_r is a solution of equation (14.29), then also $n_r' = n_r + m2\pi c/\omega\ell$ (m is an integer) is a solution of equation (14.29). To avoid this ambiguity, we must repeat the above analysis for slabs of various lengths and plot the solution of equation (14.29) as a function of the slab thickness.

As an example of this procedure, let us solve the simplest inversion problem for the homogeneous dielectric slab with permittivity $\varepsilon = 9$ and permeability $\mu = 1$. First, we use the transfer matrix given by equations (10.17)–(10.20) to calculate the transmission and reflection amplitudes as a function of slab thickness ℓ. Numerical data for the transmission and reflection amplitudes are shown in figure 14.2. Then we substitute these data into equations (14.25) and (14.26) and calculate the refractive index and the impedance. Results are shown in figure 14.3.

The retrieval procedure is an important tool for estimation of the effective permittivity ε_{eff} and effective permeability μ_{eff} of nonhomogeneous materials and of metamaterials, composed of various constituents. Example of such metamaterials are man-made left-handed materials. If we assume that the wavelength of the propagating electromagnetic wave is much larger than the typical size of the inhomogeneity of the composite material, we can consider the material to be homogeneous. Thus, we can assume that

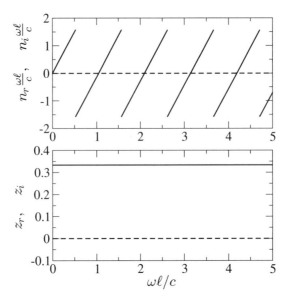

FIGURE 14.3. Upper panel: $n_r \omega \ell / c$ as a function of $\omega \ell / c$. Different branches are due to the periodicity of the function $\cos(x + 2\pi) = \cos x$. The slope of the linear dependence determines $n = 3$. The imaginary part of the refractive index is also shown by the dashed line. As expected, $n_i = 0$. Lower panel: Real and imaginary parts of the impedance z. As expected, z does not depend on the system length; $z_r = 1/3$ and $z_i \equiv 0$.

the material possesses an effective refractive index n_{eff} and effective impedance z_{eff}, and that experimental or numerical data for the transmission and reflection amplitudes can satisfy equations (14.23) and (14.24) with n and z replaced by n_{eff} and z_{eff}. Then, following the method described in this section, we can calculate the effective electromagnetic parameters of the material.

We remind the reader that, in order to determine the effective refractive index, we need both the transmission and the reflection *amplitudes*. It is not sufficient to have only transmission and reflection coefficients. This means that we also need information about the increase of the phase of the electric field *inside* the slab.

14.3 Alternating Layers with Negative Permittivity and Negative Permeability

An interesting example of a periodic layered material is a structure in which layer a has negative permeability and $\varepsilon_a \equiv 1$, and layer b has negative permittivity and $\mu_b \equiv 1$ (figure 14.4). Neither material a or b is transparent, since the refractive indices n_a and n_b are complex. Nevertheless, we find that there is an interval of frequencies, in which the periodic layered medium composed of layers a and b allows propagation of electromagnetic waves.

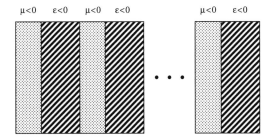

FIGURE 14.4. A periodic layered medium composed of alternating layers with negative permittivity and negative permeability.

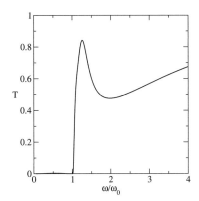

FIGURE 14.5. Transmission coefficient for $N = 100$ alternating layers of negative permittivity and negative permittivity. Layer a has negative permeability and layer b has negative permittivity.

We demonstrate this possibility on a simple model example with

$$\mu_a = 1 - \frac{\omega_m^2 - \omega^2}{\omega^2 - \omega_0^2 + i\gamma_m\omega} \tag{14.30}$$

and

$$\varepsilon_b = 1 - \frac{\omega_p^2}{\omega^2 + i\gamma_e\omega}, \tag{14.31}$$

with parameters $\nu_0 = \omega_0/2\pi = 10\,\text{GHz}$, $\omega_m = 1.5\omega_0$, $\omega_p = 4\omega_0$, $\gamma_m = 0.02\omega_0$, and $\gamma_e = 0.005\omega_0$. The thickness of slab a is $\ell_a = 0.001\lambda_0$ and the thickness of slab b is $\ell_b = 0.00066\lambda_0$. Here, λ_0 is the wavelength of the electromagnetic wave with the frequency ω_0, $\lambda_0 = 2\pi/\omega_0 = 3\,\text{cm}$.

We use the transfer matrix, and calculate the transmission and the reflection amplitudes of the electromagnetic waves through the described structure. The data obtained are shown in figure 14.5. We see that the layered structure possesses a transmission peak at frequencies $\omega > \omega_0$, which is a frequency region where both ε_b and μ_a are negative. To understand this result, we calculate the effective permittivity and permeability of the investigated material. Since both ℓ_a and ℓ_b are much smaller than the wavelength of the wave at frequency ω_0, we can apply either the method from section 14.1 or the retrieval procedure explained in section 14.2.

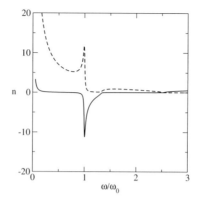

FIGURE 14.6. Effective refractive index for a negative-permittivity and negative-permittivity layered structure obtained from numerical data for the transmission and the absorption amplitudes for the layered structure shown in figure 14.4.

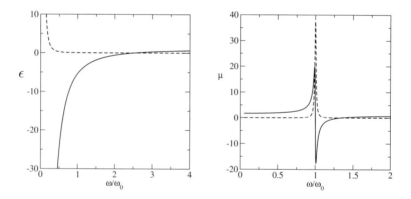

FIGURE 14.7. Effective permittivity (left panel) and effective permeability (right panel) of a negative-permittivity and negative-permittivity layered structure.

The retrieved refractive index is shown in figure 14.6. We see that the structure indeed possesses a transmission band with a negative real part of the refractive index, $n_r < 0$. Figure 14.7 shows the retrieval effective permittivity ε_{eff} and permeability μ_{eff} of the system.

The data shown in figure 14.7 are identical, with effective parameters

$$\varepsilon_{\text{eff}} = \frac{\varepsilon_a \ell_a + \varepsilon_b \ell_b}{\ell_a + \ell_b} \quad \text{and} \quad \mu_{\text{eff}} = \frac{\mu_a \ell_a + \mu_b \ell_b}{\ell_a + \ell_b}, \tag{14.32}$$

calculated in section 14.1, with $\varepsilon_a = \mu_b = 1$ and μ_a and ε_b given by equations (14.30) and (14.31), respectively. This is not a surprise, since both methods should lead to the same results in this particular case.

14.4 Problem

Problem 14.1

Using equations (14.4) and (14.13), calculate the effective permittivity and permeability of the layered medium discussed in section 14.3. Compare your results with those shown in figure 14.7.

15 Wave Propagation in Nonlinear Structures

In this chapter, we will investigate the nonlinear response of wave propagation in one-dimensional structures.

When dielectric materials are arranged periodically, electromagnetic waves at some frequencies are forbidden to propagate. This was discussed in detail in chapter 13. Most of the interest in multilayer structures focuses on the *linear* regime in which the dielectric constant is independent of the field strength. However, the presence of optical *nonlinearity* in a system leads to a much richer and more complex response to radiation. We will see that the transmission coefficient is a function of the intensity of the incoming electromagnetic wave. This leads to an interesting phenomenon, known as *bistability*, in which two possible output states exist for a simple input. Bistability promises interesting applications in ultrafast optical switches.

We first solve the problem of a single nonlinear δ-function impurity [96]. The bistability will be demonstrated and analyzed. Then we discuss the nonlinear Kronig-Penney δ-function model, which captures most of the essential features of the nonlinear response in layered structures [95, 96, 98].

15.1 Single δ-Function Layer of a Nonlinear Dielectric

In this section we will consider a very thin nonlinear layer. The nonlinearity is expressed in first order by an intensity-dependent term in the electric permittivity $\varepsilon(z)$,

$$\varepsilon(z) = \varepsilon_0 \left[1 + \Lambda |E(z)|^2\right] \delta(z). \tag{15.1}$$

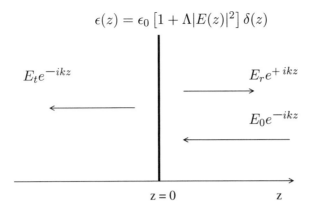

$$\epsilon(z) = \epsilon_0 \left[1 + \Lambda |E(z)|^2\right] \delta(z)$$

$E_t e^{-ikz}$

$E_r e^{+ikz}$

$E_0 e^{-ikz}$

$z = 0$

z

FIGURE 15.1. A plane wave of amplitude E_0 strikes a nonlinear δ-function dielectric layer from the right, giving rise to a reflected and a transmitted wave.

Here, ε_0 is defined as the limit of $n^2 \ell \to \varepsilon_0$ with $n \to \infty$ and $\ell \to 0$ of a finite-width layer of extent ℓ and index of refraction n (see section 13.6 for details). The parameter $\varepsilon_0 \Lambda$ is the corresponding nonlinear *Kerr coefficient*. The limit $\Lambda = 0$ corresponds to the linear problem. Both $\Lambda > 0$ and $\Lambda < 0$ are of interest. In the particular case of a single δ-function layer, only negative values of Λ give us bistable behavior of the nonlinear system.

Consider the propagation of the electromagnetic wave through a single δ-function impurity with an intensity-dependent permittivity as given by equation (15.1). This is schematically shown in figure 15.1. Our interest lies in the steady-state response of the system. Let a plane wave $E_0 e^{-ikz}$ with a wave vector $k = \omega c$ be incident from the right. This gives rise to a reflected wave $E_r e^{+ikz}$ ($z > 0$), as well as to a transmitted wave $E_t e^{-ikz}$ ($z < 0$). We need to solve the wave equation

$$\frac{\partial^2 E(z)}{\partial z^2} + \frac{\omega^2}{c^2} \varepsilon(z) E(z) = 0. \tag{15.2}$$

The boundary conditions at the position of the δ-function layer, $z = 0$, were derived in section 3.1. The electric field must be continuous at $z = 0$,

$$E(z = 0^-) = E(z = 0^+), \tag{15.3}$$

and the first derivative of the field must satisfy the following relationship:

$$\left.\frac{\partial E}{\partial z}\right|_{z=0^+} - \left.\frac{\partial E}{\partial z}\right|_{z=0^-} = \frac{\omega^2}{c^2} \varepsilon_0 \left(1 + \Lambda |E(z)|^2\right) E(z = 0), \tag{15.4}$$

in agreement with equation (3.11), derived for the case of the δ-function quantum potential.

Inserting the electric field shown in figure 15.1 into equation (15.4), we obtain from equation (15.3)

$$E_t = E_0 + E_r, \tag{15.5}$$

and from equation (15.4),

$$-ik E_t + ik E_0 - ik E_r = \frac{\omega^2}{c^2}\left(1 + \Lambda |E_t|^2\right) E_t. \tag{15.6}$$

From equation (15.5) we have $E_r = E_t - E_0$. Inserting this into equation (15.6) we obtain a relationship between the amplitude of the incident field E_0 and the amplitude of the transmitted field E_t,

$$E_0 = E_t\left[1 - \frac{1}{2}\frac{\omega^2}{ik}\left(1 + \Lambda |E_t|^2\right)\right]. \tag{15.7}$$

Now we define $X = |E_t|^2$ and $Y = |E_0|^2$, which have the physical meaning of the squares of the amplitudes of the transmitted and incoming waves, respectively. Then we obtain from equation (15.7) the following nonlinear relation:

$$Y = X\left[1 + \tau(1 + \Lambda X)^2\right]. \tag{15.8}$$

In equation (15.8) we introduced a parameter τ,

$$\tau = \left(\frac{\varepsilon_0 \omega^2}{2c^2 k}\right)^2 = \frac{\varepsilon_0^2 k^2}{4}. \tag{15.9}$$

To obtain the expression for τ, we used $\omega = ck$. When nonlinearity is absent, $\Lambda = 0$, we have that the transmission coefficient T_0 through the single δ-function layer $(\omega^2/c^2)\varepsilon_0$ is given by equation (3.21):

$$T_0 = \frac{|E_t|^2}{|E_0|^2} = \frac{4k^2}{4k^2 + \left(\frac{\omega^2}{c^2}\varepsilon_0\right)^2} = \frac{1}{1+\tau}. \tag{15.10}$$

We can express τ in terms of T_0 as follows:

$$\tau = \frac{1 - T_0}{T_0}. \tag{15.11}$$

Note that for $\Lambda < 0$ equation (15.8) always has a solution

$$X = Y = \frac{1}{|\Lambda|}, \tag{15.12}$$

for which the system exhibits resonant transmission $T = X/Y = 1$. Clearly, the larger is $|\Lambda|$, the smaller are the switching intensities $(X = |E_0|^2)$. The nonlinear system will exhibit bistability if we can have more than one output for a given input. This will happen if Y is a nonmonotonic function of X, so that the function $Y = Y(X)$ has a local maximum or minimum. To determine the condition for instability, we calculate from equation (15.8) the first derivative

$$\frac{\partial Y}{\partial X} = 1 + \tau + 4\Lambda\tau X + 3\Lambda^2\tau X^2, \tag{15.13}$$

and look for physical solutions of

$$\frac{\partial Y}{\partial X} = 0. \tag{15.14}$$

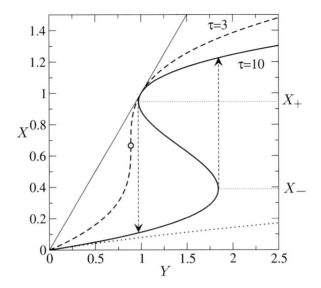

FIGURE 15.2. Transmitted intensity Y as a function of the incident intensity X for a nonlinear δ-function layer. The solid (dashed) line corresponds to $\Lambda = -1$ and $\tau = 10$, ($\tau = 3$). The dotted line corresponds to $\Lambda = +1$, $\tau = 10$. The thin line is $X = Y$. The circle shows critical parameters $X_+ = X_- = 2/3$ for $\Lambda = -1$ and $\tau = 3$.

Since $\tau > 0$ and $X > 0$, we can have bistability only when $\Lambda < 0$. Then, equaiton (15.13) has the following solutions:

$$X_\pm = \frac{1}{3|\Lambda|}\left(2 \pm \sqrt{1 - 3/\tau}\right). \tag{15.15}$$

From equation (15.15), we see that bistability appears only when $\tau > 3$, or, with the use of relation (15.11), we have that $T_0 > 1/4$.

The nonlinear response of our system with negative Λ is shown in figure 15.2. The switching-up and switching-down intensities are given by $Y_\pm = Y(X_\mp)$, respectively. For $\Lambda = -1$ and $\tau = 3$, we have the onset of bi-stability, $Y_+ = Y_- = 8/9|\Lambda|$ (marked by a small circle in figure 15.2). As τ increases, we use $\sqrt{1+x} \approx 1 + x/2$ for $x \ll 1$ and obtain from equation (15.15)

$$X_+ = \frac{1}{|\Lambda|} - \frac{1}{2|\Lambda|\tau}, \quad X_- = \frac{1}{3|\Lambda|} + \frac{1}{2|\Lambda|\tau}. \tag{15.16}$$

Inserting this into equation (15.8) we obtain $Y_- \to 1/|\Lambda|$ and $Y_+ = 4\tau/(27|\Lambda|)$. For large τ, the switching-down intensity Y_- converges to the resonance intensity $1/|\Lambda|$.

Note that, in the case of resonance transmission intensity $X = Y = 1/|\Lambda|$, given by equation (15.12), the permittivity of the δ-function layer, $\varepsilon(z = 0) = 0$. Thus, the resonant transmission is due to the fact that the refractive index mismatch between the δ-function layer and the background effectively disappears.

Now we apply our studies to the more realistic problem of a finite slab of a nonlinear dielectric. The condition for the total transmission is $2\ell = m\lambda$, where ℓ is the thickness of the slab, m is an integer, and $\lambda = 2\pi c/(n\omega)$ is the wavelength of the electromagnetic wave inside the layer. If we incorporate a nonlinear coefficient in the permittivity given by equation (15.1), then we expect that for a suitable choice of parameters we can also have bistability. For example, if $\ell < \lambda/2$ and the nonlinear parameter $\Lambda > 0$, then the permittivity ε effectively increases due to the nonlinearity. This reduces the wavelength λ. We expect that for some intensity of the incident wave we can obtain the result that the resonance condition $\ell = \lambda/2$ is satisfied. Hence, an increase in the intensity of the incident field will lead to resonance transmission. Moreover, higher-order resonances ($m > 1$) are also possible in a layer of finite thickness, giving rise to *multistability*.

Similarly, for a negative nonlinear parameter $\Lambda < 0$, the nonlinearity causes an increase of the wavelength λ. The resonance is possible if the incoming wave reduces the effective refractive index of the slab so that Fabry-Pérot resonance appears inside the lower energy band. The case when the nonlinearity reduces the effective refractive index of the slab to zero corresponds to the δ-function layer case discussed above in this section.

15.2 Nonlinear Kronig-Penney δ-Function Model

Now we investigate the nonlinear response of wave propagation in a one-dimensional layered structure. The bistable response of such a system results from the modulation of the transmission by an intensity-dependent refractive index. This causes an intensity-dependent phase shift of the electromagnetic waves traveling inside the nonlinear structure. In addition to bistability and multistability, we show that additional transmission modes exist within the band gap, These modes are localized solutions, known as "gap solitons." Under appropriate conditions, they can couple with the incident wave giving rise to the resonant transmission.

15.2.1 One-Dimensional Array of δ-Function Nonlinear Layers

We first present the theoretical formulation of the wave transmission in a nonlinear one-dimensional lattice. For simplicity, we consider the one-dimensional Kronig-Penney model with δ-function nonlinear layers.

Consider a system of N δ-function layers distributed periodically at $z = n\ell$, $n = 0$, $1, \ldots, N-1$. Here, ℓ is the lattice constant. The structure is shown in figure 15.3. The electromagnetic waves outside the nonlinear structure are described by

$$E(z) = \begin{cases} E_i e^{-ikz} + E_r e^{+ikz}, & z \geq N\ell, \\ E_t e^{-kz}, & z \leq 0. \end{cases} \tag{15.17}$$

Inside the structure, the electric field satisfies the wave equation

$$\frac{\partial^2 E(z)}{\partial z^2} + \frac{\varepsilon\omega^2}{c^2} \sum_{n=0}^{N-1} \left[1 + \Lambda |E(z)|^2\right] E(z)\delta(z - n\ell) = 0. \tag{15.18}$$

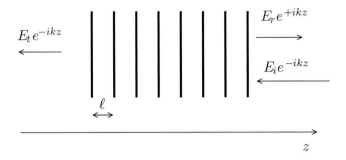

FIGURE 15.3. One-dimensional system of N nonlinear δ-function layers separated by lattice spacing ℓ. The z-dependence of the permittivity is $\varepsilon(z) = \sum_n \varepsilon_0 \left[1 + \Lambda |E(z)|^2\right] \delta(z - n\ell)$ with $n = 1, 2, \ldots, N$.

This equation is formally equivalent to the Kronig-Penney model of electrons discussed in chapter 4. Then, we can follow the arguments of problem 5.5 and rewrite equation (15.18) in the discrete form

$$E_{n+1} + E_{n-1} = \left[2 \cos k\ell - a(1 + \Lambda |E_n|^2)k\ell \sin k\ell\right] E_n, \tag{15.19}$$

where

$$a = \varepsilon. \tag{15.20}$$

The nonlinear Kerr coefficient is now Λa.

We discussed the *linear* Kronig-Penney model in section 13.6. As can be seen in figure 13.16, the linear model possesses bands and gaps in the transmission spectrum. From equation (13.46), we see that the position of the bands and gaps can be obtained by solving the equation

$$\left| \cos k\ell - \frac{a}{2} k\ell \sin k\ell \right| = 2. \tag{15.21}$$

From equation (15.19) it is clear that the effective nonlinearity is strongly modified by the wave vector. The nonlinearity is much less effective in the long-wave regime, $k\ell \ll 1$, and at the bottom of the band, where $\sin k\ell \sim 0$.

We are interested in the transmission characteristics of different frequencies and input or output intensities. To obtain the transmission coefficient numerically, we use the method derived in chapter 5. Assume that left of the structure we have electric fields $E_0 = E_t$ and $E_{-1} = E_0 e^{-ik\ell}$. Then, iterating equation (15.19), we obtain the electric fields E_N and E_{N-1}. The transmission coefficient is then given by equation (5.41),

$$T = \frac{4 \sin^2(k\ell)|E_0|^2}{|e^{-ik\ell}E_n - E_{n-1}|^2}. \tag{15.22}$$

In a linear system ($\Lambda = 0$) the transmission coefficient depends only on the wave vector. The transmission of the linear system as a function of $k\ell$ is shown in figure 13.17.

In contrast to linear systems, where T is a function of only $k\ell$, the transmission coefficient of a the nonlinear system depends also on the intensity of the transmitted

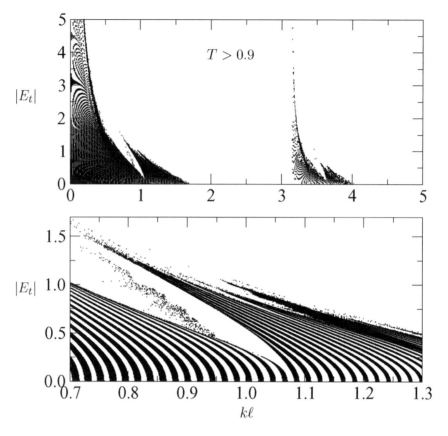

FIGURE 15.4. Plot of parameters $k\ell$ and $|E_t|$ for which the transmission coefficient T of the one-dimensional nonlinear system, is greater than 0.9. The length of the system is $L = 100\ell$. The Kerr coefficient is positive, $\Lambda = 1$, and $a = 1$. As shown in figure 13.16, the band gap spans from $1.724 < k\ell < \pi$. We see that there are no transmission states inside the gap of the linear system. The lower panel shows details of the upper panel for $0.7 < k\ell < 1.3$.

field. Figure 15.4 shows how the transmission through a system of $N = 100$ nonlinear δ-functions with positive nonlinear parameter $\Lambda = 1$ depends on $k\ell$ and $|E_t|$. We plot all points in the two-parametric plane $k\ell, |E_t|$ for which the transmission coefficient $T > 0.9$. Note that the limit $|E_t| \to 0$ corresponds to the linear system, in which transmission oscillates as a function of $k\ell$ (figure 13.17). These oscillations manifest themselves as a series of black and white strips in figure 15.4 (we remind the reader that black points denote parameters for which $T > 0.9$). For $\Lambda \equiv 0$, transmission does not depend on the field intensity. Therefore, the strips would be parallel with the horizontal axis. Due to the nonlinearity, strips bend to the left, so that the transmission possesses maximum values for smaller values of $k\ell$.

First, note that the transmission is nonzero only in the regions of transmission bands of the linear system (compare figure 15.4 with figures 13.16 and 13.17). This is easy to understand. For positive Λ, the parameter

$$a' = a\left(1 + \Lambda |E_t|^2\right) \tag{15.23}$$

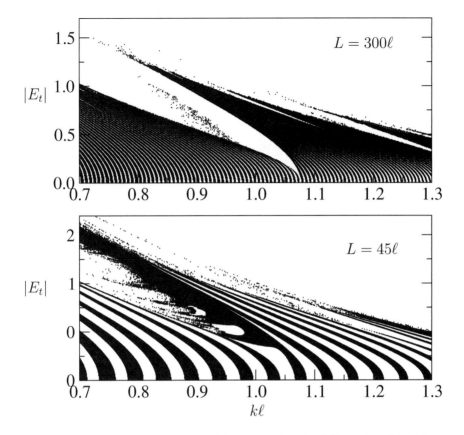

FIGURE 15.5. The same as the lower panel of figure 15.4, but for different lengths of the system: $L = 300\ell$ (upper panel) and $L = 45\ell$ (lower panel). Note how the structure of the subbands depends on the length of the system.

increases when field intensity $|E_t|$ increases. On the other hand, we see from figure 13.16 that a larger parameter a' gives broader gaps.

Inside the bands of a linear system, the transmission coefficient exhibits a highly nontrivial dependence on the wave vector $k\ell$ and the intensity $|E_t|$. Note that the transmission is also determined by the length of the system, as is clearly visible from figure 15.5, where we compare the transmission through two nonlinear systems with $N = 45$ and 300.

Contrary to the case of $\Lambda > 0$, the *negative nonlinear parameter* $\Lambda = -1$ also allows transmission *inside* the band gap. Indeed, increasing field intensity reduces the effective δ-function layer parameter a', defined by equation (15.23). This means that the band gap becomes narrower. We see in figure 15.6 that the transmission dependence on $k\ell$ and $|E_t|$ is even more complicated.

15.2.2 Gap Soliton

We have seen that systems with negative nonlinear parameter Λ possess transmission states in the band gap of the original linear system. Transmission states inside the band

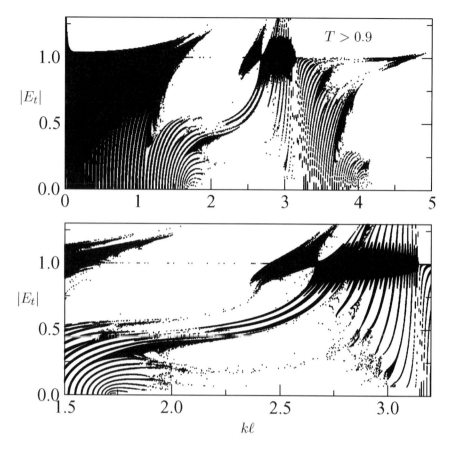

FIGURE 15.6. Plot of parameters $k\ell$ and $|E_t|$ for which the transmission coefficient T of the one-dimensional nonlinear system is greater than 0.9. In this case, the nonlinear Kerr coefficient is negative, $\Lambda = -1$, and $a = 1$. The system length is $L = 100\ell$. Note that transmission states exist inside the linear band gap, $1.724 < k\ell < \pi$. The lower panel shows details of the upper panel for $k\ell$ in the band gap.

gap can be explained by the creation of *gap solitons*. The origin of the solitons can be explain within a simple physical picture.

In the gap, the wave amplitude can either exponentially decrease or exponentially increase. In a linear system, we must reject increasing solutions because the exponential increase leads to infinite energy for sufficiently long systems, which is not physical. However, in a nonlinear medium, the intensity of the field influences the properties of the medium. This enables feedback: when the field increases above a certain limit, an exponential increase of the field amplitude stops and is followed by the exponential decrease. Therefore, there is no necessity a priori to reject all exponentially increasing solutions [97].

Consider now the case when the field increases exponentially from a small starting value, as it enters the nonlinear system from outside. We learned in section 15.1 that a larger field reduces the band gap. Consequently, the effective gap narrows and

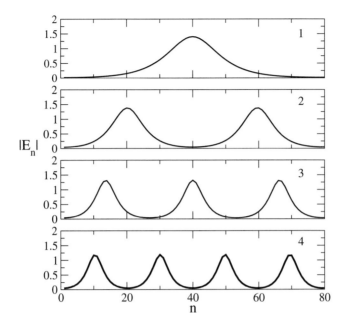

FIGURE 15.7. Spatial dependence of the electric field inside the structure for four different values of the wave vector $k\ell$, shown in figure 15.8. The used values of $k\ell$ and $|E_t|$ are highlighted in figure 15.8 by open circles (from top to bottom) (3.122, 0.1910), (3.086, 0.0488), (2.994, 0.0405), and (2.838, 0.0447). Note that $k\ell$ lies inside the gap of the linear model ($\Lambda = 0$). In all these cases, the transmission coefficient is very close to 1 (higher than 0.995). Note that the length of the nonlinear structure is an integer multiple of the width of the soliton. Note also that different transmission lines shown in figure 15.8 correspond to solutions with different numbers of solitons.

eventually closes when the field intensity is sufficiently high. The field then behaves as if it is propagating and starts to oscillate. However, as soon as the the field intensity falls to a value such that the gap opens and the frequency is once again in the gap, the fields begins to decrease exponentially. When the field decreases to the original value $E = E_i$, it starts to increase again. The overall spatial shape of the field is shown in figure 15.7.

Since such solutions appears only inside the gap of the linear system, they are called gap solitons. We show gap solitons for the nonlinear periodic structure defined by equation (15.19). In figure 15.8 we see that transmission is possible inside the *linear* band gap, $2.362 < ka < \pi$, shown in figure 13.16. Figure 15.7 shows the spatial dependence of the field inside the structure for various values of $k\ell$. Indeed, we see that the field inside the structure possesses a sharp maximum and decays exponentially on both sides of the maximum. The transmission is close to 1 if the length of the system, L, becomes an integer multiple of the spatial extent of the soliton, Δ ($L = n \times \Delta$ with $n = 1, 2, 3$, and 4 for the cases shown in figure 15.7). This explains why the position of the resonant transmission trajectories, shown in figure 15.8, depends on the length of the system.

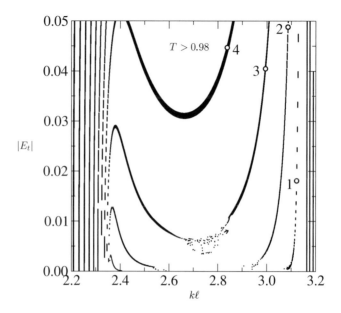

FIGURE 15.8. Resonant transmission trajectories of single and multiple solitons in the first gap, $2.362 < k\ell < \pi$, of the *linear* model. The length of the system is $L = 80\ell$. The parameters of the nonlinear system are $\Lambda = -1$ and $a = 0.35$. Black points show parameters $k\ell$ and $|E_t|$ for which the system exhibits transmission $T > 0.98$. Note the very small values of the transmitted field $|E_t|$. The parallel strips inside the transmission band (for $k\ell < 2.362$) are due to the oscillation of the transmission as a function of $k\ell$, known from the linear problem (figure 13.17). The system is almost linear in this region. In the gap, nonlinearity enables a large transmission by creating gap solitons. The open solid circles show pairs of $k\ell$ and $|E_t|$ for which we show in figure 15.7 the spatial dependence of the electric field. The integers give the number of solitons excited along the lattice.

15.3 Problems

Problem 15.1

Show that the nonlinear problem, defined by equation (15.19), is invariant with respect to the transformation

$$\Lambda \to a\Lambda \quad \text{and} \quad E \to E/a^2. \tag{15.24}$$

This symmetry enables us to restrict our analysis to values of $\Lambda = 1$ and -1.

Problem 15.2

An alternative to equation (15.22) is to use Chebyshev's formula (13.50) with transfer matrix $\mathbf{M}_1$ given by equation (13.49) to calculate the transmission coefficient of N nonlinear δ-function layers defined by the nonlinear permittivity (15.1). For $\Lambda = 1$ and $a = 1$, compare your results with those shown in figure 15.4.

Problem 15.3
Repeat the analysis shown in figure 15.8 for a few different values of N. Describe how the transmission lines change as a function of the number N of nonlinear δ-functions.

Problem 15.4
Verify that all points lying at a given resonant transmission line shown in figure 15.8 indeed correspond to solutions with a given number of solitons. Note that you must first calculate the exact position of transmission lines in order to determine accurate the values of $k\ell$ and E_t for which a soliton solution exists.

Problem 15.5
Calculate the form of the gap soliton for a few transmission lines shown in figure 15.6.

16 Left-Handed Materials

In this chapter, we summarize the electromagnetic properties of so-called left-handed materials. This name was given to man-made composites that possess, in a certain frequency region, negative real parts of both the permittivity and permeability.

We have discussed some properties of left-handed materials in previous chapters. In chapter 9, we found that an interface between a vacuum and a left-handed medium might allow perfect transmission. In chapter 11, we analyzed in detail the existence of surface electromagnetic waves localized at the interface between vacuum and a left-handed material. In chapter 13, we used left-handed materials in construction of infinite layered systems and found that such structures can possess an $\bar{n} = 0$ band gap. All the above examples show that electromagnetic properties of left-handed materials differ qualitatively from those of conventional materials, dielectric or metal.

In this chapter we discuss in more detail the basic properties of left-handed materials [13,14,107,119–121,125,134,142,147,157,158]. In the first part of this chapter, we show that negativeness of the permittivity and permeability leads to antiparallel orientation of the wave vector $\vec{k}$ and the Poynting vector $\vec{S}$. Also, from the requirement that the electromagnetic field energy be positive, we find that left-handed materials must be dispersive: both ε and μ must be functions of frequency.

Left-handed materials became famous because of the negativeness of the index of refraction [142,157], which, in turn, causes a negative refraction of the electromagnetic wave at the interface between vacuum and the left-handed medium [114,138]. An even more surprising property of left-handed materials is the ability to amplify evanescent waves [119]. This enables us to use left-handed materials for construction of a perfect lens.

It is important to note that we are not aware of the existence of such materials in nature. However, recently man-made composites were prepared, which exhibit, in a certain frequency region, negative *effective* permittivity [126,127,144] and permeability [100,128]. "Effective" means that these parameters describe the response to an external electromagnetic field with a frequency ω such that the wavelength of the field, $\lambda = 2\pi c/\omega$, is much larger than any spatial inhomogeneity of the material (see also chapter 14). We describe the structure of such composites in the second part of this chapter. We find that the left-handed medium is not homogeneous, but consists of a periodic repetition of elementary "cells." Each cell contains a so-called split-ring resonator and a thin metallic wire [145]. The structure of the elementary cell [86,138] determines the tranmission properties of the entire left-handed material.

Negative refraction of electromagnetic waves is observed also at the interface of two-dimensional photonic crystals. Since the theory of the negative refractive index in photonic crystals is rather complicated, we refer the reader to the original papers [65–67,112].

16.1 Electromagnetic Properties of Left-Handed Materials

16.1.1 Wave Equation

We start with the wave equation for an electromagnetic wave in a medium with permittivity ε and μ. In Appendix B.35, we derived equation (B.38),

$$\left[k^2 - \varepsilon\mu \frac{\omega^2}{c^2} \right] \vec{E} = 0, \tag{16.1}$$

from which it follows that the plane wave

$$\vec{E}(\vec{r}, t) = \vec{E}_0 e^{i(\vec{k}\cdot\vec{r} - \omega t)} \tag{16.2}$$

can propagate in the medium if the product $\mu\varepsilon$ is positive. This is possible if either both ε and μ are positive, or *both* parameters are *negative*. Thus we conclude that propagation of electromagnetic waves is possible in a left-handed medium.

16.1.2 Left-Handed Rule

As shown in Appendix B, in regular materials, the vectors $\vec{E}$, $\vec{H}$ and $\vec{k}$ follow the right-handed triplet of vectors, and the Poynting vector $\vec{S} \parallel \vec{k}$. However, if both ε and μ are *negative*, we obtain from Maxwell's equations that

$$\begin{aligned} \vec{k} \times \vec{E} &= -\frac{|\mu|\omega}{c} \vec{H}, \\ \vec{k} \times \vec{H} &= +\frac{|\varepsilon|\omega}{c} \vec{E}. \end{aligned} \tag{16.3}$$

It is clear that the vectors $\vec{E}$, $\vec{H}$, and $\vec{k}$ follow now the left-handed triplet of vectors, and vectors $\vec{k}$ and $\vec{S}$ have *opposite* orientation as shown in figure 16.1.

$$\epsilon > 0 \quad \mu > 0 \qquad\qquad\qquad \epsilon < 0 \quad \mu < 0$$

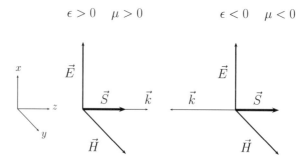

FIGURE 16.1. Orientation of the electric and magnetic vectors of the intensity of electromagnetic fields, $\vec{E}$ and $\vec{H}$, with respect to the wave vector $\vec{k}$ and Poynting vector $\vec{S}$.

Note that, the direction of the propagation of the electromagnetic wave is determined by the orientation of the Poynting vector $\vec{S}$, and *not* by the orientation of the wave vector $\vec{k}$. The opposite sign of $\vec{k}$ only means that the phase velocity of the wave inside the left-handed medium is negative.

16.1.3 Dispersion

From the expression for the energy of the electromagnetic field [26]

$$U = \frac{1}{8\pi} \left\{ \frac{\partial(\varepsilon\omega)}{\partial\omega} E^2 + \frac{\partial(\mu\omega)}{\partial\omega} H^2 \right\}, \tag{16.4}$$

we immediately see that both permittivity ε and permeability μ must depend on the frequency. Otherwise, the energy U reduces to the expression $U = (\varepsilon E^2 + \mu H^2)/(8\pi)$, which would be negative for both ε and μ negative. On the other hand, the energy U given by equation (16.4) is always positive, because the permittivity and permeability must satisfy the inequalities [26]

$$\frac{\partial\omega\varepsilon(\omega)}{\partial\omega} > 0 \quad \text{and} \quad \frac{\partial\omega\mu(\omega)}{\partial\omega} > 0. \tag{16.5}$$

The frequency dependence of ε and μ means, due to the Kramers-Kronig relations [26], that both the permittivity and permeability must be complex. The nonzero imaginary parts of the permittivity and permeability lead unavoidably to absorption losses of the electromagnetic wave propagating through the left-handed medium.

16.1.4 Index of Refraction and Snell's Law

One of the most important properties of left-handed materials is that they also possess a negative index of refraction. Since $n = \sqrt{\varepsilon\mu}$, it seems that n might be positive, even if $\varepsilon < 0$ and $\mu < 0$. To find the correct sign in front of the square root, we use the physical constraint for the imaginary part of the refractive index n_i derived in appendix B. We remind the reader that n_i must be always positive in passive materials

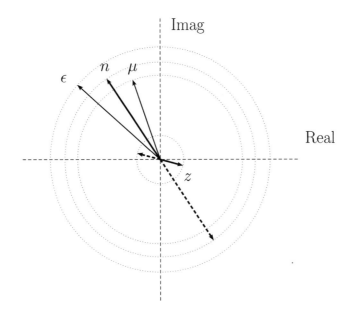

FIGURE 16.2. Graphical explanation of the negative real part of the refractive index. The arrows show the position of the permittivity ε and the permeability μ in the complex plane. There are two solutions for n given by equation (16.8), which correspond to $m = 0$ (full line) and 1 (dashed line). Only one of them—that with $n_r < 0$— satisfies the requirement that $n_i > 0$ [equation (B.86)] and is, therefore, a true physical solution for the refractive index. We show also the orientation of the impedance $z = \sqrt{\mu/\varepsilon}$. Again, the physical requirement $z_r > 0$ [equation (B.65)] determines the sign of the square root.

[equation (B.86)] since negative n_i would cause an exponential increase of the amplitude of the electromagnetic field, which is not physical.

Let us consider both ε and μ complex and write them in the form

$$\varepsilon = |\varepsilon|e^{i\phi_\varepsilon}, \quad \mu = |\mu|e^{i\phi_\mu}. \tag{16.6}$$

Then, the refractive index n is also complex, $n = n_r + in_i$, and can be written as follows:

$$n = |n|e^{i\phi_n}. \tag{16.7}$$

All three phases are confined by $0 \leq \phi_\varepsilon \leq \pi$ and $0 \leq \phi_\mu \leq \pi$ and $0 \leq \phi_n \leq \pi$, in order to have the electromagnetic losses, given by equation (B.73), positive. From the relation $n = \sqrt{\varepsilon\mu}$, we see that

$$|n| = \sqrt{|\varepsilon\mu|} \quad \text{and} \quad \phi_n = \frac{1}{2}(\phi_\varepsilon + \phi_\mu + 2\pi m), \tag{16.8}$$

where $m = 0$ or 1. Of these two solutions, only that with $n_i > 0$ is physical. From figure 16.2, we easily see that this solution corresponds to

$$n_r < 0, \tag{16.9}$$

so that the real part of the index of refraction of a left-handed medium is indeed negative.

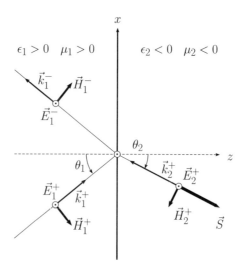

FIGURE 16.3. Negative refraction of the electromagnetic plane wave on the interface of a "right-handed" and a "left-handed" material. The wave has TE polarization. In the left-handed material, the wave propagates in the direction of negative angle. Note that the directions of the Poynting vector $\vec{S}$ and $\vec{k}$ are opposite.

In the same way, we determine the proper sign of the impedance $z = \sqrt{\mu/\varepsilon}$ using the requirement, given by equation (B.65), that the real part of z must be positive, $z_r > 0$.

16.1.5 Electromagnetic Waves at a Vacuum–Left-Handed Medium Interface

Figure 16.3 shows the propagation of the TE wave through the interface between the dielectric and the left-handed medium. With the use of Maxwell's equations we estimate the orientation of the vectors of electric and magnetic intensities and of the wave vector.

Since the index of refraction of the left-handed medium is negative,

$$n < 0, \tag{16.10}$$

Snell's law tells us that the angle of refracted waves is also negative, as is shown in figure 16.3. In contrast to the situation in a dielectric material, the Poynting vector and the wave vector have opposite orientations in a left-handed medium.

The boundary conditions for the electric and the magnetic fields, equations (9.15) and (9.16), as well as the condition of the continuity of tangential component of the wave vector,

$$k_{2x} = k_{1x}, \tag{16.11}$$

must be satisfied in the present case, too. The z-component of the wave vector is negative:

$$k_{2z} = -k_{1z} |n_2|. \tag{16.12}$$

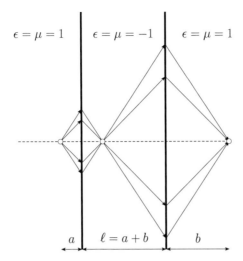

$\epsilon = \mu = 1$ $\epsilon = \mu = -1$ $\epsilon = \mu = 1$

a $\ell = a + b$ b

FIGURE 16.4. Transmission of an electromagnetic wave through a planar slab with $\varepsilon = -1$ and $\mu = -1$. All the waves emitted from a source at a distance a left of the slab propagate toward a common crossing point on the right side of the slab. The distance b of the image from the planar slab is $b = \ell - a$. Note that there is also a second crossing point of all the waves inside the slab [157].

Keeping in mind the orientation of the wave vector, we can calculate the transmission and reflection coefficients through an interface between the vacuum and the left-handed material with the help of general formulas derived in section 9.2. We will not repeat them here.

16.2 Transmission through a Slab of Left-Handed Material

A slab constructed from a left-handed material with negative permittivity and permeability can be treated in the same way as a dielectric slab, discussed in chapter 10. It is, however, worthwhile to emphasize two special properties of the left-handed slab. First, negative refraction at the interface of the vacuum and left-handed medium enables us to create a planar lens. Second, we show that the left-handed slab is able to amplify incoming evanescent waves. This property makes left-handed materials different in principle from any other known material, and offers various applications of the left-handed materials, for instance, the construction of so-called perfect lenses.

16.2.1 Focusing

Negative refraction at the interface between a vacuum and a left-handed medium enables the construction of a slab lens, shown in figure 16.4.

Consider first the ideal case of

$$\varepsilon_2 = -1 \quad \text{and} \quad \mu = -1. \tag{16.13}$$

Then also $n = -1$ and from equation (16.12), we obtain

$$k_{2z} = -k_{1z}. \tag{16.14}$$

In this very simple case, the transfer matrix of a planar left-handed slab, given by equations (10.17) and (10.20), simplifies to

$$M = \begin{pmatrix} e^{ik_{2z}\ell} & 0 \\ 0 & e^{-ik_{2z}\ell} \end{pmatrix}. \tag{16.15}$$

This means that we have no reflection at any interface, in agreement with our analysis of perfect transmission in section 9.5.2.

Consider now a point source located at distance a from the left interface of the left-handed slab (figure 16.4). We easily see that all the waves emitted from the source focus on one point (image) on the other side of the slab. If ℓ is the thickness of the slab, then the distance of the image from the slab is $b = \ell - a$. Moreover, we see that the distance each ray travels in the slab is exactly the same as it travels in a vacuum. Since the wave vector in the left-handed slab has opposite sign to the wave vector outside the slab, but both vectors are equal in magnitude, we conclude that the phase difference between point source and image is zero. We see that not only does the left-handed slab focus all outgoing waves into one point, but these waves arrive at the image with the same phase they had in the source. The phase coherence of the incoming waves is very important for the construction of an image.

Unfortunately, all said above is valid only when conditions (16.13) are satisfied. Consider now a left-handed material with $n \neq -1$. Then, a wave with an incident angle θ_1 propagates in the left-handed slab in the direction given by angle θ_2 ($\sin \theta_1 = n \sin \theta_2$). It is easy to see that focusing of all the waves requires that the condition

$$(a + b)\tan \theta_1 = -\ell \tan \theta_2 \tag{16.16}$$

holds for all incidence angles θ_1. This is possible only when $\theta_1 = -\theta_2$, which means only when the refractive index $n = -1$.

We can try to construct a focusing left-handed slab with some other combinations of negative permittivity ε and permeability μ, which keep $n = -1$. This is, however, of little use, since in this case the surface impedance $z_{21} = \sqrt{\mu/\varepsilon} \neq 1$. This means that the boundary of the left-handed slab is no longer perfect. Some portion of the incident wave is reflected back. This is easy to see from equation (10.23), because

$$\frac{r}{t} = M_{12}^{(s)} = \frac{i}{2}\left[z_{21} - \frac{1}{z_{21}}\right]\sin k_{2z}\ell \tag{16.17}$$

possesses nonzero values for $\mu_2 \neq -1$, so that the reflection $r \neq 0$.

16.2.2 Evanescent Waves in a Left-Handed Slab

Another fascinating property of the left-handed slab is that it amplifies incoming evane-scent waves [119]. To understand this phenomenon, let us study first the "transmission" of the evanescent wave through a single interface located at $z = 0$ between a vacuum and a left-handed material with refractive index n.

An incoming electromagnetic field decreases exponentially along the z axis,

$$E = |E|e^{ik_x x - \kappa_1 z}, \qquad z < 0, \tag{16.18}$$

where we used that $k_{1z} = i\kappa_1$ and where

$$\kappa_1 = \sqrt{k_x^2 - \frac{\omega^2}{c^2}} > 0. \tag{16.19}$$

Right of the interface, in the left-handed medium, we have the wave

$$E_2 = |E_2|e^{ik_x x - \kappa_2 z}, \qquad z > 0, \tag{16.20}$$

where

$$\kappa_2 = +\sqrt{k_x^2 - \frac{\omega^2}{c^2} n^2}. \tag{16.21}$$

We have to use a positive sign of the square root in order to prevent an exponential increase of the amplitude of the electromagnetic wave, which is not allowed in a semi-infinite medium.

Now we calculate the transmission coefficient of the evanescent wave, given by equation (16.18), through a left-handed slab of final thickness ℓ. For simplicity consider $n = -1$.

First, we insert $k_{1z} = i\kappa_1$ in the expression for the matrix element M_{22} given by equation (10.19), and obtain

$$M_{22}^{(s)} = \cos k_{2z}\ell - \frac{1}{2}\left[\frac{\mu_1}{\mu_2}\frac{k_{2z}}{\kappa_1} - \frac{\mu_2}{\mu_1}\frac{\kappa_1}{k_{2z}}\right]\sin k_{2z}\ell. \tag{16.22}$$

Then, we insert $k_{2z} = i\kappa_2$, and use that $k_{2z}\sin k_{2z} = -\kappa_{2z}\sinh\kappa_{2z}$ and $(\sin \kappa_{2z})/k_{2z} = (\sinh\kappa_{2z})/\kappa_{2z}$. Using these relations, we obtain that

$$M_{22}^{(s)} = \cosh \kappa_2\ell + \frac{1}{2}\left[\frac{\mu_1}{\mu_2}\frac{\kappa_2}{\kappa_1} + \frac{\mu_2}{\mu_1}\frac{\kappa_1}{\kappa_2}\right]\sinh \kappa_2\ell. \tag{16.23}$$

Keeping in mind that $\mu_2/\mu_1 = -1$ and $\kappa_2 = \kappa_1$ (for $n = -1$), we finally obtain that

$$M_{22} = e^{-\kappa_2\ell}, \tag{16.24}$$

so that the transmission amplitude

$$t = \frac{1}{M_{22}} = e^{\kappa_2\ell} \tag{16.25}$$

increases exponentially inside the left-handed slab.

In the same way, we find that the off-diagonal elements of the transfer matrix $M_{12} = M_{21} = 0$. This means there is no reflection of the evanescent waves at the interface.

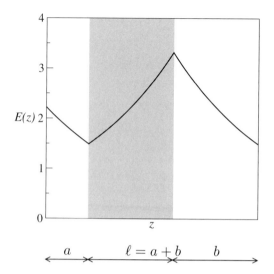

FIGURE 16.5. Transmission of an evanescent electromagnetic wave through a planar slab with $\varepsilon = -1$ and $\mu = -1$. An evanescent wave emitted from the source propagates toward the left-handed slab and decreases as $e^{-\kappa z}$ ($\kappa \ell = 4$ in this case). When the wave reaches the surface of the left-handed slab, it starts to increase exponentially as $e^{+\kappa z}$. After crossing the interface on the opposite side of the left-handed slab, the evanescent wave again decreases exponentially. Note that the amplitude of the evanescent wave at the distance b from the second boundary is the same as at the position of the source [119, 166].

Finally, we calculate $M_{11} = e^{+\kappa_2 \ell}$. Since $M_{11} = t' - r'r/t$, we see that the transmission amplitude $t' = e^{\kappa_2 \ell}$ also increases exponentially with the slab thickness. The z-dependence of an evanescent wave propagating through a left-handed slab with $\varepsilon = -1$ and $\mu = -1$ is shown in figure 16.5.

To obtain the numerical data shown in figure 16.5, we first consider complex permittivity and permeability $\varepsilon = -1 + i\gamma$ and $\mu = -1 + i\gamma$. The small imaginary part γ helps us to choose the correct sign of the wave vector k_2, such that the imaginary part of k_2 is positive (see problem 16.4).

Amplification of evanescent waves is a generic property of left-handed materials. Although this is perfect only for the ideal case (16.13), it exists also in materials with any values of *negative* ε and μ. However, the transfer matrix is, in general, not diagonal, so that part of the incident evanescent wave is reflected back. As an example, we show in figure 16.6 the amplification of the evanescent waves inside the left-handed slab with $n = -1$, but with surface impedance $z \neq 1$ (which corresponds to values $\mu = 1/\varepsilon$). Since $n = -1$, we have $\kappa_1 = \kappa_2 = \kappa$. Details of the calculation are given in problem 16.5. Here we note only that, because of the reflection at both interfaces, the electromagnetic field at the left side of the left-handed slab consists of two terms,

$$Ae^{-\kappa z} + Be^{+\kappa z}. \tag{16.26}$$

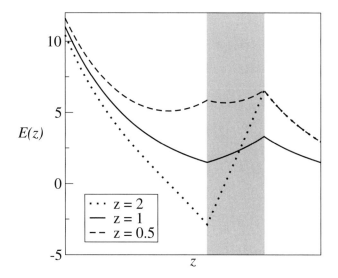

FIGURE 16.6. Transmission of an evanescent wave through a left-handed slab for various values of the surface impedance z for $\kappa\ell = 4$. When the surface impedance $z \neq 1$, part of the incoming wave is reflected at the left interface. Therefore, in addition to the incoming wave $E^+ = \exp[-\kappa z]$, we observe on the left-hand side of the slab also a reflected wave $B\exp[+\kappa z]$. This is the reason that we observe a local maximum of the field also at the left interface.

An incoming wave is proportional to A and a reflected wave is proportional to B. This is the reason that a field possesses local maxima at both boundaries of the slab. Although the field distribution looks similar to that of surface waves, discussed in chapter 11, we want to stress the different physical origins of both phenomena. We remind the reader that surface waves can be excited only for very specific values of wave vector and frequency, given by the dispersion relation $\omega = \omega(k_x)$. No such restriction exists for the amplification of evanescent waves in the left-handed slab.

16.2.3 Perfect Lens

Amplification of evanescent waves opens the possibility, at least in principle, to create a lens able to reconstruct infinitesimally small details of an object.

We know from optics that the best resolution of a conventional lens is of the order of the wavelength λ of the electromagnetic wave. The reason for this constraint can be explained as follows. Consider two points at the distance Δ, as shown in figure 16.7. In order to distinguish these two points in the image plane right of the lens, we must transfer through the lens electromagnetic waves with the x-component of the wave vector, $k_x > 2\pi/\Delta$. However, an electromagnetic wave propagates in the z-direction, and the two components of the wave vector, k_x and k_z, are related by the dispersion relation

$$\frac{\omega^2}{c^2} = k_x^2 + k_z^2. \tag{16.27}$$

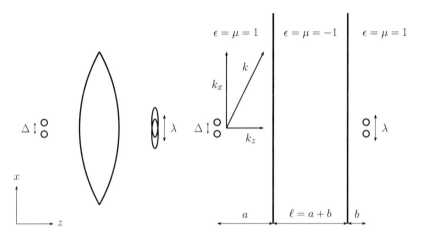

Figure 16.7. Left: Resolution of a conventional lens for a given frequency. The resolution is limited by the wavelength of the incident electromagnetic wave. Two points at a distance Δ smaller than the wavelength λ (left of the lens) are not distinguishable in the image (right). Right: Resolution of the ideal planar left-handed lens with $\varepsilon = -1$ and $\mu = -1$. Any two points of the source are distinguished at the image. This is due to the ability of the left-handed slab to transfer evanescent waves with large transverse components of the wave vector k_x. These waves contain information about spatial details of the source that are smaller than the wavelength λ [119, 166].

Since k_z must be real for propagating waves, we obtain the result that the maximum possible value of k_x of the propagating electromagnetic wave with frequency ω is given by

$$k_x < k_x^{\max} = \frac{\omega}{c} = \frac{2\pi}{\lambda}. \tag{16.28}$$

Components of the electromagnetic field with *larger* values of k_x cannot propagate along the z-axis. Although waves with $k_x > k_x^{\max}$ exist, they decay exponentially along the z-direction as $e^{-\kappa z}$, where $\kappa = \sqrt{k_x^2 - \omega^2/c^2}$. No such exponentially decaying wave can be used for the reconstruction of the image at the opposite side of the conventional lens. Therefore, information about tiny details of the object is lost. The best resolution of a conventional lens, defined as the smallest distance between two points that can be distinguished from each other in the image, is given by

$$\Delta = \frac{2\pi}{k_x^{\max}} = \lambda. \tag{16.29}$$

Consider now a planar slab of a left-handed material with permittivity $\varepsilon = -1$ and permeability $\mu = -1$. In section 16.2.1 we have seen that such a left-handed slab focuses all propagating waves, emitted from a point source, to one image point at the opposite side of the interface. Thus, the left-handed slab acts as a lens. Next, in section 16.2.2 we showed that the left-handed slab also enhances all the incident evanescent waves. This means that the evanescent wave with $k_x > k_{\max}$, emitted from a source at distance a from the slab, can be amplified inside the left-handed slab so that at the image point at the distance b on the other side of the left-handed slab it possesses the same magnitude as it had at the source, and can be used for the reconstruction of the source. Consequently, there is no resolution limit of a planar slab made from the left-handed material. In principle,

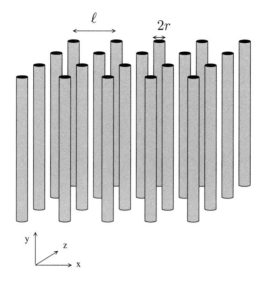

FIGURE 16.8. Periodic array of thin metallic wires. This structure behaves as a medium with negative effective permittivity ε given by equation (16.31), if the electromagnetic wave propagates with $\vec{E}$ parallel to the wires. The typical radius of wires is much smaller than the distance between neighboring wires [126, 127, 139].

this enables reconstruction of all details of the object. We conclude that a planar slab can work as a *perfect* lens [119]. A perfect planar lens constructed from a left-handed material is shown schematically on the right side of figure 16.7.

The scenario described concerns the ideal case of a left-handed material with $\varepsilon = -1$ and $\mu - 1$. However, we know that no such ideal material can be constructed. That is, real left-handed materials have frequency-dependent permittivity and permeability. Even if we succeed in having a refractive index n close to -1 in a given frequency region, the impedance $z = \sqrt{\mu/\varepsilon}$ will differ from 1 at these frequencies. Therefore, we can never obtain perfect imaging. Nevertheless, even when the refractive index n and the impedance z differ from their ideal values, the slab of a left-handed medium can transfer evanescent waves and enable us to obtain an image with a resolution much smaller than the wavelength.

16.3 Structure of Left-Handed Materials

16.3.1 Array of Thin Metallic Wires

In section 13.5.1, we learned that a periodic layered medium consisting of thin metallic layers separated by air behaves effectively as a medium with reduced plasma frequency

$$\omega_p^{\text{red}} = \omega_1 = \frac{\pi c}{\ell_b n_b} \tag{16.30}$$

where ℓ_b is the distance between metallic layers. Consider now the structure that consists of a periodic array of very thin metallic wires, shown in figure 16.8. In analogy with the one-dimensional metallic problem, we assume that the wire structure possesses a band gap below some effective plasma frequency ω_p. A more detailed analysis [106, 126, 127, 133, 139, 144] show that if the radius r of the wire is much smaller than the distance between neighboring wires ℓ, then the wire medium behaves as a medium with effective plasmalike permittivity

$$\varepsilon(\omega) = 1 - \frac{\omega_p^2}{\omega^2 + i\omega\gamma}, \tag{16.31}$$

where the effective plasma frequency ω_p is given by the structure of the material,

$$\nu_p = \frac{\omega_p}{2\pi} = \frac{c}{\ell\sqrt{2\pi\ln(\ell/r)}}. \tag{16.32}$$

We see that the plasma frequency decreases with increasing distance between wires as $\nu_p \sim \ell^{-1}$, and is a logarithmic function of the ratio ℓ/r.

Of course, the relation (16.31) holds only for the electromagnetic wave polarized with $\vec{E}$ parallel to the wires. For the other polarization, with the electric field perpendicular to the wires, the permittivity $\varepsilon \equiv 1$.

It is worth mentioning that the above formulas were derived under the assumption that metal is a very good conductor, i.e., the frequency ω is much smaller than the plasma frequency of the metal. The choice of the plasma frequency determines the length parameters of the wire array. To obtain $\nu_p \sim 10\,\text{GHz}$, we need the period of the wire array, ℓ, to be of the order of a few millimeters. For instance, the choice $\ell = 5\,\text{mm}$ and $r = 0.003\ell$ gives $\nu_p \approx 10.6\,\text{GHz}$.

Figure 16.9 shows the effective permittivity of a periodic array of thin metallic wires. To obtain these data, we first calculate transmission and reflection amplitudes, and then apply the retrieval procedure, described in section 14.2. Note that the data obtained agree with our estimation of the plasma frequency. In the inset of figure 16.9 we show the absorption, which exhibits a maximum at the plasma frequency ω_p. This is qualitatively similar to the frequency dependence of the absorption of an electromagnetic wave propagating through thin metallic films shown in figure 10.10. Thus, we conclude that the electromagnetic response of an array of thin metallic wires is qualitatively very similar to the electromagnetic response of a metal, but with an effective reduced plasma frequency ω_p.

16.3.2 Resonant Magnetic Response

In the previous section, we found that an array of thin metallic wires behaves as a medium with a negative effective permittivity. Now, we need to find a similar medium that possesses a negative magnetic permeability. The simplest element that has a magnetic response is a simple metallic ring. However, although an array of metallic rings can create a diamagnetic medium with effective permeability $\mu < 1$, it is not possible to reduce the permeability to negative values. This would be possible if the medium consisted of resonant elements. The simplest element that possesses a resonant magnetic response is a split metallic ring [128], shown in figure 16.10.

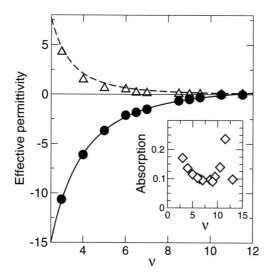

FIGURE 16.9. Effective permittivity for a lattice of thin metallic wires calculated numerically versus frequency v, measured in GHz. The wire radius is $r = 15\,\mu m$ and the lattice constant is $\ell = 5$ mm. The metallic permittivity is $\varepsilon_m = -2000 + 10^6 i$, typical for the GHz region. Solid and dashed lines are numerical fits to equation (16.31) with plasma frequency $v_p = 11.1$ GHz and loss factor $\gamma = 1.2$ GHz. Inset: Numerically calculated absorption as a function of frequency [106].

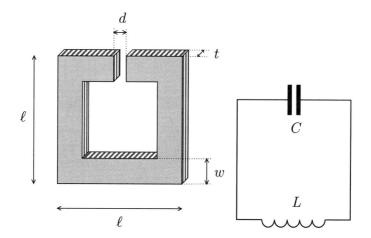

FIGURE 16.10. Left: Simple split-ring resonator is given by a cut metallic ring deposited on a dielectric substrate (not shown) [128]. Right: An equivalent electric circuit. The split is replaced by a capacitor $C = \varepsilon_c(wt/d)/4\pi$ and the loop is replaced by the inductance L, given by equation (16.34) [100].

The magnetic response of the split ring can be estimated on the basis of the analogy with an LC circuit as shown in figure 16.10. The capacitance C is due to the gap. It can be estimated as

$$C = \frac{\varepsilon_c}{4\pi}\frac{wt}{d} \tag{16.33}$$

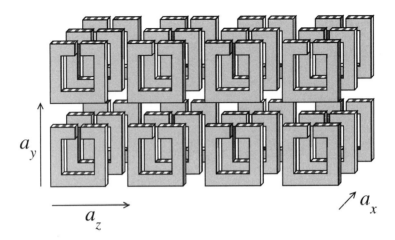

FIGURE 16.11. A periodic array of split-ring resonators [128, 145].

(ε_c is the permittivity of the material that fills the gap). The inductance L is proportional to the ring area ℓ^2, and inverse to the ring thickness t,

$$L = \frac{4\pi}{c^2} \frac{\ell^2}{t},$$

(16.34)

as follows from the formula for a coil with $N = 1$ windings.

Thus, we expect that the split ring behaves as a resonator with resonant frequency

$$\omega_{LC} = \frac{1}{\sqrt{LC}}.$$

(16.35)

By substituting C and L in equation (16.35) we obtain

$$\omega_{LC} = \frac{c}{\ell \sqrt{\varepsilon_c}} \sqrt{\frac{d}{w}}.$$

(16.36)

The corresponding free-space wavelength

$$\lambda_{LC} = \frac{2\pi \ell \sqrt{\varepsilon_c}}{c} \sqrt{\frac{w}{d}}$$

(16.37)

is proportional to the size ℓ of the split-ring resonator. In principle, λ_{LC} can be much larger than ℓ. This is a fundamental difference between left-handed materials and photonic crystals. As we mentioned in chapter 13, the frequency of a operation of a photonic crystal corresponds to a wavelength of the same order of magnitude as the size of the unit cell. Since in left-handed materials the operation wavelength λ_{LC} can be much larger than the size of the unit cell, a left-handed material can be considered as a homogeneous effective medium and described using effective medium theories, which simplify its description.

Consider now a regular three-dimensional array of split-ring resonators, shown in figure 16.11. We divide the space into small unit cells of size a_x, a_y and a_z in the direction

x, y and z, respectively. Each unit cell contains one resonator so that the volume V contains

$$N = \frac{V}{a_x a_y a_z} \qquad (16.38)$$

resonators. We need to estimate the magnetic response of this structure to the external magnetic field

$$H_x = H_{0x} e^{i\omega t}. \qquad (16.39)$$

We start with the estimation of the voltage in each split ring. The external magnetic-field-induced voltage U_{ind} is compensated by the self-induction voltage V_L and by the voltage drop over the capacitance V_C,

$$U_L + U_c = U_{ind}, \qquad (16.40)$$

which can be written as

$$L \frac{\partial I}{\partial t} + \frac{1}{C} \int dt\, I = -\frac{1}{c} \frac{\partial \phi}{\partial t}, \qquad (16.41)$$

with external magnetic flux ϕ given by

$$\phi = \ell^2 H_z. \qquad (16.42)$$

Taking the time derivative of equation (16.41), we obtain

$$\frac{\partial^2 I}{\partial t^2} + \frac{1}{LC} I = -\frac{\ell^2}{cL} \frac{\partial^2 H_x}{\partial^2 t}. \qquad (16.43)$$

Now, we express the current I as

$$I = I_0 e^{i\omega t}, \qquad (16.44)$$

and obtain

$$\left(-\omega^2 + \omega_{LC}^2\right) I_0 = \frac{\ell^2 \omega^2}{c} H_{0x}. \qquad (16.45)$$

The magnetization M is a function of the current,

$$M = M_{0x} e^{i\omega t}, \qquad M_0 = \frac{N}{V} \frac{\ell^2}{c} I_0. \qquad (16.46)$$

Inserting these two expressions into equation (16.43), we obtain the resonant frequency dependence of the magnetization

$$M_{0x} = \frac{1}{\omega_{LC}^2 - \omega^2} \frac{\ell^4 \omega^2}{c^2 L} \frac{N}{V} H_{0x}. \qquad (16.47)$$

Now we use the expression for the inductance L [equation (16.34)], for the ratio N/V [equation (16.38)] and the relation (B.12),

$$M_{0x} = \frac{\chi}{4\pi} H_{0x}, \qquad (16.48)$$

and obtain

$$\mu(\omega) = 1 + \chi = 1 + \frac{F\omega^2}{\omega_{LC}^2 - \omega^2}. \tag{16.49}$$

In equation (16.49) we introduced the filling factor,

$$F = \frac{\ell^2 t}{a_x a_y a_z}, \tag{16.50}$$

which represents the portion of the volume occupied by the split rings.

The magnetic permeability diverges to infinity when $\omega \to \omega_0 = \omega_{LC}$, and changes sign when ω crosses ω_{LC}. Thus, due to the resonant magnetic response, we have constructed a medium that possesses negative magnetic permeability for frequencies $\omega > \omega_{LC}$.

The parameter F is given as the ratio of the smallest possible volume of the elementary cell, $\ell^2 t$, to the volume of a given unit cell. It is clear that

$$0 \leq F \leq 1. \tag{16.51}$$

The larger F, the stronger the magnetic resonance of the system.

The relationship (16.49) plays a crucial role in the construction of structures with negative permeability. Thanks to the resonance, we obtain an interval of frequencies $\omega_0 < \omega < \omega_0/\sqrt{1-F}$ where the permeability is *negative*.

Note that, in the limit of $d \to 0$, the gap closes. The resonance frequency $\omega_0 \to 0$ and the magnetic permeability $\mu(\omega) \to 1 - F$ is always positive. An array of closed metallic rings creates only an effective diamagnetic medium. The split of the ring is essential for the magnetic resonance and the negative permeability.

The resonant frequency of ω_{LC} given by equation (16.36) increases when the size of the split ring, ℓ, decreases. In the first experiments [145] the size of the unit cell of ~ 3 cm, and the resonant frequency was in the gigahertz region. Since the electromagnetic properties of a metal depend on the frequency, there is no simple scaling relation between the size of the structure and the frequency, similar to that discussed in chapter 13 [165]. Materials with resonant frequency in the terahertz region were prepared only recently [55, 56, 99, 148].

16.3.3 Design of Left-Handed Materials

By combining an array of thin metallic wires with an array of split-ring resonators, we obtain a structure that possesses, in a given frequency region, both negative permittivity and permeability [145]. A simple unit cell of such a left-handed material is shown in figure 16.12.

However, keep in mind that the structure obtained exhibits left-handed electromagnetic proprieties only for one direction of propagation of the electromagnetic waves. Indeed, to have negative permeability, the magnetic field must be oriented perpendicularly to the plane of split rings. Also, negative permittivity requires that the electric field is oriented parallel to the metallic wires. Therefore, the only possible direction of propagation of an electromagnetic wave is along the z-axis in figure 16.12. Even in this case, only one

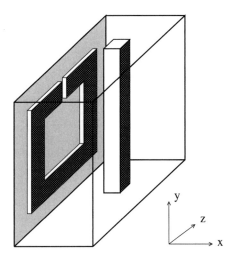

FIGURE 16.12. A schematic description of one unit cell of a left-handed medium. The structure is given by periodic repetition of the unit cell in each direction. Thin wires create a medium with negative permittivity, and an array of split-ring resonators creates a medium with negative magnetic permeability. The electromagnetic wave propagates in the z direction and is polarized with $\vec{E} \parallel y$ and $\vec{H} \parallel x$ [138, 145].

polarization, with $\vec{E}_0 \parallel y$ and $\vec{H} \parallel x$, propagates in a left-handed medium. The other polarization, with $\vec{E} \parallel x$ and $\vec{H} \parallel y$, propagates through the system without any interaction with the structural components. The left-handed structures prepared in laboratories confirm the negative refraction of electromagnetic waves [114, 138].

We summarize the dispersion relations for a left-handed medium. The permittivity is given by the relation (16.31),

$$\varepsilon(\omega) = 1 - \frac{\omega_p^2}{\omega^2 + i\gamma_e \omega}, \tag{16.52}$$

which is formally equivalent to the Drude formula for the permittivity of a metal. The main difference between these two formulas is in the value of the plasma frequency ω_p, which is $\sim 2000\,\text{THz}$ in the case of a metal, but only $\sim 10\,\text{GHz}$ for an array of metallic wires separated by a distance of the order of millimeters. The frequency dependence of the permittivity shown in figure 16.9 is formally identical with that of a metal, shown in figure B.3.

The magnetic permeability is given by the relationship

$$\mu(\omega) = 1 + \frac{F\omega^2}{\omega^2 - \omega_0^2 + i\gamma_m \omega}, \tag{16.53}$$

which is relation (16.49) with an additional loss term γ_m.

The resonant frequency dependence of the magnetic permeability is shown in figure 16.13. We see that the permeability is negative in the frequency interval

$$\omega_0 < \omega < \frac{\omega_0}{\sqrt{1 - F}}. \tag{16.54}$$

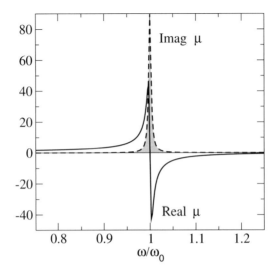

FIGURE 16.13. Effective permeability $\mu(\omega)$ given by equation (16.53) for a left-handed material with damping factor $\gamma_m = 0.00625\omega_0$.

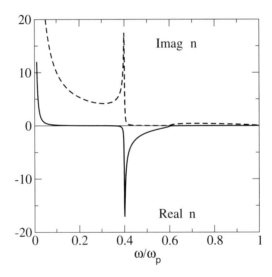

FIGURE 16.14. Frequency dependence of the refractive index of a left-handed medium with permittivity and permeability given by equations (16.31) and (16.53), respectively. $\omega_0 = 0.4\omega_p$, $F = 0.56$, and $\gamma_e = \gamma_m = 0.0025\omega_p$. ω_p is the plasma frequency of the wire medium.

In figure 16.14 we plot the frequency dependence of the refractive index $n = \sqrt{\varepsilon\mu}$ with the permittivity and permeability given by equations (16.52) and (16.53), respectively. The sign of the square root is chosen such that the imaginary part of the refractive index is always positive. Indeed, the real part of n is negative.

As discussed above, both the permittivity and permeability are complex quantities. Therefore, we always have a complex refractive index. This means that losses are

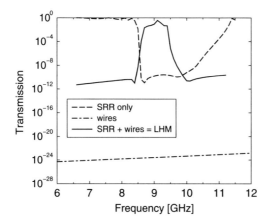

FIGURE 16.15. Simple transmission experiments that indicate the existence of a left-handed transmission band. Shown are the transmissions through three different structures. For an array of thin metallic wires with plasma frequency $\omega_p = 20$ GHz, we expect no transmission, since $\omega < \omega_p$. The transmission through an array of split rings exhibits a deep minimum in the frequency interval where we expect the magnetic permeability to be negative. Finally, the combination of both structures creates a material that has both the permittivity and the permeability negative in the frequency interval 8–10 GHz. We see that the transmission through this structure is high (close to 1) exactly in the frequency region where transmission was very small for an array of split-ring resonators [107, 145].

unavoidable in left-handed materials. Fortunately, it is possible to suppress the imaginary part of the refractive index so that losses are almost negligible, and a left-handed material can be considered as a transparent medium.

If $\omega_0 < \omega_p$, we expect left-handed transmission bands to appear in the spectrum. We can verify the negativeness of the refractive index by transmission experiments, as explained in the caption for figure 16.15.

Because of the complicated spatial structure of the left-handed materials, most of the relevant data for propagation of electromagnetic waves are obtained numerically. Details about numerical programs based on the transfer matrix technique can be found in [105, 118, 122, 123, 160].

16.4 Problems

Problems with Solutions

Problem 16.1

Using the dispersion relations for permittivity and permeability given by equations (16.31) and (16.53), respectively, estimate the frequency at which $\varepsilon = -1$ and $\mu = -1$. Consider the case of a lossless medium ($\gamma_e = \gamma_m = 0$).

Solution. From the dispersion relation (16.31), we find that $\varepsilon = -1$ at the frequency

$$\omega_\varepsilon = \frac{\omega_p}{\sqrt{2}}. \tag{16.55}$$

Similarly, from equation (16.53), we find that $\mu = -1$ at the frequency

$$\omega_\mu = \frac{2}{\sqrt{2-F}} \, \omega_0. \tag{16.56}$$

Problem 16.2

Find the frequencies at which $n = -1$ and $z = 1$.

Solution. $n = -1$ means that $\varepsilon\mu = 1$. Inserting equations (16.31) and (16.53), we obtain a quadratic equation for the frequency ω_n^2,

$$F\omega_n^4 + (1 - F)\omega_p^2\omega_n^2 - \omega_p^2\omega_n^2 = 0. \tag{16.57}$$

The impedance $z = 1$ if $\mu(\omega_z) = \varepsilon(\omega_z)$. This leads to the quadratic equations for ω_z^2,

$$F\omega_z^4 - \omega_p^2\omega_z^2 + \omega_p^2\omega_0^2 = 0. \tag{16.58}$$

Inserting $F = 0.56$ and $\omega_0 = 0.4\omega_p$, we get

$$\omega_n = 0.52\omega_p \quad \text{and} \quad \omega_z = 0.4215\omega_p. \tag{16.59}$$

Problem 16.3

Consider a complex permittivity and permeability $\varepsilon = \varepsilon_r + i\varepsilon_i$, and $\mu = \mu_r + i\mu_i$. Show that, if $\varepsilon_r < 0$ and $\mu_r < 0$, then the real part of the refractive index, n_r, is also negative.

Solution. From the relationship $n = \sqrt{\varepsilon\mu}$ we obtain

$$n^2 = n_r^2 - n_i^2 + 2in_r n_i = (\varepsilon_r\mu_r - \varepsilon_i\mu_i) + i\,[\varepsilon_r\mu_i + \mu_r\varepsilon_i]. \tag{16.60}$$

We compare the imaginary parts of both sides of equation (16.60):

$$2n_r n_i = \varepsilon_r\mu_i + \mu_r\varepsilon_i. \tag{16.61}$$

Since $\varepsilon_i > 0$ and $\mu_i > 0$, the right-hand side of equation (16.61) is negative, so that $n_i n_r < 0$. Since the imaginary part of the refractive index must be positive, we obtain $n_r < 0$.

Problem 16.4

Discuss the sign of the wave vector inside the left-handed medium.

Solution. Consider a left-handed material with complex refractive index $n = n_r + in_i$.
The wave vector k_2 satisfies the dispersion relation

$$\frac{\omega^2}{c^2}(n_r + in_i)^2 = k_2^2 = k_{2x}^2 + k_{2z}^2. \tag{16.62}$$

Due to the nonzero imaginary part of the refractive index, the wave vector k_2 is complex too. Since $k_{2x} = k_{1x} = k_x$ (the component of the wave vector parallel to the interface is conserved), we have that k_{2z} is complex,

$$k_{2z} = k_r + i k_i. \tag{16.63}$$

From equation (16.62) we find that k_r and k_i must satisfy the following equations:

$$k_r^2 - k_i^2 = -k_x^2 + \frac{\omega^2}{c^2}(n_r^2 - n_i^2) \tag{16.64}$$

and

$$k_r k_i = \frac{\omega^2}{c^2} n_r n_i. \tag{16.65}$$

If k_x is small, then the wave propagates inside the left-handed material, but is absorbed because of the nonzero imaginary part of the refractive index. The imaginary part of the wave vector, k_i, must be positive [equations (B.84) and (B.86)]. Then we find from equation (16.65) that $k_r < 0$. This result remains valid also in the limit of $n_i \to 0$, and gives an additional proof of equation (16.12).

When k_x increases, $|k_r|$ decreases and k_i increases, but the signs of k_r and k_i do not change: k_r is always negative and k_i is always positive. Similarly, the sign of the wave vector cannot change if the imaginary part of the refractive index, n_i, becomes infinitesimally small. We conclude that the signs of both real and imaginary parts are unambiguously determined.

Problem 16.5
Calculate the transmission of evanescent waves through a left-handed slab of thickness ℓ. The left-handed medium has refractive index $n = -1$, and negative permittivity and permeability, $\varepsilon = 1/\mu$.

Solution. Consider that the slab is located in the xy plane with interfaces at $z = -a$ and $z = +a$ ($\ell = 2a$). The waves in the three regions are

$$
\begin{aligned}
A e^{-\kappa_1 z} + B e^{+\kappa_1 z}, &\quad z \leq -a, \\
C e^{-\kappa_2 z} + D e^{+\kappa_2 z}, &\quad -a \leq z \leq a, \\
F e^{-\kappa_1 z}, &\quad a \leq z.
\end{aligned}
\tag{16.66}
$$

The coefficients C and D can be obtained from the relation

$$
\begin{pmatrix} C e^{-\kappa_2 a} \\ D e^{+\kappa_2 a} \end{pmatrix} = \mathbf{M} \begin{pmatrix} F e^{-\kappa_1 a} \\ 0 \end{pmatrix}, \tag{16.67}
$$

where $\mathbf{M}$ is the transfer matrix for a single interface. For TE polarization, $\mathbf{M}$ is given by equation (9.22). In our case, we have $k_1 = i\kappa_1$ and $k_2 = i\kappa_2$, so that

$$
\mathbf{M} = \frac{1}{2} \begin{pmatrix} 1 + \mu \dfrac{\kappa_1}{\kappa_2} & 1 - \mu \dfrac{\kappa_1}{\kappa_2} \\ 1 - \mu \dfrac{\kappa_1}{\kappa_2} & 1 + \mu \dfrac{\kappa_1}{\kappa_2} \end{pmatrix}. \tag{16.68}
$$

The coefficients A and B could be found either from the transfer matrix for the slab, given by equations (10.17)–(10.20), or by using the known coefficients C and D:

$$\begin{pmatrix} Ae^{\kappa_1 a} \\ Be^{-\kappa_1 a} \end{pmatrix} = \frac{1}{2} \begin{pmatrix} 1 + \dfrac{1}{\mu}\dfrac{\kappa_2}{\kappa_1} & 1 - \dfrac{1}{\mu}\dfrac{\kappa_2}{\kappa_1} \\ 1 - \dfrac{1}{\mu}\dfrac{\kappa_2}{\kappa_1} & 1 + \dfrac{1}{\mu}\dfrac{\kappa_2}{\kappa_1} \end{pmatrix} \cdot \begin{pmatrix} Ce^{+\kappa_2 a} \\ De^{-\kappa_2 a} \end{pmatrix}.$$

(16.69)

Results are shown in figure 16.6.

Problem 16.6

Consider the structure shown in figure 16.12. Discuss propagation of electromagnetic waves in the directions x and y.

Solution. Electromagnetic waves propagating along the y direction do not feel any interaction with the metallic wires, because the electric field is perpendicular to the wires. The magnetic response depends on the orientation of the magnetic field. No propagation is possible for polarization with $\vec{H} \parallel x$, because the magnetic permeability is negative due to the response of the split rings. However, if $\vec{H} \parallel z$, then both $\varepsilon \approx 1$ and $\mu \approx 1$, and the wave can propagate through the structure.

In the same way, if an electromagnetic wave propagates in the x direction, then we find that the magnetic response of the split rings is negligible. The wave polarized with $\vec{E} \parallel y$ decays because ε is negative and μ is positive, but the wave polarized with $\vec{E} \parallel z$ propagates through the structure.

Problem without Solution

Problem 16.7

Plot the refractive index $n(\omega)$ as a function of ω for various values of the loss parameters γ_e and γ_m. Study how the frequencies ω_n and ω_z depend on losses.

Appendix A   Matrix Operations

We present here some very simple properties of matrices. Since only matrices of size 2×2 are used in this book, we restrict ourselves to this particular case.

A.1 The Determinant and the Trace of the Matrix

Consider a 2×2 matrix $\mathbf{A}$,

$$\mathbf{A} = \begin{pmatrix} A_{11} & A_{12} \\ A_{21} & A_{22} \end{pmatrix}. \tag{A.1}$$

The determinant of the matrix $\mathbf{A}$ is defined as

$$\det \mathbf{A} = A_{11} A_{22} - A_{12} A_{21}. \tag{A.2}$$

The trace is given by the sum of the diagonal elements of the matrix:

$$\mathrm{Tr}\, \mathbf{A} = A_{11} + A_{22}. \tag{A.3}$$

If the matrix $\mathbf{A}$ is *diagonal*,

$$\mathbf{A} = \begin{pmatrix} \lambda_1 & 0 \\ 0 & \lambda_2 \end{pmatrix}, \tag{A.4}$$

then its elements are equal to the eigenvalues λ_1 and λ_2 of the matrix $\mathbf{A}$. In this particular case the determinant is given by the product of two eigenvalues,

$$\det \mathbf{A} = \lambda_1 \lambda_2, \tag{A.5}$$

and the trace is given by the sum of two eigenvalues,

$$\mathrm{Tr}\, \mathbf{A} = \lambda_1 + \lambda_2. \tag{A.6}$$

When det $\mathbf{A}$=1, then it follows from equation (A.5) that $\lambda_2 = \lambda_1^{-1}$. From equation (A.6) we have that

$$\mathrm{Tr}\, \mathbf{A} = \lambda_1 + \frac{1}{\lambda_1}. \tag{A.7}$$

Note that, if $\lambda_1 = e^{ix}$ and x is real, we obtain from equation (A.7) that $\mathrm{Tr}\, \mathbf{A} = 2\cos x$ so that $|\mathrm{Tr}\, \mathbf{A}| \leq 2$. On the other hand, if $\lambda_1 = \pm e^{\kappa}$ and κ is real, then $|\mathrm{Tr}\, \mathbf{A}| > 2$.

A.2 Inverse, Transpose, and Unitary Matrices

For a given matrix $\mathbf{A}$ we define the *inverse matrix* $\mathbf{A}^{-1}$ by the relation

$$\mathbf{A}^{-1}\mathbf{A} = 1. \tag{A.8}$$

It is easy to verify by direct matrix multiplication of both matrices $\mathbf{A}$ and $\mathbf{A}^{-1}$ that the inverse matrix has the form

$$\mathbf{A}^{-1} = \frac{1}{\det \mathbf{A}} \begin{pmatrix} A_{22} & -A_{21} \\ -A_{12} & A_{11} \end{pmatrix}. \tag{A.9}$$

Note that the inverse matrix exists only if the matrix $\mathbf{A}$ has a nonzero determinant.

We can also define the *complex conjugate* matrix $\mathbf{A}^*$ by the relation

$$\mathbf{A}^* = \begin{pmatrix} A_{11}^* & A_{12}^* \\ A_{21}^* & A_{22}^* \end{pmatrix} \tag{A.10}$$

and the transpose matrix of $\mathbf{A}$ is defined by

$$\mathbf{A}^T = \begin{pmatrix} A_{11} & A_{21} \\ A_{12} & A_{22} \end{pmatrix}. \tag{A.11}$$

The *Hermitian conjugate* matrix $\mathbf{A}^\dagger$ of $\mathbf{A}$ is the complex conjugate of the transpose matrix and is given by

$$\mathbf{A}^\dagger = \begin{pmatrix} A_{11}^* & A_{21}^* \\ A_{12}^* & A_{22}^* \end{pmatrix} = (\mathbf{A}^T)^*. \tag{A.12}$$

The matrix $\mathbf{A}$ is called *Hermitian* if it satisfies the condition

$$\mathbf{A}^{\dagger} = \mathbf{A}. \tag{A.13}$$

For instance, the Hamiltonian of the quantum particle is a Hermitian matrix.
  Matrix $\mathbf{A}$ is called *unitary* if it satisfies the relation

$$\mathbf{A}^{\dagger}\mathbf{A} = 1. \tag{A.14}$$

We will prove that the scattering matrix $\mathbf{S}$ is unitary.
  We can easily verify the following relation for the inverse of the product of two matrices:

$$[\mathbf{AB}]^{-1} = \mathbf{B}^{-1}\mathbf{A}^{-1} \tag{A.15}$$

and a similar relation for the Hermitian conjugation of the product of two matrices:

$$[\mathbf{AB}]^{\dagger} = \mathbf{B}^{\dagger}\mathbf{A}^{\dagger}. \tag{A.16}$$

A given vector u with two components u_1 and u_2 can be written as a column vector

$$u = \begin{pmatrix} u_1 \\ u_2 \end{pmatrix}. \tag{A.17}$$

We can obtain the transpose vector u^T,

$$u^T = (u_1 \quad u_2), \tag{A.18}$$

and the Hermitian conjugate vector,

$$u^{\dagger} = (u_1^* \quad u_2^*). \tag{A.19}$$

Then

$$u^{\dagger}v = u_1^* v_1 + u_2^* v_2 \tag{A.20}$$

is the scalar product of two vectors. Note also that

$$(\mathbf{A}u)^T = u^T \mathbf{A}^T \tag{A.21}$$

In terms of matrix elements,

$$u^T \mathbf{A}^T = (u_1 \ u_2) \begin{pmatrix} A_{11} & A_{21} \\ A_{12} & A_{22} \end{pmatrix} = (A_{11}u_1 + A_{12}u_2 \ \ A_{21}u_1 + A_{22}u_2). \tag{A.22}$$

In the same way we obtain for the Hermitian conjugate that

$$(u\mathbf{A})^{\dagger} = u^{\dagger}\mathbf{A}^{\dagger} = (u_1^* \ u_2^*) \begin{pmatrix} A_{11}^* & A_{21}^* \\ A_{12}^* & A_{22}^* \end{pmatrix} \tag{A.23}$$

$$= (A_{11}^* u_1^* + A_{12}^* u_2^* \ \ A_{21}^* u_1^* + A_{22}^* u_2^*).$$

A.3 Eigenvalues and Eigenvectors

The eigenvalues λ_1, λ_2 of the matrix $\mathbf{A}$ are solutions of the following equation:

$$\det[\mathbf{A} - \lambda] = 0. \tag{A.24}$$

Since the matrix $\mathbf{A}$ is a 2×2 matrix, we can find the eigenvalues explicitly:

$$\lambda_{1,2} = \frac{1}{2}\left[\mathrm{Tr}\mathbf{A} \pm \sqrt{\mathrm{Tr}^2\mathbf{A} - 4\det\mathbf{A}}\right], \tag{A.25}$$

which in terms of matrix elements reads

$$\lambda_{1,2} = \frac{A_{11} + A_{22}}{2} \pm \sqrt{\frac{(A_{11} - A_{22})^2}{4} + A_{12}A_{21}}. \tag{A.26}$$

The *right* eigenvectors u of the matrix $\mathbf{A}$ are defined by the relation

$$\mathbf{A}u = \lambda u, \tag{A.27}$$

where λ is the eigenvalue. The elements of the eigenvector u can be found from the system of linear equations $(\mathbf{A} - \lambda)u = 0$,

$$\begin{aligned}(A_{11} - \lambda)u_1 \quad &+ A_{12}u_2 \quad = 0, \\ A_{21}u_1 \quad &+ (A_{22} - \lambda)u_2 \quad = 0.\end{aligned} \tag{A.28}$$

Since the determinant of the matrix $(\mathbf{A} - \lambda)$ is by definition equal to zero, the two equations (A.28) are equivalent to each other. We therefore need another relation for the elements of the vector u. This is given by the normalization condition of the vector u,

$$|u_1|^2 + |u_2|^2 = 1 \tag{A.29}$$

In the same way we can define two *left* eigenvectors v by the relation

$$\lambda v = v\mathbf{A}. \tag{A.30}$$

From this definition it follows that the elements of the left eigenvectors are given by the solution of the system of equations $v(\mathbf{A} - \lambda) = 0$,

$$\begin{aligned}(A_{11} - \lambda) \quad &v_1 + A_{21}v_2 \quad = 0, \\ A_{12} \quad &v_1 + (A_{22} - \lambda)v_2 \quad = 0,\end{aligned} \tag{A.31}$$

with the additional constraint $|v_1|^2 + |v_2|^2 = 1$.

A.4 Similarity Transformations

The above equations for left and right eigenvectors can be written in the matrix form as

$$\mathbf{A} = \mathbf{R}\Lambda\mathbf{L} \tag{A.32}$$

where Λ is a diagonal matrix

$$\Lambda = \begin{pmatrix} \lambda_1 & 0 \\ 0 & \lambda_2 \end{pmatrix} \tag{A.33}$$

and the matrix $\mathbf{R}$ contains in its *columns* the vectors u and the matrix $\mathbf{L}$ contains in its *rows* the vectors v. It is easy to prove the relations

$$\mathbf{LR} = 1 \tag{A.34}$$

and

$$\mathbf{AR} = \mathbf{R}\Lambda \quad \text{and} \quad \mathbf{LA} = \Lambda\mathbf{L}. \tag{A.35}$$

If the matrix $\mathbf{A}$ is Hermitian, then $\mathbf{R}$ is a unitary matrix and $\mathbf{L} = \mathbf{R}^\dagger$. In this particular case we see that

$$\mathbf{A} = \mathbf{R}\Lambda\mathbf{R}^\dagger. \tag{A.36}$$

In this book, we deal mostly with transfer matrices $\mathbf{A}$ that are *not* Hermitian. Therefore we have to keep in mind that we need to distinguish the left eigenvectors from the right ones.

Note also that with the help of the matrices $\mathbf{R}$ and $\mathbf{L}$ we can diagonalize the matrix $\mathbf{A}$:

$$\mathbf{R}^{-1}\mathbf{AR} = \Lambda. \tag{A.37}$$

It is easy to prove that the matrices $\mathbf{A}$ and $\mathbf{R}\Lambda\mathbf{L}$ have the same trace and the same determinant.

A.5 Degeneracy

It often happens that the eigenvalues of the matrix $\mathbf{A}$ are the same. We call such a matrix a *degenerate* matrix:

$$\mathbf{A}_0 = \begin{pmatrix} \lambda_0 & 0 \\ 0 & \lambda_0 \end{pmatrix}. \tag{A.38}$$

The eigenvectors of the degenerate matrix are simply

$$u_1 = \begin{pmatrix} 1 \\ 0 \end{pmatrix} \quad \text{and} \quad u_2 = \begin{pmatrix} 0 \\ 1 \end{pmatrix}. \tag{A.39}$$

Also, any linear combination of these two vectors is again an eigenvector of matrix $\mathbf{A}_0$.

Consider now the matrix

$$\mathbf{V} = \begin{pmatrix} 0 & \varepsilon \\ \varepsilon & 0 \end{pmatrix}. \tag{A.40}$$

We need to find the eigenvalues of the matrix

$$\mathbf{A} = \mathbf{A}_0 + \mathbf{V} = \begin{pmatrix} \lambda_0 & \varepsilon \\ \varepsilon & \lambda_0 \end{pmatrix}. \tag{A.41}$$

From equation (A.24) we see that the eigenvalues are given by the solution of the equation $\det(\mathbf{A} + \mathbf{V} - \lambda) = 0$, which in our case has the form

$$(\lambda_0 - \lambda)^2 - \varepsilon^2 = 0. \tag{A.42}$$

It is easy to show that the eigenvalues are of the form

$$\lambda_1 = \lambda_0 + \varepsilon \tag{A.43}$$

and

$$\lambda_2 = \lambda_0 - \varepsilon. \tag{A.44}$$

We come to the very important conclusion that any small perturbation of the degenerate system removes the degeneracy of the eigenvalues.

Appendix B   Summary of Electrodynamics Formulas

Electrodynamics is the theory of fields and forces acting on stationary and moving charges. The classical theory is fully described by Maxwell's equations. In this appendix we briefly summarize basic concepts and definitions of the most important quantities of the electromagnetic theory. We do not intend to give here the detailed electromagnetic theory, only to summarize the most important formulas used in the text. For detailed studies, we refer readers to standard textbooks on the subject [2, 20, 26, 37, 39].

B.1 Maxwell's Equations

All the analysis of wave propagation is based on Maxwell's equations, which govern the time dependence of the intensity of the electric and magnetic field $\vec{E}$ and $\vec{H}$, respectively. These two fields vectors are often used to describe the electromagnetic field.

In the cgs system of units, Maxwell's equations can be written as follows: Faraday's law,

$$\nabla \times \vec{E} = -\frac{1}{c}\frac{\partial \vec{B}}{\partial t};$$

(B.1)

Ampere's law with Maxwell's correction,

$$\nabla \times \vec{H} = +\frac{1}{c}\frac{\partial \vec{D}}{\partial t} + \frac{4\pi}{c}\vec{j};$$

(B.2)

Gauss's law,

$$\nabla \cdot \vec{D} = 4\pi\rho;$$

(B.3)

and

$$\nabla \cdot \vec{B} = 0. \tag{B.4}$$

The quantities $\vec{D}$ and $\vec{B}$ are called the *electric displacement* and the *magnetic induction*, respectively. The quantities ρ and $\vec{j}$ are the electric charge density and the current density, respectively.

The parameter c is the velocity of light in vacuum,

$$c = 2.99795 \times 10^8 \, \text{m/s}. \tag{B.5}$$

In a medium with no free charges ($\rho = 0$) and no currents ($\vec{j} = 0$), Maxwell's equations are simpler and given by

$$\nabla \times \vec{E} = -\frac{1}{c} \frac{\partial \vec{B}}{\partial t}, \tag{B.6}$$

$$\nabla \times \vec{H} = +\frac{1}{c} \frac{\partial \vec{D}}{\partial t}, \tag{B.7}$$

$$\nabla \cdot \vec{E} = 0, \tag{B.8}$$

$$\nabla \cdot \vec{B} = 0. \tag{B.9}$$

We can also introduce the electric permittivity ε and the magnetic permeability μ. These two parameters characterize the response of the material to an external electric and magnetic field,

$$\vec{D} = \varepsilon \vec{E} = \vec{E} + 4\pi \vec{P}, \tag{B.10}$$

$$\vec{B} = \mu \vec{H} = \vec{H} + 4\pi \vec{M}, \tag{B.11}$$

where $\vec{P}$ and $\vec{M}$ are the electric and magnetic polarizations, respectively. When an electromagnetic field is present in a material, the electric field can perturb the motion of electrons and produce a dipole polarization $\vec{P}$. The application of a magnetic field to a material can induce a magnetization $\vec{M}$.

Equations (B.10) and (B.11) are assumed to be linear. This is correct only if the electric and magnetic fields $\vec{E}$ and $\vec{H}$ are not too strong.

In isotropic materials we have the following relations:

$$4\pi \vec{P} = \chi_e \vec{E}, \tag{B.12}$$

$$4\pi \vec{M} = \chi_m \vec{H}, \tag{B.13}$$

where χ_e (χ_m) is the electric (magnetic) susceptibility. By substituting equations (B.12) and (B.13) into equations (B.10) and (B.11), we obtain

$$\varepsilon = 1 + \chi_e, \tag{B.14}$$

$$\mu = 1 + \chi_m. \tag{B.15}$$

These expressions hold in the cgs system of units.

B.1.1 Maxwell's Equations for a Plane Wave

For a plane wave, the electric and magnetic fields can be expressed as

$$\vec{E}(\vec{r}, t) = \vec{E}_0 e^{i[\vec{k}\cdot\vec{r}-\omega t]},$$
$$\vec{H}(\vec{r}, t) = \vec{H}_0 e^{i[\vec{k}\cdot\vec{r}-\omega t]},$$

(B.16)

where ω is the frequency of the field and $\vec{k}$ is the wave vector, and $\vec{E}_0$ and $\vec{H}_0$ define the amplitude and the direction of the vectors $\vec{E}$ and $\vec{H}$, respectively.

Now, we substitute expressions (B.10) and (B.11) into Maxwell's equations, given by equations (B.6)–(B.9). With the use of relations (B.16), Maxwell's equations become

$$\vec{k} \times \vec{E} = +\frac{\mu\omega}{c} \vec{H},$$

(B.17)

$$\vec{k} \times \vec{H} = -\frac{\varepsilon\omega}{c} \vec{E},$$

(B.18)

$$\vec{k} \cdot \vec{E} = 0,$$

(B.19)

$$\vec{k} \cdot \vec{H} = 0.$$

(B.20)

Equations (B.19) and (B.20) show that both the electric field vector $\vec{E}$ and the magnetic field vector $\vec{H}$ are perpendicular to the wave vector $\vec{k}$,

$$\vec{E} \perp \vec{k} \quad \text{and} \quad \vec{H} \perp \vec{k}.$$

(B.21)

Also, from equations (B.17) and (B.18) we see that the electric and magnetic field vectors are perpendicular to each other,

$$\vec{E} \perp \vec{H}.$$

(B.22)

Thus, we obtain that the three vectors $\vec{E}$, $\vec{H}$, and $\vec{k}$ form a triplet of mutually perpendicular vectors.

By multiplying both sides of equation (B.17) by $\vec{k} \times \vec{E}$, we obtain

$$k^2 E^2 = \left(\frac{\mu\omega}{c}\right)^2 H^2,$$

(B.23)

where $E = |\vec{E}| = |\vec{E}_0|$ and $H = |\vec{H}| = |\vec{H}_0|$.

Similarly, by multiplying both sides of equation (B.18) by $\vec{k} \times \vec{H}$, we obtain

$$k^2 H^2 = \left(\frac{\varepsilon\omega}{c}\right)^2 E^2.$$

(B.24)

Dividing equation (B.23) by equation (B.24), we obtain a relation between the absolute values of the intensities of the electric and magnetic fields:

$$E = \sqrt{\frac{\mu}{\varepsilon}} H.$$

(B.25)

B.1.2 Maxwell's Equations in the SI System of Units

For completeness, we present also Maxwell's equations in the SI system of units:

$$\nabla \times \vec{E} = -\frac{\partial \vec{B}}{\partial t}, \tag{B.26}$$

$$\nabla \times \vec{H} = \vec{j} + \frac{\partial \vec{D}}{\partial t}, \tag{B.27}$$

$$\nabla \cdot \vec{D} = \rho, \tag{B.28}$$

$$\nabla \cdot \vec{B} = 0. \tag{B.29}$$

Equations (B.10) are written as

$$\vec{D} = \varepsilon \vec{E} = \vec{E} + \vec{P}, \tag{B.30}$$

$$\vec{B} = \mu \vec{H} = \vec{H} + \vec{M}. \tag{B.31}$$

In the SI, we use the vacuum permittivity $\varepsilon_0 = 10^7/(4\pi c^2) = 8.854 \times 10^{-12}$ F/m and vacuum permeability $\mu_0 = 4\pi \times 10^{-7}$ H/m. The light velocity is then given by

$$c = \frac{1}{\sqrt{\varepsilon_0 \mu_0}}. \tag{B.32}$$

B.2 Wave Equation

From equations (B.6), (B.7), (B.10), and (B.11), we can derive the wave equation for the electric and magnetic fields. Applying the operator $\nabla \times$ to both sides of equation (B.6) we get

$$\nabla \times (\nabla \times E) = -\frac{\mu}{c}\frac{\partial}{\partial t}\nabla \times \vec{H}. \tag{B.33}$$

The left-hand side of equation (B.33) can be simplified with use of the identity

$$\nabla \times (\nabla \times \vec{E}) = \nabla(\nabla \cdot \vec{E}) - \nabla^2 \vec{E}. \tag{B.34}$$

The right-hand side of equation (B.33) can be rewritten using equation (B.7), and finally we obtain the equation

$$\nabla^2 \vec{E} = \frac{\varepsilon\mu}{c^2}\frac{\partial^2 \vec{E}}{\partial t^2}, \tag{B.35}$$

which is called the *wave equation*. A similar equation can be obtained for the magnetic field $\vec{H}$. The wave equation is telling us that the fields $\vec{E}$ and $\vec{H}$ propagate through empty space ($\varepsilon = \mu = 1$) with a speed equal to the velocity of light, c. However, in a medium with ε and μ different from 1, the electromagnetic fields propagate with a velocity

$$v = \frac{c}{\sqrt{\varepsilon\mu}} = \frac{c}{n}, \tag{B.36}$$

where $n = \sqrt{\varepsilon\mu}$ is the *index of refraction*.

Equation (B.35) has the solution

$$\vec{E}(\vec{r}, t) = \vec{E}_0 e^{i(\vec{k}\cdot\vec{r} - \omega t)}, \tag{B.37}$$

identical with the plane wave, equation (B.16).

Inserting (B.37) into (B.35) we obtain

$$\left[k^2 - \varepsilon\mu \frac{\omega^2}{c^2} \right] \vec{E} = 0, \tag{B.38}$$

which determines the dispersion relation of a plane wave propagating in a medium with permittivity ε and permeability μ,

$$k^2 = \varepsilon\mu \frac{\omega^2}{c^2} = n^2 \frac{\omega^2}{c^2}. \tag{B.39}$$

We will see in appendix B.6 that both the permittivity and permeability could depend on the frequency ω:

$$\varepsilon = \varepsilon(\omega) \quad \text{and} \quad \mu = \mu(\omega), \tag{B.40}$$

so that the dispersion relations express in the general case a rather complicated nonlinear relation between frequency and wave vector.

The wave vector determines the wavelength of the wave,

$$\lambda = \frac{2\pi}{k}. \tag{B.41}$$

In terms of the frequency,

$$\lambda = \frac{2\pi c}{\omega} = \frac{c}{\nu}, \tag{B.42}$$

since the frequency ν is related to the angular frequency ω by

$$\nu = \omega/2\pi. \tag{B.43}$$

B.3 Group Velocity and Phase Velocity

Consider now two waves

$$\vec{E}_+ = E \cos\left[(k + \delta k)x - (\omega + \delta\omega)t \right] \tag{B.44}$$

and

$$\vec{E}_- = E \cos\left[(k - \delta k)x - (\omega - \delta\omega)t \right], \tag{B.45}$$

where $\delta k \ll k$ and $\delta\omega \ll \omega$. The sum

$$\vec{E} = \vec{E}_+ + \vec{E}_- = 2E \cos\left[\delta k x - \delta\omega t\right] \cos(kx - \omega t) \tag{B.46}$$

is shown in the upper panel of figure B.1. We distinguish two velocities: the phase velocity

$$v_{\text{ph}} = \frac{\omega}{k}, \tag{B.47}$$

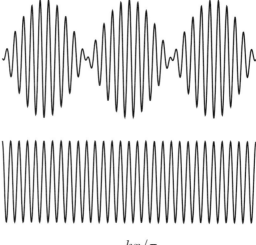

$$kx/\pi$$

FIGURE B.1. Upper panel: Space dependence of the wave $E(x, t)$, given by equation (B.46) for a given time t. We used $\delta k/k = 0.05$. Lower panel: Wave $E\cos(kx - \omega t)$.

which determines the change of the *phase* of the electric field, and the group velocity

$$v_\mathrm{g} = \frac{\delta\omega}{\delta k}, \tag{B.48}$$

which determines the spatial change of the *envelope*. While the magnitude of the phase velocity is not restricted by any physical law (it might even be negative), the group velocity must be positive and less than c since it can be interpreted as the velocity of energy transfer in the medium. However, note that the last statement is true only in isotropic nondispersive media. In the case of strong dispersion, the wave packet shown in figure B.1 is not stable, and the group velocity has no clear physical interpretation.

In the limit of $\delta k \to 0$ and $\delta\omega \to 0$, we come back to a monochromatic wave propagating with the group velocity

$$v_\mathrm{g} = \frac{\partial\omega}{\partial k}. \tag{B.49}$$

Inserting $k = n\omega/c$, we obtain

$$v_\mathrm{g}^{-1} = \frac{1}{c}\frac{\partial}{\partial\omega}\omega n(\omega). \tag{B.50}$$

The requirement of positiveness of the group velocity can be written as

$$\frac{\partial}{\partial\omega}\omega n(\omega) = n(\omega) + \frac{\partial n(\omega)}{\partial\omega} > 0. \tag{B.51}$$

B.4 Poynting Vector

Poynting's theorem states that the time rate of flow of the electromagnetic energy per unit area is given by the vector $\vec{S}$, called the Poynting vector, defined as the cross product of the electric and magnetic fields,

$$\vec{S} = \frac{c}{4\pi} \vec{E} \times \vec{H}. \tag{B.52}$$

Since $\vec{S}$ has the physical meaning of an energy flow, it must be real. Therefore, in equation (B.52), only the real parts of all three vectors $\vec{E}$, $\vec{H}$, and $\vec{S}$ should be considered:

$$\vec{S} = \frac{c}{4\pi} \operatorname{Re} \vec{E} \times \operatorname{Re} \vec{H}. \tag{B.53}$$

In this book, we are interested only in the propagation of electromagnetic waves where both the electric and the magnetic fields oscillate as $e^{i\omega t}$. Therefore, it is more convenient to average the Poynting vector $\vec{S}$ over one period

$$T = \frac{2\pi}{\omega} \tag{B.54}$$

of the oscillation of electric and magnetic fields,

$$\vec{S} = \frac{c}{4\pi} \frac{1}{T} \int_0^T dt \left[\operatorname{Re} \vec{E} \times \operatorname{Re} \vec{H} \right]. \tag{B.55}$$

Inserting $\vec{E} = \vec{E}_0 e^{i(\phi_E + \omega t)}$ and $\vec{H} = \vec{H}_0 e^{i(\phi_H + \omega t)}$, we have that

$$\operatorname{Re} \vec{E} = \vec{E}_0 \cos(\phi_E + \omega t) \quad \text{and} \quad \operatorname{Re} \vec{H} = \vec{H}_0 \cos(\phi_H + \omega t). \tag{B.56}$$

After substitution of these two expressions into the integral on the r.h.s. of equation (B.55), we can integrate over time. With the use of the relation $\cos a \cos \beta = \frac{1}{2} \left[\cos(a + \beta) + \cos(a - \beta) \right]$, we obtain

$$\operatorname{Re} \vec{E} \times \operatorname{Re} \vec{H} = \frac{1}{2} \vec{E}_0 \times \vec{H}_0 \left\{ \cos(2\omega t + \phi_E + \phi_H) + \cos(\phi_E - \phi_H) \right\}. \tag{B.57}$$

The time integral of the first term, $\cos(\phi_E + \phi_H + 2\omega t)$, is zero, because

$$\frac{1}{T} \int_0^T dt \, \cos(2\omega t + a) = \frac{1}{2\pi} \int_0^{2\pi} dx \, \cos(2x + a) = 0 \tag{B.58}$$

[we substitute $x = \omega t$ and used the definition of the period T given by equation (B.54)]. The second term on the r.h.s. of equation (B.57), $\cos(\phi_E - \phi_H)$, does not depend on t, so that the integration is trivial. We obtain the following expression for the time-averaged Poynting vector:

$$\vec{S} = \frac{c}{8\pi} \vec{E}_0 \times \vec{H}_0 \cos(\phi_E - \phi_H), \tag{B.59}$$

which can be expressed in terms of the complex fields $\vec{E}$ and $\vec{H}$ in the form [39]

$$\vec{S} = \frac{c}{8\pi} \operatorname{Re} \left[\vec{E} \times \vec{H}^* \right]. \tag{B.60}$$

We can use equation (B.17) to express $\vec{H}$ in terms of $\vec{E}$ and express the Poynting vector in the form

$$\vec{S} = \frac{c}{8\pi} \frac{c}{\omega} \mathrm{Re} \left[\frac{\vec{k}}{\mu} |E|^2 \right]. \tag{B.61}$$

Equations (B.60) and (B.61) are often used in this book in the analysis of an transmission of an electromagnetic wave through various structures. From equation (B.61) we see that, if μ is real, then $\vec{S}$ is nonzero only when the real part of the wave vector $\vec{k}$ is nonzero. Consider, for instance, the so-called *evanescent* waves, which are waves decaying exponentially in space,

$$\vec{E} = \vec{E}_0 e^{-\kappa \vec{r} - i\omega t}, \qquad \vec{H} = \vec{H}_0 e^{-\vec{\kappa} \vec{r} - i\omega t}. \tag{B.62}$$

With $\vec{\kappa} = i\vec{k}$, these waves can be considered as plane waves with an imaginary wave vector. Inserting (B.62) into equation (B.61), we immediately see that evanescent waves do not carry any energy, since $\vec{S} = 0$.

For the plane wave $\vec{E} = \vec{E}_0 e^{i(\vec{k}\cdot\vec{r}-\omega t)}$, $\vec{H} = \vec{H}_0 e^{i(\vec{k}\cdot\vec{r}-\omega t)}$, we also use another expression for the Poynting vector. First, we express the wave vector in the form $\vec{k} = k\vec{s} = (\omega/c)n\vec{s}$, where $\vec{s}$ is a unit vector in the direction of $\vec{k}$. Inserting this into equation (B.60), we obtain

$$\vec{S} = \frac{c}{8\pi} \mathrm{Re} \left[\frac{1}{z} \right] |E|^2 \vec{s} = \frac{c}{8\pi} \mathrm{Re}[z] |H|^2 \vec{s}, \tag{B.63}$$

where we introduce the *impedance z* by the relation

$$z = \sqrt{\frac{\mu}{\varepsilon}}. \tag{B.64}$$

Since the energy must flow in the direction of propagation, we obtain an important condition for the real part of the impedance, which must be positive,

$$z_r > 0. \tag{B.65}$$

B.5 Boundary Condition at an Interface

At the interface between two media, we need to satisfy the following boundary conditions:

$$\vec{n} \cdot (\vec{B}_2 - \vec{B}_1) = 0, \tag{B.66}$$
$$\vec{n} \cdot (\vec{D}_2 - \vec{D}_1) = 0, \tag{B.67}$$
$$\vec{n} \times (\vec{E}_2 - \vec{E}_1) = 0, \tag{B.68}$$
$$\vec{n} \times (\vec{H}_2 - \vec{H}_1) = 0. \tag{B.69}$$

Here, $\vec{n}$ is the unit vector perpendicular to the plane surface and indices 1 (2) refer to the first (second) medium. The relations (B.66)–(B.69) follow directly from Maxwell's equations [2, 20]. They play a key role in our analysis of transmission of electromagnetic wave through various structures.

From equations (B.68) and (B.69), it follows that the tangential components of both $\vec{E}$ and $\vec{H}$ are continuous at the interface:

$$\vec{E}_{1t} = \vec{E}_{2t} \quad \text{and} \quad \vec{H}_{1t} = \vec{H}_{2t}. \tag{B.70}$$

Similarly, by using equations (B.68) and (B.69) we obtain that the *normal* components of the electric displacement and the magnetic inductance are continuous at the interface,

$$\vec{D}_{1n} = \vec{D}_{2n} \quad \text{and} \quad \vec{B}_{1n} = \vec{B}_{2n}. \tag{B.71}$$

The continuity conditions, given by equations (B.70) and (B.71), are very important in the analysis of the propagation of electromagnetic waves through the interface between two media.

B.6 Permittivity and Permeability

In section B.1 we introduced the electric permittivity ε and the magnetic permeability μ. These parameters characterize the electric and magnetic responses of a given material to an applied electromagnetic field. Since the permittivity and the permeability are the most important parameters of the material, it is worth studying their properties for different materials.

First, we want to emphasize that both these parameters are in general complex,

$$\varepsilon = \varepsilon_r + i\varepsilon_i \quad \text{and} \quad \mu = \mu_r + i\mu_i. \tag{B.72}$$

From the wave equation (B.35) we see that the imaginary parts cause electromagnetic losses in the material. The losses in the material are given by the formula [26]

$$Q = \frac{1}{2\pi} \int d\omega \, \omega \left[\varepsilon_i |E|^2 + \mu_i |H|^2 \right]. \tag{B.73}$$

For a passive material, the losses must be positive. This means that the imaginary parts of both the permittivity and the permeability must be positive,

$$\varepsilon_i > 0, \quad \mu_i > 0. \tag{B.74}$$

The real and imaginary parts of the permittivity are not independent of each other. They are related by the Kramer-Kronig relations [20, 26].

The imaginary part of the permittivity is closely related to the current inside the material,

$$\vec{j} = \sigma \vec{E}, \tag{B.75}$$

where σ is the electrical conductivity. Indeed, in the case of nonzero electric current, we obtain from Maxwell's equations the following wave equation:

$$\nabla^2 \vec{E} = \frac{\varepsilon \mu}{c^2} \frac{\partial^2 \vec{E}}{\partial t^2} + \frac{4\pi \mu \sigma}{c^2} \frac{\partial E}{\partial t}, \tag{B.76}$$

and equation (B.38) has the form

$$\left[k^2 - \mu \frac{\omega^2}{c^2} \left(\varepsilon + i \frac{4\pi\sigma}{\omega} \right) \right] \vec{E} = 0. \tag{B.77}$$

Equation (B.76) can be interpreted as a wave equation with a complex permittivity,

$$\varepsilon = \varepsilon_r + i \frac{4\pi\sigma}{\omega}. \tag{B.78}$$

This relationship between the imaginary part of the electric permittivity and the current explains the main difference between the electric responses of metals and dielectrics. Metals are usually very good conductors, so that the imaginary part ε_i is large. On the other hand, dielectrics have very small electrical conductivity and the imaginary part of the permittivity is very small. The ratio $\varepsilon_i / \varepsilon_r$ is called the tangent loss. Typical values of $\varepsilon_i / \varepsilon_r$ in a dielectric are $10^{-2}-10^{-3}$. For such small losses, we can neglect the imaginary part, $\varepsilon_i \approx 0$, and to a good approximation consider dielectrics to be lossless materials.

If the permittivity ε is complex, then also the index of refraction n and the impedance z,

$$n = \sqrt{\varepsilon\mu}, \tag{B.79}$$

$$z = \sqrt{\frac{\mu}{\varepsilon}} \tag{B.80}$$

are complex,

$$n = n_r + i n_i \quad \text{and} \quad z = z_r + i z_i. \tag{B.81}$$

From equation (B.77) we also have that

$$k = n \frac{\omega}{c}. \tag{B.82}$$

Inserting k in the expression for the electric field, $\vec{E} = \vec{E}_0 e^{i(\vec{k}\cdot\vec{r}-\omega t)}$, we see that the imaginary part of the index of refraction determines an exponential decrease of the plane wave in space,

$$\vec{E} \sim e^{i\vec{k}_r \cdot \vec{r} - \vec{k}_i \cdot \vec{r}}, \tag{B.83}$$

where

$$k_r = \frac{\omega}{c} n_r \tag{B.84}$$

and

$$k_i = \frac{\omega}{c} n_i. \tag{B.85}$$

For passive materials we require

$$n_i > 0. \tag{B.86}$$

The condition (B.86) guarantees that the amplitude of the electromagnetic wave decreases exponentially in an absorbing material.

Table B.1. Plasma frequency v_p and γ_0 for some metals. Values are given in THz $= 10^{12}$ Hz. Note that parameters used in the Drude formula (B.87) are $\omega_p = 2\pi v_p$ and $\gamma = 2\pi\gamma_0$ [61]. Visible light spans the frequencies $430 < v < 750$ THz.

Metal	v_p	γ_0
Al	3570	19.40
Cu	1914	8.34
Au	2175	6.50
Ag	2175	4.35

Equations (B.74) define the positiveness of the imaginary parts of the permittivity and permeability. There is no such relation for the real parts of ε and μ. In fact, ε_r and μ_r can possess any values, positive or negative. In the case of dielectrics, $\mu_r = 1$ and the permittivity ε_r is positive. Both μ_r and ε_r are almost independent of the frequency of the electromagnetic field. In contrast to this, the electric response of metals strongly depends on the frequency. Also, the real part of the permittivity is negative. We will discuss the electric response of metals in section B.7.

The magnetic response of materials is mostly weaker than the electric response. For nonmagnetic materials, the magnetic permeability $\mu = 1$. In most magnetic materials, μ can be either larger or smaller than 1, but is always positive. There are magnetic materials that have either ferromagnetic or antiferromagnetic resonance, which in principle can cause the real part of the permeability, μ_r, to be negative in certain frequency intervals. A second class of materials, the so-called *left-handed materials*, have the real part of the magnetic permeability negative, $\mu_r < 0$. If also $\varepsilon_r < 0$, these materials allow wave propagation with a negative index of refraction. We discuss the electromagnetic properties of left-handed materials in chapter 16.

B.7 Metals

In nonmagnetic *metals* we have that $\mu = 1$, but the permittivity ε depends strongly on the frequency, as given by the Drude formula

$$\varepsilon_m = 1 - \frac{\omega_p^2}{\omega^2 + i\gamma\omega}. \tag{B.87}$$

The relationship (B.87) expresses the qualitative behavior of the metallic permittivity from microwaves ($\omega \sim$ GHz) through the visible range ($\omega \sim 10^2$ THz) up to frequencies of 10^4 THz.

The frequency ω_p is called the *plasma frequency*. In the literature, the values of plasma frequency are usually given as $v_p = \omega_p/(2\pi)$. Typical values of v_p are $v_p \sim 2000$ THz and $\gamma_0 = \gamma/2\pi \sim 5-10$ THz [61]. The inverse of γ, γ^{-1}, represents the relaxation time of electrons. Typical values of the plasma frequency for a few metals are given in table B.1.

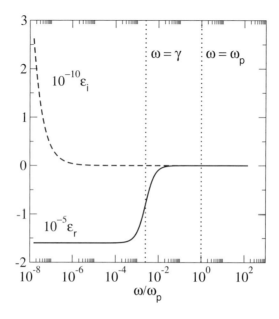

FIGURE B.2. Real and imaginary parts of the metallic permittivity given by the Drude formula (B.87) with plasma frequency $v_p = \omega_p/(2\pi) = 2000\,\text{THz}$ and $\gamma_0 = 5\,\text{THz}$. Note that the real part of the permittivity is re-scaled by a factor of 10^{-5} and the imaginary part is re-scaled by a factor of 10^{-10} in order to fit the data into the same figure. Note also the logarithmic scale on the x axis. Since $v_p \sim 10^3\,\text{THz}$, values of $\omega/\omega_p \sim 10^{-6}$ correspond to the GHz region. Visible light lies approximately in the frequency region 400–700 THz, which corresponds to $v_p/5 - v_p/3$.

From equation (B.87), we can express the real and imaginary parts of the permittivity. The real part is given by

$$\varepsilon_r = 1 - \frac{\omega_p^2}{\omega^2 + \gamma^2}, \tag{B.88}$$

and the imaginary part is given by

$$\varepsilon_i = \frac{\omega_p^2 \gamma}{\omega(\omega^2 + \gamma^2)}. \tag{B.89}$$

Note that both real and imaginary parts of the electric permittivity strongly depend on the frequency, as is shown in figures B.2 and B.3. We discuss three frequency regions: the low-frequency limit $\omega \ll \gamma$, the high-frequency limit $\gamma < \omega < \omega_p$, and, finally, the limit of $\omega > \omega_p$.

B.7.1 Low-Frequency Limit $\omega \ll \gamma$

For small frequencies

$$\omega \ll \gamma, \tag{B.90}$$

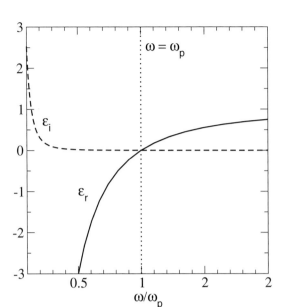

FIGURE B.3. Real and imaginary parts of the metallic permittivity given by the Drude formula (B.87) for frequencies close to the plasma frequency ω_p.

the formulas (B.88) and (B.89) for the real and imaginary parts of the permittivity reduce to

$$\varepsilon_r \approx -\frac{\omega_p^2}{\gamma^2} \tag{B.91}$$

and

$$\varepsilon_i \approx \frac{\omega_p^2}{\gamma\omega}. \tag{B.92}$$

The real part ε_r does not depend on the frequency, and is negative. Note that the imaginary part is much larger than the real part,

$$\frac{\varepsilon_i}{|\varepsilon_r|} = \frac{\gamma}{\omega} \gg 1, \tag{B.93}$$

so to a good approximation the permittivity can be considered as purely imaginary,

$$\varepsilon_m \approx i\frac{\omega_p^2}{\omega\gamma}. \tag{B.94}$$

The imaginary part ε_i can be expressed in terms of the electrical conductivity:

$$\varepsilon_i = \frac{4\pi\sigma}{\omega}, \tag{B.95}$$

where we have used equation (B.78). By using a typical value of the plasma frequency, $\nu_p \approx 2000\,\text{THz} \sim 10^{15}\,\text{Hz}$, we obtain that the imaginary part of the permittivity, ε_i, possesses huge values. For instance, $\varepsilon_i \sim 10^9$ for $\nu = 1\,\text{GHz}$.

B.7.2 High-Frequency Limit $\gamma \ll \omega < \omega_p$

In the frequency interval of

$$\gamma \ll \omega < \omega_p, \tag{B.96}$$

we can neglect the loss factor γ in the denominator of equations (B.88) and (B.89) for the real and imaginary parts of the permittivity. We obtain

$$\varepsilon_r \approx 1 - \frac{\omega_p^2}{\omega^2} \tag{B.97}$$

and

$$\varepsilon_i \approx \frac{\omega_p^2 \gamma}{\omega^3}. \tag{B.98}$$

With the help of the relationship (B.96) we see that in this frequency interval the imaginary part ε_i is much less than the absolute value of the real part $|\varepsilon_r|$,

$$\frac{\varepsilon_i}{|\varepsilon_r|} = \frac{\gamma}{\omega} \ll 1. \tag{B.99}$$

The metal can be considered as a lossless material with negative permittivity. From the values of the plasma frequency given in table B.1, we see that this approximation is good when the frequency ω is of the order of 1000 THz.

B.7.3 Frequency Region $\omega > \omega_p$

For frequencies

$$\omega \gg \omega_p, \tag{B.100}$$

the real part of the permittivity of a metal becomes positive, and the electromagnetic response of metals is similar to that of lossy dielectrics.

Bibliography

Textbooks and Monographs

[1] Ashcroft, N. W., and N. D. Mermin, *Solid State Physics*. Thomson Learning (1976).

[2] Born, M., and E. Wolf, *Principles of Optics*, 7th ed. Cambridge University Press, Cambridge (1999).

[3] Brillouin, L., *Wave Propagation in Periodic Structures*. Dover, New York (1953).

[4] Burns, G., *Solid State Physics*. Academic, New York (1985).

[5] Caloz, C., and T. Itoh, *Electromagnetic Metamaterials: Transmission Line Theory and Microwave Applications*. Wiley IEEE Press, New York (2005).

[6] Christman, J. R., *Fundamentals of Solid State Physics*. John Wiley & Sons, New York (1988).

[7] Cohen-Tannoudji, C., B. Diu, and F. Lalce, *Quantum Mechanics*. John Wiley & Sons, New York (1977).

[8] Datta, S., *Electronic Transport in Mesoscopic Systems*, 6th ed. Cambridge University Press, Cambridge (2005).

[9] Datta, S., *Quantum Transport: Atom to Transistor*. Cambridge University Press, Cambridge (2005).

[10] Dragoman, D., and M. Dragoman, *Quantum-Classical Analogies*. Springer, Berlin (2006).

[11] Economou, E. N., *Green's Functions in Quantum Physics*, 2nd ed. Springer, Berlin (1979).

[12] Efetov, K. B., *Supersymmetry in Disorder and Chaos*. Cambridge University Press, Cambridge (1997).

[13] Eleftheriades, G. V., and K. G. Balmain, *Negative-Refractive Metamaterials: Fundamental Principles and Applications*. Wiley IEEE Press, New York (2005).

[14] Engheta, N., and R. W. Ziolkowski (editors), *Metamaterials: Physics and Engineering Explorations*. J. Wiley & Sons, New York (2006).

[15] Flügge, S., *Practical Quantum Mechanics*. Springer, Berlin (1994, reprinted 1999).

[16] Fowles, G. R., *Introduction to Modern Optics*, 2nd ed. Holt, Reinhart and Winston, New York (1975).

[17] Gasiorowicz, S., *Quantum Physics*, 3rd. ed. J. Wiley & Sons, New York (2003).

[18] Grimvall, G., *Thermophysical Properties of Materials*. Elsevier, Amsterdam (1999), chapter 17.

[19] Imry, Y., *Introduction to Mesoscopic Physics*. Oxford University Press, Oxford (1977).

[20] Jackson, J. D., *Classical Electrodynamics*, 3rd. ed. J. Wiley & Sons, New York (1999).

[21] Joannopoulos, J. D., R. D. Mead, and J. N. Winn, *Photonic Crystals: Molding the Flow of Light*. Princeton University Press, Princeton (1995).

[22] Johnson, S. G., and J. D. Joannopoulos, *Photonic Crystals: The Road from Theory to Practice*. Kluwer Academic, Dordrecht (2002).

[23] Kittel, C., *Introduction to Solid State Physics*, 8th ed. J. Wiley & Sons, New York (2005).

[24] Kivshar, Yu. S., and G. P. Agrawal, *Optical Solitons*. Academic, San Diego (2003).

[25] Landau, L. D., and E. M. Lifshitz, *Quantum Mechanics*. Pergamon, Oxford (1984).

[26] Landau, L. D., E. M. Lifshitz, and L. P. Pitaevskii, *Electrodynamics of Continuous Media*. Pergamon, Oxford (1984).

[27] Londergan, J. T., J. P. Carini, and D. P. Murdock, *Binding and Scattering in Two-Dimensional Systems: Application to Quantum Wires, Waveguides and Photonic Crystals*. Springer, New York (1999).

[28] Lourtioz, J. M., H. Benisty, V. Berger, and J. M. Gerard, *Photonic Crystals: Towards Nanoscale Photonic Devices*. Springer, Berlin (1999).

[29] Mello, P. A., and N. Kumar, *Quantum Transport in Mesoscopic Systems: Complexity and Statistical Fluctuations*. Oxford University Press, Oxford (2004).

[30] Merzbacher, E., *Quantum Mechanics*, 3rd. ed. John Wiley & Sons, New York (1998).

[31] Mott, N. F., and E. A. Davis, *Electron Processes in Non-Crystalline Materials*. Clarendon, Oxford (1979).

[32] Press, W. H., S. A. Teukolsky, V. T. Vetterling, and B. P. Flannery, *Numerical Recipes*. Cambridge University Press, Cambridge (1992).

[33] Sakoda, K., *Optical Properties of Photonic Crystals*. Springer, Berlin (2005).

[34] Sheng, P., *Introduction to Wave Scattering, Localization, and Mesoscopic Phenomena*, 2nd ed. Academic, San Diego (2000).

[35] Soukoulis, C. M. (editor), *Photonic Band Gap Materials*. Kluwer, Dordrecht (1996).

[36] Soukoulis, C. M. (editor), *Photonic Crystals and Light Localization in the 21st Century*. Kluwer, Dordrecht (2001).

[37] Stratton, J. A., *Electromagnetic Theory*. John Wiley & Sons, New York (2007).

[38] Taflove, A., *Computational Electrodynamics*. Artech House, New York (1995).

[39] Yeh, P., *Optical Waves in Layered Media*, 2nd ed. Wiley Series in Pure and Applied Optics. John Wiley & Sons, New York (2005).

Papers and Review Articles

[40] Abrahams, E., P. W. Anderson, D. C. Licciardello, and T. V. Ramakrishnan, Scaling theory of localization: Absence of quantum diffusion in two dimensions. *Phys. Rev. Lett.* **42**, 673 (1979).

[41] Abrikosov, A., The paradox with the static conductivity of a one-dimensional metal. *Solid State Commun.* **37**, 997 (1981).

[42] Agio, M., and C. M. Soukoulis, Mini-stop bands in photonic crystals. *Phys. Rev. E* **64**, 055603(R) (2001).

[43] Altshuler, B. L., V. E. Kravtsov, and I. V. Lerner, Statistical properties of mesoscopic fluctuations and similarity theory. *JETP Lett.* **43**, 441 (1986).

[44] Anderson, P. W., Absence of diffusion in certain random lattices. *Phys. Rev.* **109**, 1492 (1958).

[45] Anderson, P. W., D. J. Thouless, E. Abrahams, and S. D. Fisher, New method for a scaling theory of localization. *Phys. Rev. B* **22**, 3519 (1980).

[46] Ando, T., Numerical study of symmetry effects on localization in two dimensions. *Phys. Rev. B* **40**, 5325 (1989).

[47] Ando, T., Quantum point contacts in magnetic fields. *Phys. Rev. B* **44**, 8017 (1991).

[48] Aydin, K., *et al.*, Experimental observation of true left-handed transmission peak in meta-materials. *Opt. Lett.* **29**, 2623 (2004).

[49] Beenakker, C. W. J., Random-matrix theory of quantum transport. *Rev. Mod. Phys.* **69**, 731 (1997).

[50] Chan, C. T., S. Datta, K. M. Ho, and C. M. Soukoulis, The A-7 structure: A family of photonic crystals. *Phys. Rev. B* **50**, 1988 (1994).

[51] Derrida, B., and E. Gardner, Lyapounov exponent of the one dimensional Anderson model: weak disorder expansions. *J. Phys. (Paris)* **45**, 1283–1295 (1984).

[52] Derrida, B., K. Mecheri, and J.-L. Pichard, Lyapounov exponents of products of random matrices: weak disorder expansion. Application to localization. *J. Phys. (Paris)* **48**, 733 (1987).

[53] Deubel, M., *et al.*, Direct laser writing of 3D photonic crystal templates for photonic bandgaps at 1.5 micrometers. *Nat. Mater.* **3**, 444 (2004).

[54] Deych, L. I., A. A. Lisyansky, and B. L. Altshuler, Single parameter scaling in one-dimensional localization revisited. *Phys. Rev. Lett.* **84**, 2678 (2000).

[55] Dolling, G., *et al.*, Observation of simultaneous negative phase and group velocity of light. *Science* **312**, 892 (2006).

[56] Dolling, G., *et al.*, Negative-index metamaterial at 780 nm wavelength. *Opt. Lett.* **32**, 53 (2007).

[57] Dorokhov, O. N., Localization and transmission coefficient for two coupled metal chains with disorder. *Solid State Commun.* **44**, 915 (1982).

[58] Economou, E. N., Surface plasmons in thin films. *Phys. Rev.* **182**, 539 (1969).

[59] Economou, E. N. and C. M. Soukoulis, Static conductance and scaling theory of localization in one dimension. *Phys. Rev. Lett.* **46**, 618 (1981); *Phys. Rev. Lett.* **47**, 973 (1981).

[60] Economou, E. N., C. M. Soukoulis, and A. D. Zdetsis, Localized states in disordered systems as bound states in potential wells. *Phys. Rev. B* **30**, 1686 (1984).

[61] El-Kady, I., *et al.*, Metallic photonic crystals at optical wavelengths. *Phys. Rev. B* **62**, 15299 (2000).

[62] Erdös, P., and R. C. Herdnon, Theories of electrons in one-dimensional disordered systems. *Adv. Phys.* **31**, 65 (1982).

[63] Evangelou, E. N., and T. Ziman, The Anderson transition in two dimensions in the presence of spin-orbit coupling. *J. Phys. C: Solid State Phys.* **20**, L235 (1987).

[64] Feiertag, G., *et al.*, Fabrication of photonic crystals by deep X-ray lithography. *Appl. Phys. Lett.* **71**, 1441 (1997).

[65] Foteinopoulou, S., E. N. Economou, and C. M. Soukoulis, Refraction at media with negative refractive index. *Phys. Rev. Lett.* **90**, 107402 (2003).

[66] Foteinopoulou, S., and C. M. Soukoulis, Negative refraction and left-handed behavior in 2D photonic crystals. *Phys. Rev. B* **67**, 235107 (2003).

[67] Foteinopoulou, S., and C. M. Soukoulis, Electromagnetic wave propagation in 2D photonic crystals: A study of anomalous refractive effects. *Phys. Rev. B* **72**, 165112 (2005).

[68] Fowler, A. B., A. Harstein, and R. A. Webb, Conductance in restricted-dimensionality accumulation layers. *Phys. Rev. Lett.* **48**, 196 (1982).

[69] Garcia, N., and A. Z. Genack, Anomalous photon diffusion at the threshold of the Anderson localization transition. *Phys. Rev. Lett.* **66**, 1850 (1991).

[70] Genack, A. Z., and N. Garcia, Observation of photon localization in a three-dimensional disordered system. *Phys. Rev. Lett.* **66**, 2064 (1991).

[71] Grigorenko, A. N., *et al.*, Nanofabricated media with negative permeability at visible frequencies. *Nature* **438**, 335 (2005).

[72] Ho, K. M., C. T. Chan, and C. M. Soukoulis, Existence of a photonic gap in periodic dielectric structures. *Phys. Rev. Lett.* **65**, 3152 (1990).

[73] Ho, K. M., *et al.*, Photonic band gaps in three dimensions: New layer-by-layer periodic structures. *Solid State Commun.* **89**, 413 (1994).

[74] Hückestein, B., Scaling theory of the integer quantum Hall effect. *Rev. Mod. Phys.* **67**, 357 (1995).

[75] Ishii, K., Localization of eigenstates and transport phenomena in one-dimensional disordered systems. *Prog. Theor. Phys. Suppl.* **53**, 77 (1973).

[76] Izrailev, F. M., and A. A. Krokhin, Localization and the mobility edge in one-dimensional potentials with correlated disorder. *Phys. Rev. Lett.* **82**, 4062 (1999).

[77] Izrailev, F. M., and N. M. Makarov, Anomalous transport in low-dimensional systems with correlated disorder. *J. Phys. A* **49**, 10613 (2005).

[78] Janßen, M., Statistics and scaling in disordered mesoscopic electron systems. *Phys. Rep.* **295**, 1 (1998).

[79] Kappus, M., and F. Wegner, Anomaly in the band centre of the one-dimensional Anderson model. *Z. Phys. B* **45**, 15 (1981).

[80] Katsarakis, N., *et al.*, Electric coupling to the magnetic resonance of split ring resonators. *Appl. Phys. Lett.* **84**, 2943 (2004).

[81] Kirkman, P. D., and J. B. Pendry, The statistics of one-dimensional resistances. *J. Phys. C: Solid State Phys.* **17**, 4327 (1984).

[82] Kirkman, P. D., and J. B. Pendry, The statistics of the conductance of one-dimensional disordered chains. *J. Phys. C: Solid State Phys.* **17**, 5707 (1984).

[83] Kogan, E., and M. Kumar, Random-matrix-theory approach to the intensity distributions of waves propagating in a random medium. *Phys. Rev. B* **52**, R3813 (1995).

[84] Koschny, Th., P. Markoš, E. N. Economou, D. R. Smith, D. Vier, and C. M. Soukoulis, Impact of the inherent periodic structure on the effective medium description of left-handed and related meta-materials. *Phys. Rev. B* **71**, 245105 (2005).

[85] Koschny, Th., R. Moussa, and C. M. Soukoulis, Limits on the amplification of evanescent waves of left-handed materials. *J. Opt. Soc. Am. B* **23**, 485 (2006).

[86] Koschny, Th., L. Zhang, and C. M. Soukoulis, Isotropic 3D left-handed and related meta-materials of the split-ring resonator and wire type. *Phys. Rev. B* **71**, 036617 (2005).

[87] Kramer, B., and A. MacKinnon, Localization—theory and experiment. *Rep. Prog. Phys.* **56**, 1469 (1993).

[88] Kuske, R., *et al.*, Schrödinger's equation on a one-dimensional lattice with weak disorder. *SIAM J. Appl. Math.* **53**, 1210 (1993).

[89] Landauer, R., Electrical resistance of disordered one-dimensional lattices. *Philos. Mag.* **21**, 863 (1970).

[90] Landauer, R., Spatial variation of currents and fields due to localized scatterers in metallic conduction. *IBM J. Res. Dev.* **1**, 223 (1957), reprinted **44**, 253 (2000).

[91] Larose, E., *et al.*, Weak localization of seismic waves. *Phys. Rev. Lett.* **93**, 048501 (2004).

[92] Lee, P. A., and T. V. Ramakrishnan, *Disordered electronic systems.* Rev. Mod. Phys. **57**, 287 (1985).

[93] Lee, P. A., A. D. Stone, and H. Fukuyama, Universal conductance fluctuations in metals: Effects of finite temperature, interactions, and magnetic field, *Phys. Rev. B* **35**, 1039 (1987).

[94] Li, J., L. Zhou, C. T. Chan, and P. Sheng, Photonic band gap from a stack of positive and negative index materials. *Phys. Rev. Lett.* **90**, 083901 (2003).

[95] Li, Q., *et al.*, Wave propagation in nonlinear photonic band-gap materials. *Phys. Rev. B* **53**, 15577 (1996).

[96] Lidorikis, E., *et al.*, Optical nonlinear response of a single nonlinear dielectric layer sandwiched between two linear dielectric structures. *Phys. Rev. B* **56**, 15090 (1997).

[97] Lidorikis, E., Q. Li, and C. M. Soukoulis, Wave propagation in nonlinear multilayer structures. *Phys. Rev. B* **54**, 10249 (1996).

[98] Lidorikis, E., and C. M. Soukoulis, Pulse-driven switching in one-dimensional nonlinear photonic band-gap materials: A numerical study. *Phys. Rev. E* **61**, 5825 (2000).

[99]   Linden, S., *et al.*, Magnetic response in metamaterials at 100 THz. *Science* **306**, 1351 (2004).

[100] Linden, S., *et al.*, Photonic metamaterials: Magnetism at optical frequencies. *IEEE J. Sel. Top. Quantum Electron.* **12**, 1097 (2006).

[101] MacKinnon, A., and B. Kramer, One-parameter scaling of localization length and conductance in disordered systems. *Phys. Rev. Lett.* **47**, 1546 (1981).

[102] Maldovan, M., and E. L. Thomas, Diamond-structured photonic crystals. *Nat. Mater.* **3**, 593 (2004).

[103] Markoš, P., Numerical analysis of the Anderson localization. *Acta Phys. Slovaca* **51**, 581 (2006).

[104] Markoš, P., and B. Kramer, Statistical properties of Lyapunov exponents and of quantum conductance of random systems in the regime of hopping transport. *Ann. Phys.* **2**, 339 (1993).

[105] Markoš, P., and C. M. Soukoulis, Numerical studies of left-handed materials and arrays of split ring resonators. *Phys. Rev. E* **65**, 036622 (2002).

[106] Markoš, P., and C. M. Soukoulis, Absorption losses in periodic arrays of thin metallic wires. *Opt. Lett.* **28**, 846 (2003).

[107] Markoš, P., and C. M. Soukoulis, Left-handed materials, in *Wave Scattering in Complex Media: From Theory to Applications*, B. A. van Tiggelen, and S. E. Skipetrov (editors), Kluwer Academic, Dordrecht (2003).

[108] Mello, P. A., P. Pereyra, and N. Kumar, Macroscopic approach to multichannel disordered conductors. *Ann. Phys. (NY)* **181**, 290 (1988).

[109] Mirlin, A., Statistics of energy levels and eigenfunctions in disordered systems. *Phys. Rep.* **326**, 259 (2000).

[110] Moško, M., *et al.*, Coherent metallic resistance and medium localization in a disordered one-dimensional insulator. *Phys. Rev. Lett.* **91**, 136803 (2003).

[111] Mott, N. F., and W. D. Twose, The theory of impurity conduction. *Adv. Phys.* **10**, 107 (1961).

[112] Notomi, M., Theory of light propagation in strongly modulated photonic crystals: Refraction like behavior in the vicinity of the photonic band gap. *Phys. Rev. B* **62**, 10696 (2000).

[113] Özbay, E., *et al.*, Measurement of three-dimensional band gap in new crystal structure made of dielectric rods. *Phys. Rev. B* **50**, 1945 (1999).

[114] Parazzoli, C. G., *et al.*, Experimental verification and simulation of negative index of refraction using Snell's law. *Phys. Rev. Lett.* **90**, 107401 (2003).

[115] Pendry, J. B., The evolution of waves in disordered media. *J. Phys. C: Solid State Phys.* **15**, 3493 (1982).

[116] Pendry, J. B., A transfer matrix approach to localization in 3D. *J. Phys. C: Solid State Phys.* **17**, 5317 (1984).

[117] Pendry, J. B., Quasi-extended electron states in strongly disordered systems. *J. Phys. C: Solid State Phys.* **20**, 733 (1987).

[118] Pendry, J. B., Photonic band structures. *J. Mod. Opt.* **41**, 209 (1994).

[119] Pendry, J. B., Negative refraction makes a perfect lens. *Phys. Rev. Lett.* **85**, 3966 (2000).

[120] Pendry, J. B., Electromagnetic materials enter the negative age. *Phys. World* **14**, 47 (2001).

[121] Pendry, J. B., Negative refraction. *Contemp. Phys.* **45**, 191 (2004).

[122] Pendry, J. B., and P. M. Bell, Transfer matrix techniques for electromagnetic waves, in *Photonic Band Gap Materials* (Ref. [36]), p. 203.

[123] Pendry, J. B., and A. MacKinnon, Calculation of photon dispersion relations. *Phys. Rev. Lett.* **69**, 2772 (1992).

[124] Pendry, J. B., A. MacKinnon, and P. J. Roberts, Universality classes and fluctuations in disordered systems. *Proc. R. Soc. London, Ser. A* **437**, 67 (1992).

[125] Pendry, J. B., *et al.*, Extremely low frequency plasmons in metallic mesostructures. *Phys. Rev. Lett.* **76**, 4773 (1996).

[126] Pendry, J. B., *et al.*, Low frequency plasmons in thin-wire structures. *J. Phys.: Condens. Matter* **10**, 4785 (1998).

[127] Pendry, J. B., *et al.*, Magnetism from conductors and enhanced non-linear phenomena. *IEEE Trans. Microwave Theory Tech.* **47**, 2075 (1999).

[128] Pendry, J. B., D. Schurig, and D. R. Smith, Controlling electromagnetic fields. *Science* **312**, 1780 (2006).

[129] Pichard, J. L., Random transfer matrix theory and conductance fluctuations, in *Quantum Coherence in Mesoscopic Systems*, NATO ASI Ser. B: Physics Vol. **254**, B. Kramer (editor), Plenum, New York (1991).

[130] Pichard, J. L., and G. Sarma, Finite size approach to Anderson localization. *J. Phys. C: Solid State Phys.* **14**, L127 (1981).

[131] Pichard, J. L., The one-dimensional Anderson model: Scaling and resonances revisited. *J. Phys. C.: Solid State Phys.* **19**, 1519 (1986).

[132] Pichard, J. L., N. Zanon, Y. Imry, and A. D. Stone, Theory of random multiplicative transfer matrices and its implications for quantum transport. *J. Phys. (Paris)* **51**, 587 (1990).

[133] Pokrovsky, A. L., and A. L. Efros, Electrodynamics of metallic photonic crystals and the problem of left-handed materials. *Phys. Rev. Lett.* **89**, 093901 (2002).

[134] Ramakrishna, S. A., Physics of negative refractive index materials. *Phys. Rep.* **68**, 449 (2005).

[135] Robertson, W. M., *et al.*, Observation of surface photons on periodic dielectric arrays. *Opt. Lett.* **18**, 528 (1993).

[136] Ruppin, R., Surface polaritons of a left-handed medium. *Phys. Lett. A* **277**, 61 (2000).

[137] Ruppin, R., Surface polaritons in a left-handed material slab. *J. Phys.: Condens. Matter* **13**, 1811 (2001).

[138] Shelby, R. A., D. R. Smith, and S. Schultz, Experimental verification of a negative index of refraction. *Science* **292**, 77 (2001).

[139] Sigalas, M. M., C. T. Chan, K. M. Ho, and C. M. Soukoulis, Metallic photonic band-gap materials. *Phys. Rev. B* **52**, 11744 (1995).

[140] Slevin, K., P. Markoš, and T. Ohtsuki, Reconciling conductance fluctuations and the scaling theory of localization. *Phys. Rev. Lett.* **86**, 3594 (2001).

[141] Slevin, K., and T. Nagao, Maximum-entropy models of transport in mesoscopic wires. *Int. J. Mod. Phys B* **2**, 103 (1995).

[142] Smith, D. R., and N. Kroll, Negative refractive index in left-handed materials. *Phys. Rev. Lett.* **85**, 2933–2936 (2000).

[143] Smith, D. R., S. Schultz, P. Markoš, and C. M. Soukoulis, Determination of effective permittivity and permeability of metamaterials from reflection and transmission coefficients. *Phys. Rev. B* **65**, 195104 (2002).

[144] Smith, D. R., *et al.*, Experimental and theoretical results for a 2D metal photonic band gap cavity. *Appl. Phys. Lett.* **65**, 645 (1994).

[145] Smith, D. R., *et al.*, Composite medium with simultaneously negative permeability and permittivity. *Phys. Rev. Lett.* **84**, 4184 (2000).

[146] Smith, D. R., *et al.*, Electromagnetic parameter retrieval from inhomogeneous metamaterials. *Phys. Rev. E* **71**, 121103 (2005).

[147] Soukoulis, C. M., M. Kafesaki, and E. N. Economou, Negative-index materials: New frontiers in optics. *Adv. Mater.* **18**, 1941 (2006).

[148] Soukoulis, C. M., S. Linden, and M. Wegener, Negative refractive index at optical wavelengths. *Science* **315**, 47 (2007).

[149] Soukoulis, C. M., M. J. Velgakis, and E. N. Economou, One-dimensional localization with correlated disorder. *Phys. Rev. B* **50**, 5110 (1994).

[150] Stamp, A. P., and G. C. McIntosh, A time-dependent study of resonant tunneling through a double barrier. *Am. J. Phys.* **64**, 264 (1996).

[151] Stoychev, M., and A. Z. Genack, Measurement of the probability distribution of total transmission in random waveguides. *Phys. Rev. Lett.* **79**, 309 (1997).

[152] Szmulowitz, F., Analytic, graphical, and geometric solutions for photonic band gaps. *Am. J. Phys.* **72**, 1392 (2004).

[153] Thouless, D. J., A relation between the density of states and range of localization for one dimensional random systems. *J. Phys. C: Solid State Phys.* **5**, 77 (1972).

[154] Thouless, D. J., Electrons in disordered systems and the theory of localization. *Phys. Rep.* **13**, 93 (1974).

[155] Vagner, P., *et al.*, Coherent resistance of a disordered one-dimensional wire: Expressions for all moments and evidence for non-Gaussian distribution. *Phys. Rev. B* **67**, 165316 (2003).

[156] van Rossum, M. C. W., and Th. M. Nieuwenhuizen, Multiple scattering of classical waves: microscopy, mesoscopy, and diffusion. *Rev. Mod. Phys.* **71**, 313 (1999).

[157] Veselago, V. G., The electrodynamics of substancies with simultaneously negative values of ε and μ. *Sov. Phys. Usp.* **10**, 509 (1968).

[158] Veselago, V. G., Electrodynamics of materials with negative index of refraction. *Phys. Usp.* **46**, 764 (2003).

[159] von Klitzing, K., The quantized Hall effect. *Rev. Mod. Phys.* **58**, 519 (1986).

[160] Ward, A. J., and J. B. Pendry, Refraction and geometry in Maxwell's equations. *J. Mod. Opt.* **43**, 773 (1996).

[161] Washburn, S., and R. A. Webb, Aharonov-Bohm effect in normal metal quantum coherence and transport. *Adv. Phys.* **35**, 375 (1986).

[162] Winn, J. N., Y. Fink, S. Fan, and J. D. Joannopoulos, Omnidirectional reflection from a one-dimensional photonic crystal. *Opt. Lett.* **23**, 1573 (1998).

[163] Yablonovitch, E., Inhibited spontaneous emission in solid-state physics and electronics. *Phys. Rev. Lett.* **58**, 2059 (1987).

[164] Yablonovitch, E., T. J. Gmitter, and K. M. Leung, Photonic band structure: The face-centered-cubic case employing nonspherical atoms. *Phys. Rev. Lett.* **67**, 2295 (1991).

[165] Yeh, P., A. Yariv, and C. S. Hong, Electromagnetic propagation in periodic stratified media. I. General theory. *J. Opt. Soc. Am.* **67**, 423 (1977).

[166] Zhen, Ye, Optical transmission and reflection of perfect lenses by left handed materials. *Phys. Rev. B* **67**, 193106 (2003).

[167] Zhou, J., *et al.*, *Limit of the negative magnetic response of split-ring resonators at optical frequencies.* *Phys. Rev. Lett.* **95**, 223902 (2005).

[168] Ziolkowski, R. W., and E. Heyman, Wave propagation in media having negative permittivity and permeability. *Phys. Rev. E* **64**, 056 625 (2001).

Index